D1033925

Comparative Biochemistry of Parasites

Proceedings of an International Symposium organized by the Janssen Research Foundation and held at Janssen Pharmaceutica Beerse, Belgium, September 1-3, 1971

Comparative Biochemistry of Parasites

edited by

H. Van den Bossche

Department of Comparative Biochemistry
Janssen Pharmaceutica
Beerse, Belgium

Ⓐ Ⓟ **Academic Press**
New York and London
1972

ACADEMIC PRESS, INC.
111 Fifth Avenue, New York, New York 10003

United Kingdom Edition published by
ACADEMIC PRESS, INC. (LONDON) LTD.
24/28 Oval Road, London NW1

LIBRARY OF CONGRESS CATALOG CARD NUMBER: 72-187249

PRINTED IN THE UNITED STATES OF AMERICA

CONTENTS

CONTENTS

PARTICIPANTS
*Asterisk denotes chairmen

Bafort, J. M., Instituut voor Tropische Geneeskunde, 2000 Antwerp, Belgium

Beames, Calvin G., Jr., * Department of Physiology, Oklahoma State University, Stillwater, Oklahoma 74074

Bone, G. J., Rue Baron de Castro 7, 1040 Brussels, Belgium

Borgers, M., Janssen Pharmaceutica, 2340 Beerse, Belgium

Borst, P., Laboratorium voor Biochemie, Jan Swammerdam Instituut, Amsterdam, The Netherlands

Bout, D. T. J., Service d'Immunologie et de Biologie Parasitaire — Université de Lille, Lille, France

Bowman, I. B. R., Department of Biochemistry, University of Edinburgh Medical School, Edinburgh EH 8 9AG, Scotland

Bryant, C., * Department of Zoology, Box 4 G.P.O., Canberra A.C.T. 2600, Australia

Bueding, Ernest, * The Johns Hopkins University, Baltimore, Maryland 21205

Cheah, K. S., A.R.C. Meat Research Institute, Langford, Bristol BS18 7DY, England

Chowdhury, N., Laboratorium voor Zoöfysiologie, Rijksuniversiteit Gent, 9000 Gent, Belgium

Coles, G. C., MRC Biochemical Parasitology Unit, The Molteno Institute, University of Cambridge, Cambridge CB2 — 3EE, England

Davey, K. G., Institute of Parasitology, MacDonald College 800, Province of Quebec, Canada

De Zwaan, A., Laboratorium voor Scheikunde en Dierfysiologie, Universiteit Utrecht, Utrecht, The Netherlands

Dierick, W. S. H., Laboratorium voor Biochemie, RUCA—2020 Antwerp, Belgium.

Eckert, J., Institut für Parasitologie, Universität Zürich, Zürich, Switzerland

Eeckhout, Yves, Laboratoire de Chimie Physiologique, Université de Louvain, 3000 Leuven, Belgium

Gutteridge, W. E., Biological Laboratory, University of Kent, Canterbury, Kent, England

Haese, W. H., The Johns Hopkins Hospital, Baltimore, Maryland 21205

Hill, George C., MRC Biochemical Parasitology Unit, The Molteno Institute, Cambridge, CB2-3EE, England

Howells, R. E., Liverpool School of Tropical Medicine, Liverpool L3 5QA, England

Jaffe, Julian J., University of Vermont College of Medicine, Given Building, Burlington, Vermont 05401

Janssen, Paul A. J.,* Janssen Pharmaceutica, 2340 Beerse, Belgium

Janssens, P. G., Instituut voor Tropische Geneeskunde, 2000 Antwerp, Belgium

Kaba, A., Laboratoire de Pharmacodynamie général, Université de Louvain, 3000 Leuven, Belgium

Kohler, P., Institut für Parasitologie, Universität Zürich, Zürich, Switzerland

Lee, D. L.,* Department of Pure and Applied Zoology, University of Leeds, Leeds LS2 9JT, England

Le Ray, D., Instituut voor Tropische Geneeskunde, 2000 Antwerp, Belgium

Meyer, Franz, Upstate Medical Center, Syracuse, New York 13210

Meyer, Haruko, Upstate Medical Center, Syracuse, New York 13210

Moors, A., Instituut voor Tropische Geneeskunde, 2000 Antwerp, Belgium

Mortelmans, J., Instituut voor Tropische Geneeskunde, 2000 Antwerp, Belgium

Newton, B.A.,* MRC Biochemical Parasitology Unit, The Molteno Institute, University of Cambridge, Cambridge, CB2-3EE, England

Prins, R. A., Laboratorium voor Veterinaire Biochemie, Universiteit Utrecht, Utrecht, The Netherlands

Reeves, R. E., Louisiana School of Medicine, New Orleans, Louisiana 70112

Riou, G., Institut Gustave, Roussy, Villejuif 94, France

Rombouts, W. A. J. A., Medische Fakulteit, Universiteit Louvain, 3000 Louvain, Belgium

Ruitenberg, E. J., Rijksinstituut voor de Volksgezondheid, Utrecht, The Netherlands

Ryley, J. F., Pharmaceutical Division, I.C.I., Macclesfield, Cheshire SK10 4TG, England

*Saz, Howard J.,** College of Science, University of Notre Dame, Notre Dame, Indiana 46556

Steinert, M., Faculté des Sciences, ULB, 1640 St. Genesius-Rode, Belgium

Tollenaere, J. P., Janssen Pharmaceutica, 2340 Beerse, Belgium

*Trager, W.,** The Rockefeller University, New York, New York 10021

Van den Bossche, H., Department of Comparative Biochemistry, Janssen Pharmaceutica, 2340 Beerse, Belgium

Van de Vijver, G. H., Instituut voor Zoöfysiologie, Universiteit Gent, 9000 Gent, Belgium

Van Meirvenne, N., Instituut voor Tropische Geneeskunde, 2000 Antwerp, Belgium

Van Miert, A., Laboratorium voor Veterinaire Farmacologie, Universiteit Utrecht, Utrecht, The Netherlands

Van Nueten, J. M., Janssen Pharmaceutica, 2340 Beerse, Belgium

Veenendaal, G. H., Laboratorium voor Veterinaire Farmacologie, Universiteit Utrecht, Utrecht, The Netherlands

Vercauteren, R. E., Fakulteit voor de Diergeneeskunde, Universiteit Gent, 9000 Gent, Belgium *von Brand, T. C.,* *8606 Hempstead Avenue, Bethesda, Maryland 20034

Wattiaux, R., Facultés Universitaires Notre Dame de la Paix, 5000 Namur, Belgium

Weinbach, Eugene C., Laboratory of Parasitic Diseases, National Institutes of Health, Bethesda, Maryland 20014

PREFACE

In the last two decades a tremendous amount of progress has been made in the field of parasite physiology and biochemistry. Studies on the intermediate metabolisms, on the characterization of nucleic acids, on transport mechanisms, and on the mode of action of chemotherapeutic agents have contributed to a better understanding of the biochemistry and physiology of protozoa and helminths. The main object of the Symposium on the Comparative Biochemistry of Parasites, which was held at Janssen Pharmaceutica from September 1 to 3, 1971, was to bring together experts on various aspects of protozoa and helminth biochemistry and physiology in order to evaluate present knowledge, to stimulate further progress in this field, and to find new approaches for the rational design of new chemotherapeutic agents.

It is always difficult to ascertain whether a symposium is organized at an appropriate date. The most important reason for organizing this symposium in September 1971 was that Dr. Theodor von Brand was able to participate then. Dr. von Brand has worked in the field of parasite physiology and biochemistry for nearly half a century and, as Professor Weinstein [*J. Parasitol.* **56** (1970), 625] pointed out recently, has acted as a leavening agent in this discipline. This book, which is comprised of the papers presented at the symposium, is dedicated to his interest and honor.

It is a pleasure to thank the participants and the chairmen, and particularly Dr. Paul A. J. Janssen for providing the opportunity to organize this symposium and for his continued encouragement. I am pleased to acknowledge the advice of E. Bueding, D. Fairbairn, H. J. Saz, D. Thienpont, and T. von Brand in the preparation of the symposium, and the efficient help of a great number of my colleagues at Janssen Pharmaceutica.

I sincerely hope that this book fulfills its purpose as a tool for future research.

H. Van den Bossche

Comparative Biochemistry of Parasites

GLIMPSES AT THE EARLY DAYS
OF PARASITE BIOCHEMISTRY.

Theodor von Brand

8606 Hempstead Avenue
Bethesda, Maryland 20034, U.S.A.

Parasitology, constitutes only a tiny segment of natural science or medicine. No wonder then that parasitology is mentioned only in a very cursory manner in books dealing with the history of science or medicine. Many newer books on these subjects exist and I looked at a few of them taken at random from the library shelves. This is what I found. The index to the two volumes of Packard's (1963) "History of medicine in the United States" does not contain the words parasites or parasitic diseases. The book gives, however, an account of an early 17th century case of a large uterine hydatid cyst in an unfortunate woman. Lloyd's (1968) "A hundred years of medicine" has a 4 1/2 pages long chapter entitled "Some larger parasites". In Major's (1954) two volumes of "A history of medicine" and in Singer and Underwood's (1962) "A short history of medicine" one finds brief discussions of various tropical diseases, such as malaria, trypanosomiasis, ancylostomiasis, and others. Needless to say the physiology or biochemistry of parasites is not discussed.

If one looks for books dealing specifically with the history of parasitology one is struck by the fact that only very few have been written. Of very great interest are the scholarly works of Hoeppli "Parasites and parasitic infections in early medicine and science" (1959) and "Parasitic diseases in Africa and the Western hemisphere" (1969). Both make for fascinating reading, but for obvious reasons they do not contain and cannot be expected to contain a discussion of the development of parasite biochemistry. As far as I am aware only one other relevant book exists. It is Foster's (1965) "A history of parasitology". Foster traces the development of our knowledge concerning certain parasites and groups of parasites without, however, mentioning any data on parasite biochemistry.

Interesting historical data abound of course in biographies and autobiographies of such men as Sir Patrick Manson (Manson-Bahr and Alcock, 1927), Sir Ronald Ross (Ross, 1923), Geheimrat Bernhard Nocht (Martini, 1957) and others. But these men were not parasite physiologists and therefore the topics parasite physiology or biochemistry have no place in such accounts. In short, we just do not have as yet a history of parasite physiology and biochemistry.

I am not a trained historian and I do not pretend that I can fill this gap adequately, especially not in the confines of a brief lecture. I have, however, been active in the field for nearly half a century, have been interested for a long time in some of the old findings and am therefore to some extent familiar with historical developments. I am also familiar with the reasons why old findings frequently are forgotten. Younger investigators often discount old data as not being relevant to modern approaches and they seem reluctant to read the old literature. I think a survey of parasite physiologists asking them how many have read in the original Weinland's (1901b) classical paper "Uber Kohlenhydratzersetzung ohne Sauerstoffaufnahme bei *Ascaris,* einen tierischen Gärungsprozess" would be quite illuminating. Furthermore practically all journals discourage or reject long historical introductions and I certainly do not advocate that the literature citations of each research paper should go back to what amounts to be prehistoric times. I do want, however, to draw your attention to the fact that one can find in the old literature significant data which, if known to subsequent workers, could have accelerated later developments and I would not be too surprised if future historians would find more recent examples of the same type.

A case in point is the cuticle of *Ascaris.* Grube (1850) stated that it consisted of chitin and the terms chitin or chitinous were later often used in connection with it and other structures of the nematode body, such as the lining of the stoma and esophagus. And this despite the fact that Lassaigne had proven in 1843 that the cuticle of *Ascaris,* in contrast to the chitin of insects, was solubilized by potassium hydroxyde and despite the fact that he stated unequivocally that both do not have the same structure. In fairness to Grube it should be mentioned that Lassaigne's observations on *Ascaris* were contained in a paper entitled "Sur le tissue tégumentaire des insectes de differents ordres", that is in a paper where one would not expect to find data on worms.

Another quite instructive example concerns the polysaccharides of asca-

2

rids. The great French physiologist Claude Bernard established as early as 1859 that glycogen occurs not only in the liver of vertebrates, but that it can be demonstrated readily by qualitative chemical methods and also histochemically by means of the iodine reaction in such parasites as *Ascaris, Fasciola, Taenia,* and larval cestodes. These findings were immediately forgotten as indicated by the fact that the English physiologist Sir Michael Foster published in 1865 a paper in which he states explicitly that, although glycogen had been reported from numerous invertebrates, nobody so far noticed its occurrence in parasitic worms. He himself demonstrated it qualitatively in an undetermined tapeworm and quantitatively in *Ascaris lumbricoides,* but his findings also made no lasting impression on his contemporaries.

Gustav von Bunge, a scion of an old aristocratic Baltic family who worked first in Dorpat, Estonia and later took the newly established chair of physiological chemistry at Basel, Switzerland, did not know about these old observations. He had first become interested in the oxygen requirements of intestinal worms because he believed on theoretical grounds that the helminths of warm-blooded hosts would have only minimal needs for oxygen. This view was based on the assumption that fermentative processes were the source of muscular activity and that oxidative processes were related primarily to heat production. He observed experimentally that various ascarid species survived anaerobic periods rather well and that the worms produced anaerobically large amounts of CO_2 and an unidentified volatile acid (Bunge, 1883, 1889). He was however unable to hazard a guess as to the source of these metabolites. Had he been aware of the older work of Bernard or Foster, he undoubtedly would have put two and two together and would thus have been in a position to anticipate Weinland's later findings by more than a decade.

Up to the time of Bunge the chemical studies on helminths were confined to observations on their chemical composition; he was the first to carry out experimental metabolic studies. There are two reasons why Bunge did not pursue this work further. First he found it difficult to secure sufficient fresh worms at Basel and secondly he immersed himself in his later years completely in the temperance movement. A biography of Bunge written years after his death by an admirer of his antialcoholic activities (Graeter, 1952) does not mention even with a single word his work on free-living and parasitic invertebrates.

Before turning to Weinland's achievements, I would like to emphasize that quite a few additional chemical data can be found in the old helminthological literature. Time does not permit to review them in any detail, but a few examples may illustrate the type of data one can unearth. I mention first the calcareous corpuscles of cestodes. In an abstract of a paper presented by Doyère (1840) before the Société philomatique de Paris one finds the statement that those of *Echinococcus* consist of calcium carbonate. At the same time Gulliver (1840, 1841) studied the corpuscles of *Cysticercus* which he considered as the worms's eggs. He observed that they dissolve quickly upon treatment with hydrochloric ("muriatic") and acetic acid with plentiful evolution of a gas. When the solution was treated with sulfuric or oxalic acid a white precipitate was obtained. Küchenmeister (1851) did not share Gulliver's view concerning the significance of these structures. He remarked that not all corpuscles consist of calcium carbonate, since he found that those of some species dissolve in acid without the evolvement of gas, indicating that they consist of calcium phosphate. Leuckart (1863) mentions in the first edition of his famous book "Die menschlichen Parasiten und die von ihnen herrührenden Krankheiten" that a Dr. Naumann had investigated on his suggestion the inorganic substances of the *Taenia marginata* body which he thought could be referred largely to calcareous corpuscles. Dr. Naumann found mainly calcium salts accompanied by small amounts of magnesium, ironoxyde, sodium and potassium which were bound to carbonic, phosphoric, hydrochloric, and sulfuric acid. I must admit that I was totally unaware of these old observations when I first isolated cestode calcareous corpuscles some 90 years after Gulliver's and Doyère's papers appeared.

Quite a few additional old data can also be located for the cycts of *Echinococcus,* of which I will mention only two. Heintz showed as early as 1850 that the hydatid fluid contained succinic acid in the form of sodium succinate, a finding never mentioned when modern investigators discuss the production of succinic acid by cestodes. Some of the old investigators were also aware that the hydatid fluid contains sugar. This was shown first for human infections by Bernard and Axenfeld in 1857 and confirmed in 1860 by Lücke who also showed that sugar could be liberated from the membranes by treatment with acid. Incidentally, the Bernard just mentioned is not the famous Claude Bernard, but an unknown Charles Bernard. The note by Bernard and Axenfeld seemingly was not prepared by themselves, but appears to be a summary of a talk written by the Secretary of the Société de Biologie de Paris. It was perhaps the latter who pointed out that the master himself, Claude Bernard, had already found previously sugar in the hydatid fluid of sheep.

With the exception of Bunge's studies the above findings had no significant influence on subsequent developments in the field of helminth physiology. Bunge's studies are the one exception because they stimulated Weinland to take up his justly famous *Ascaris* investigations. Ernst Weinland worked at the turn of the century at the Physiological Institute of the University Munich. It is not surprising that he was interested in this aspect of comparative physiology since as a son of the well known helminthologist David Weinland he had been exposed early to parasitology and since he had not only an MD degree, but had also earned a Ph.D degree in zoology. He favored throughout his career the chemical approach to various life processes and relied almost exclusively on quantitative gravimetric methods, while having little faith in the then available colorimetric procedures. His interest centered largely around the metabolism of parasitic worms, but they were not confined to them. He published for instance a series of papers dealing with the metabolism of the fly *Calliphora* and one of his last papers dealt with the chemical composition of hedgehogs after various periods of hibernation (Weinland, 1925). His thorough knowledge of the field of comparative chemical physiology and his typical critical approach are well documented in a review he published in 1910 (Weinland, 1910). He had prepared a much more comprehensive treatise on the subject for Winterstein's Handbuch der vergleichenden Physiologie, when the outbreak of the first world war prevented its publication. After the war Weinland tried to bring the manuscript up to date, but because of rather poor health and his preoccupation with preparing his daily physiology lecture to medical students and his involvement with the turbulent politics of these years he never finished it. Finally Winterstein commissioned others and the result was a rather hastily prepared and incomplete account of the vast field of comparative metabolism (Kestner and Plaut, 1924).

However, on this occasion we are interested primarily in Weinland's parasitological work (Weinland, 1901 a, b, 1902 a, b, 1903, 1904, Weinland and Ritter, 1903). He started out by determining, without knowledge of the earlier comparable studies, the glycogen content of various parasitic helminths and was impressed by the enormous quantities stored by them. He then concentrated on *Ascaris lumbricoides,* establishing quantitative relationships between glycogen disappearance during starvation and the production of both carbon dioxide and volatile acids. He first thought that valeric acid was excreted almost exclusively; later on he recognized the acids as a mixture of valeric and caproic acids. Since he observed that the metabolic rate was about the same under aerobic and anaerobic conditions, he expressed the rather revolutionary idea that *Ascaris,* other intestinal worms, but also free-living mud- and swampdwellers do not require any molecular oxygen. He expressed this view perhaps most clearly in the lecture which he delivered according to German custom before the assembled faculty shortly after assuming the Professorship of Physiology at the University of Erlangen ("Akademische Antrittsrede", Weinland, 1913). He stated then: In these animals a supply of and a requirement for molecular oxygen is unnecessary for the functioning of their vital processes ("Bei diesen Tieren ist eine Zufuhr und ein Bedarf von elementarem Sauerstoff für den Ablauf der Lebensvorgänge nicht erforderlich").

Weinland's approach to the *Ascaris* problem was comprehensive. Besides studying the overall carbohydrate metabolism, he also showed that the worms do not utilize lipids for energy production, that they do decompose nitrogenous substances, that they produce antienzymes, and that cell-free preparations show still a fermentative metabolism. This work established him as the foremost authority on parasite metabolism of his day and his work stimulated some other investigators to take up studies in the field. I mention here only a few of them: Schimmelpfennig (1903) studied the chemical composition of *Parascaris,* von Kemnitz (1912) published a beautiful paper on the histochemical distribution of glycogen in *Ascaris* tissues, Ortner-Schönbach (1913) performed a similar study on trematodes and cestodes, and Krummacher (1918) investigated the heat production of *Ascaris*. In these same years a few papers were also published that were only indirectly influenced by Weinland's work, or had no connection with it at all. The best known examples are the investigations of the German pharmacologist Flury (1912) on the chemistry and toxicology of *Ascaris* and the outstanding study of the French biologist Fauré-Fremiet (1913) on the chemistry of the *Parascaris* reproductive cycle. Fauré-Fremiet, incidentally, is the only one of the early group of workers still alive and still active in research, although having left the field of parasite physiology years ago to concentrate on the fine structure of protozoa.

We will now examine briefly the reception accorded Weinland's views by the early workers. Not too surprisingly they were not accepted immediately and fully, but were questioned on various grounds. The first point challenged concerned the production of volatile acids by an animal. Fischer (1924) and Slater (1925, 1928) maintained that the true metabolic endproduct of *Parascaris* and *Ascaris* was lactic acid, just as it is in the anaerobic metabolism of vertebrate tissues. The former found this to be true when he studied minced worms and in retrospect one can say that his observation was correct, but his interpretation was not. Saz and Lescure (1969) showed recently that the shift in metabolism observed in minced or homogenized worms is due to disturbed segregation of enzymes. Slater (1925, 1928), on the other hand, thought that the volatile acids found regularly in incubates of ascarids were formed by bacterial activity. We know now that this assumption was incorrect, since Epps et al. showed in 1950 that axenized specimens of *Ascaris* produce large amounts of volatile acids.

In this respect then Weinland's views were fully vindicated. They fared less well when his thesis was challenged that intestinal worms do not consume oxygen. Various workers working independently at about the same time disproved the point for various helminths. In so far as *Ascaris* specifically is concerned, a Belgian worker, Adam, now Head of the Section of recent Invertebrates at the Institut Royal des Sciences Naturelles de Belgique and working then in Jordan's laboratory at Utrecht showed in 1932 that female and male specimens as well as homogenized muscle tissues consume appreciable amounts of oxygen. Even earlier, namely in 1931, had oxygen consumption been demonstrated for the cestode *Moniezia expansa* by the Americans Alt and Tischer and in 1932 the German Harnisch showed that the trematode *Fasciola hepatica* uses oxygen. These, I believe, are the earliest relevant papers. Since then similar observations were reported for many additional species and today nobody doubts any more that intestinal as well as tissue helminths are capable of utilizing oxygen if they have access to it.

Another controversial point was, and to some extent still is, the question whether intestinal worms lead in situ an anaerobic or an aerobic life. Weinland of course maintained that they do not require oxygen, but this view was challenged repeatedly, in the older days primarily by Slater (1925) and Davey (1938). The former had observed that electrically stimulated ascarids survived better in the presence than the absence of oxygen and the latter established that small nematodes of the sheep intestine could be kept in vitro for longer periods aerobically than anaerobically. As I pointed out a long time ago (von Brand 1938 a) these opposing views are really not mutually exclusive but can be reconciled if one assumes that large intestinal worms because of their organisation and the low oxygen concentrations in their environment lead in nature a predominantly anaerobic life while small parasites can gain in the same habitat significant amounts of oxygen.

It should be emphasized that the assumption of a predominantly anaerobic life of large helminths refers only to the mode of energy production. It is possible and even probable that the small amounts of oxygen which they can acquire under natural conditions may be quite important. Fairbairn (1970) emphasized the point in respect to the collagen formation of *Ascaris* which depends on the functioning of an oxygen requiring hydroxylase and the activity of the phenolase responsible for the tanning of the *Fasciola* egg shell may according to Moss (1970) account for quite a high percentage of the worm's oxygen consumption. But I dont want to say more about these new findings since it is today my task to review the accomplishments of the research done in days gone by.

Looking back from the vantage point of the seventies on the research in helminth physiology done in the old days, that is about to the start of the second World War, one is struck by several facts. First only relatively few species served as experimental tools. These were the nematodes *Ascaris* and *Parascaris,* the trematode *Fasciola,* and the cestode *Moniezia.* Other species were studied only in isolated instances. Mention may be made of the interesting study of the *Dioctophyme renale* hemoglobin by Aducco (1889), the study of Bondouy (1910) on the chemical composition of *Sclerostomum equinum,* Schopfer's (1932) investigations on the body fluids of various parasites or McCoy's (1930) analysis of the repiration of *Ancylostoma* larvae in dependence of temperature and other factors. During these years many more papers dealing with one or the other physiological aspect of the main experimental animals mentioned previously appeared. I list at this point only a few of them. Keilin described in 1925 the occurrence of cytochromes in *Ascaris* and Pintner (1922) and Stepanow-Grigoriew and Hoeppli (1926) voiced different views concerning the physiological basis of the nematode's larval migration through the host body.

These last studies already have the second characteristic of the old studies. I refer to the fact that usually entire animals, but occasionally also minced materials were used to investigate some phase of the overall metabolism, but that no studies on the intermediate metabolism were carried out. Examples are the studies on the respiration of *Ascaris* by Krueger (1936, 1937) or of *Triaenophorus* by Harnisch (1933), the metabolic studies on *Ascaris* by Schulte (1917), von Brand (1934) or Oesterlin (1937), the studies on the overall and respiratory metabolism of *Fasciola* by Flury and Leeb (1926) or Weinland and von Brand (1926) and of *Moniezia* by von Brand (1929, 1933 a) and Cook and Sharman (1930).

The third striking fact about the old research is that very little was done to study in dept the influence of helminths on the metabolism of the host. True enough, relevant data can be found in the old medical literature, for example data on some blood or tissue constituents of human patients infected with *Trichinella* (Fuchs, 1922) or hookworms

(Rake, 1894, Donomae, 1927) or data on the metabolism, especially the nitrogenous excretions during these infections (Padoa, 1909, Markowicz and Bock, 1931, Bohland, 1894). However, these and similar investigations did not contribute materially to an analysis of the influence of the parasites on the physiological processes of the host. The only really old experimental studies known to me are the investigations of Flury (1913) and Flury and Groll (1913) on the influence of a *Trichinella* infection on the metabolism, especially the nitrogen metabolism of the host.

If we turn now to the biochemistry of parasitic protozoa we find that its development differed in various respects from that just described for helminth biochemistry. First, only very few really old investigations of a clearly chemical nature exist. This of course is not surprising but simply a consequence of the fact that protozoa were difficult to secure in sufficient amounts and sufficient purity in the days when modern micro-methods of analysis had not yet been developed. It is the more remarkable therefore that the justly famous zoologist Bütschli was able to demonstrate in 1885 that the gregarine polysaccharide was water soluble, could be precipitated by alcohol and yielded a reducing sugar upon hydrolysis. The only other truly old studies are dated 1911 and 1913 and are due to Panzer who studied in Vienna primarily the lipids but to some extent also the proteins of the coccidian *Goussia gadi*. I mention here only that he identified cholesterol and that he established the fact that the fatty acids and glycerides of the parasites differed distinctly from those of the host.

Besides these isolated chemical studies quite a few histochemical data concerning parasitic protozoa can be located in the old literature. Examples are the demonstration of polysaccharides in gregarines by Maupas (1886), in rumen ciliates by Certes (1889) and parasitic amebas by Kuenen and Swellengrebel (1913), or the demonstration of lipid droplets and glycogen in verious sporozoa by Thélohan (1894), Cohn (1896), Brault and Loeper (1904), and others. All these studies however had no discernible influence on the subsequent developments. In contrast to what I said a moment ago about helminth physiology, physiological and biochemical studies of parasitic protozoa received their impetus from early investigations dealing with physiological and biochemical alterations sustained by parasitized hosts. These started early and in general preceded biochemical investigations on the protozoa themselves. To give an extreme example: The malarial pigment deposited in the organs of malarious patients was known for many years before the possible existence of an organism like a *Plasmodium* was even dreamed of.

The history of the malaria pigment is long and exemplifies the fact that

9

refinements in experimental technique can make older conclusions obsolete. A full discussion of this topic would require a special lecture; I can mention on this occasion only a few of the old workers in this specialized field. Discoloration of the internal organs of malarious patients has been known for a long time. The earliest references quoted in the literature, which however were not available to me for checking, are the reports by Lancisi (1717), Stoll (1797), and some others. The origin of this dark pigment was widely discussed because two opposing views were proposed during the 19th century: The theory of splenic origin promulgated by Meckel (1847) and Virchow (1849) and the theory of hematogenous origin, usually ascribed to Planer (1854).

Incidentally, it is by no means certain that all the above workers were always dealing with cases of chronic malaria. Indeed Meckel's (1847) autopsy report concerned an insane woman who had been confined for many years in a mental institution and who was not known to have suffered from malaria. This was emphasized by Meckel himself when he described a few years later (Meckel, 1850) the regular appearance of pigment in the spleen and the blood of malarious patients, but Virchow (1849) stated clearly that some of his autopsy cases had suffered from intermittens as he called malaria. Neither Meckel (1847, 1850) nor Virchow (1849) expressed definite views as to the chemical nature of the pigment. The former, however, made solubility tests and observed color changes of the pigment under the influence of acids and alkali. From reading his papers one gets the impression that he allied the pigment to the socalled melanotic pigments. For obvious reasons these old investigators took it for granted that the human body itself produced the pigment. This changed of course very soon after the malaria parasites had been detected and the view was generally accepted that the pigment was derived from the hemoglobin of the host erythrocytes and was formed within the parasites.

For years the view persisted that the malarial pigment was a melanin, that is, an iron-free pigment (e.g. Schridde, 1921), but doubts began to appear as evidenced by the fact that new names were coined for it. Ross (1910) called it plasmodin, Askanazy (1921) haemo-melanin, and eventually the now current name, hemozoin, was generally accepted. The reasons for finally abandoning the old view were on the one hand differences in solubilities and reactions to the bleaching action of oxidizing between the malarial pigment and genuine melanin (Brown, 1911, and others) and on the other hand chemical and spectroscopic data which seemed to indicate close resemblance or even identity of malaria pigment and hematin (Carbone, 1891, Ascoli, 1910, Brown, 1911, and others). All these investigations were done on pigment derived from the organs of malarious patients. Pigment isolated from the parasites themselves *(Plasmodium knowlesi)* was studied first in India (Sinton and Ghosh, 1934 a, b, Ghosh and Sinton, 1934, Ghosh and Nath, 1934) and again the conclusion was reached that hemozoin was identical with hematin. This view persisted for years, but beginning with the investigations of Deegan and Maegraith (1956 a, b) it became obvious that the appearance of hematin was due to the relatively drastic isolation procedures used by the earlier workers. Hemozoin is now considered to be a partly decomposed hemoglobin, that is, an iron porphyrin linked to a protein or polypeptide, but a further discussion of these newer developments is beyond the scope of the present discussion. As already hinted at, studies on the metabolism of the malarial parasites themselves began only many years after the first studies on malarial pigment had been published. The first relevant paper appeared in 1938 and is due to Christophers and Fulton; it initiated the modern era of investigations on this group of parasites.

The time lag between physiological investigations on hosts parasitized by pathogenic African trypanosomes and on the parasites themselves was shorter than in the case of malaria patients and plasmodia. This story has been told repeatedly (e.g. von Brand, 1938 b, 1951) and I do not propose to go into any detail on this occasion; a few data will suffice to refresh your memory. Schern, a veterinarian working then in Berlin and later in Montevideo, observed in 1911 that African trypanosomes after having lost their motility in vitro, could be revived by serum or liver extracts of normal, but not of parasitized animals. Basing himself on this initial observation, he developed over the next 25 years the theory that infected animals die because of damage induced by the sugar needs of the parasites (Schern, 1925, Schern and Artagaveytia-Allende, 1936). This now generally abandoned theory gave rise to numerous investigations on the so-called reviving phenomenon (for example by the Belgian Dubois, 1926) and on disturbances of the host metabolism (e.g. Bruynoghe et al., 1927, Krijgsman, 1933, Tubangui and Yutuc, 1931, and many others). However, it was only a decade or so after Schern's initial observation that real biochemical studies on the parasites slowly began to appear. The earliest studies were those of von Fenyvessy and Reiner (1924) on the cyanide resistant respiration of the African trypanosomes, the qualitative or semi-quantitative studies on the sugar consumption of the parasites by von Fenyvessy and Reiner (1928), Yorke et al. (1929) or Regendanz (1930) and my own comparative quantitative study of the carbohydrate consumption of various species (von Brand, 1933). Such studies laid the foundation for an objective appraisal of Schern's hypothesis which had been challenged first in 1927 by Regendanz and Tropp and which had given rise to a rather unpleasant literary feud between Schern and Regendanz (Schern, 1929, 1931, Regendanz, 1929, 1931). These old investigations then are the corner stones on which the vast edifice of contemporary biochemical research on trypanosomes was erected, the discussion of which lies however outside this presentation.

Physiological research on parasitic protozoa became more diversified in respect to species studied earlier than was the case with helminths. Thus sarcocystin, the toxic principle of sarcosporidia, was first described in 1891 by Pfeiffer and subsequently studied extensively, at least from a biological standpoint (Laveran and Mesnil, 1899, Rievel and Behrens, 1903, and others). Later, in the 1920 and 1930's not only the work on malarial parasites and trypanosomes mentioned a moment ago was started, but Cleveland's (1925) justly famous studies on the oxygen sensitivity and host-parasite relationships of termite flagellates began then, Trager (1932) investigated their cellulose digestion, Marguerite Lwoff (1933) studied the nutrition of the lower trypanosomids, André Lwoff (1934) their respiration, Cailleau (1935) presented data on the growth requirements of trichomonads and several investigators (Weineck, 1934, Westphal, 1934, and others) began studies on digestion and cultural requirements of rumen ciliates. Evidently not all these studies were strictly biochemical ones. However, I mention them here, because I find it difficult and not even desirable to draw a hard and fast line between biochemical and physiological research.

The last group of parasites to be considered are the endoparasitic arthropods. In so far as endoparasitic insects are concerned, the *Gastrophilus* larvae received some attention in the old days. The first relevant physiological observations showing that the larvae survive anaerobiosis rather well were made by the Dutch investigator Numan (1833, 1838) and the German Schwab (1858). The former observed also that the larvae produced carbon dioxide in the presence and the absence of oxygen and he concluded that this gas was either present preformed in the body of the larvae or that it was produced by a chemical combination of carbon and oxygen derived from some body constituent under the influence of a vital force. Later Vaney (1902) described the occurrence of hemoglobin in the conspicuous red organ of the larvae, but it was only in 1914 and 1916 that von Kemnitz published the first papers dealing with their metabolism. He unfortunately could not continue this work because he lost his life during a mountain-climbing excursion into the Alps around the time his 1916 paper appeared in print. The world famous British biochemist and parasitologist David Keilin began to study *Gastrophilus* in 1919 because he was interested in tracing the fate of the hemoglobin which is a characteristic constituent of the larvae but which does not occur even in traces in the adult flies. This study had a tremendous influence on general biochemistry because it led Keilin directly to the discovery of the cytochromes as he vividly describes in his posthumous book on the history of cell respiration and cytochrome (Keilin, 1966). Another relevant study done in the time period under consideration is that of the Rumanian Dinulescu (1932) who studied, partly from a biological and partly from a biochemical standpoint not only the usually employed *Gastrophilus intestinalis* but also the larvae of several other species.

A relatively large old literature deals with physiological host-parasite relationships of the *Gastrophilus* larvae, especially with the question of their possible toxicity to the host. It cannot be reviewed in detail here. I mention only that the original assumption (Seyderhelm and Seyderhelm, 1914) of a causal connection between the larvae and the pernicious anemia of horses proved to be incorrect. Toxic symptoms produced in horses by injection of larval extracts were subsequently recognized as anaphylactic reactions (van Es and Schalk, 1918, Cameron, 1922, and others), although it is questionable whether similar symptoms elicited in small laboratory animals can be explained on the same basis (De Kock, 1919, Roubaud and Pérard, 1924). Other endoparastic insects were hardly considered in the old days; one exception is the larva of *Cordylobia anthropophaga* for which several relevant date (digestive enzymes, tyrosinase, respiration, resistance to anaerobiosis) were reported by Blacklock et al. (1930).

Whether parasitic crustacea, especially Rizocephala, should be classified among the endoparasites may well be open to debate. I mention them here only briefly, since it is obviously impossible to review the vast literature concerning their influence on the host in any detail. Giard was the first to report in 1886 observations concerning the parasitic castration of crustacea by crustacean parasites and data on parasite-induced feminization of males or hyperfeminization of females accumulated over the next decades. This literature, however, is largely simply descriptive and not really analytical. The first biochemical data are due to Smith (1911, 1913) and Robson (1911) who thought that the blood of infected crabs was richer in lipids than that of normal specimens and who showed that the hepatopancreas of the parasitized animals had an abnormally high lipid, but low glycogen content. The interpretation proposed was that the alleged "Gargantuan appetite" of the parasites was responsible for the chemical and morphological alternations sustained by the hosts, a view that is essentially abandoned today. I cannot go deeper into the various theories developed subsequently to account for the phenomenon of sacculinization. They range from the assumption that the parasites produce female sex hormones (Biedl, 1913) to the view that castrated males correspond to neutral individuals (Lipschütz, 1924) or to intersexes (Goldschmidt, 1931), that the parasites produce toxic substances (Lévy, 1923, 1924) which might be involved in influencing the host's sexual characters (Reinhard and von Brand, 1944) and finally and this view is now in the ascendancy, that the parasites disturb the internal secretions of the host (Hartnoll, 1967).

The time allotted to this presentation is at an end. I hope that I could conveye to you the idea that contemporary research in the various aspects of parasite physiology and biochemistry has many roots that go back a fairly long time and that we all owe a measure of gratitude to many now forgotten investigators. I wish I could conveye to you the thrill I personally experienced in reading some of the old literature. I do not hesitate in stating that one of the greatest intellectual pleasures I ever had was in reading for the first time 25 years ago a very old non-parasitological paper. It was Vaucquelin's (1792) paper entitled "Observations chimiques et physiologiques sur la respiration des insectes et des vers" in which he showed experimentally that invertebrates consume and require oxygen and that CO_2 is formed during their respiration. The logic of his experimental approach and the clarity of his writing made an indelible impression on me. We can only hope that some future historian 200 years from now will say the same from some contemporary parasitological effort.

References

ADAM, W. 1932. Uber die Stoffwechselprozesse von *Ascaris suilla* Duj. I. Teil. Die Aufnahme von Sauerstoff aus der Umgebung. Z. vergl. Physiol. **16**, 229 - 251.

ADUCCO, V. 1889. La substance colorante rouge de l'*Eustrongylus gigas.* Arch. Ital. Biol. **11,** 52 - 69.

ALT, H.L. and TISCHER, O.A. 1931. Observations on the metabolism of the tapeworm, *Moniezia expansa.* Proc. Soc. Exptl. Biol. Med. **29,** 222 - 224.

ASCOLI, V. 1910. Sul pigmento malarico. Il Policlinico Sez. Med. **17,** 246 - 255.

ASKANAZY, M. 1921. Aussere Krankheitsursachen. In: Aschoff, L. (Ed.): Pathologische Anatomie. 5th ed. **1,** 58 - 307. Fischer, Jena.

BERNARD, Ch. and AXENFELD (no initial) 1857. Présence du sucre dans le liquide d'un kyste hydatique du foie. Compt. rend. Soc. Biol. Ser. 2, **3**, 90 - 91.

BERNARD, CI. 1859. De la matière glycogène chez les animaux dépourvus de foie. Compt. rend. Soc. Biol. Ser. 3, **1**, 53 - 55.

BIEDL, A. 1913. Innere Sekretion. Urban u. Schwarzenberg, Berlin.

BLACKLOCK, D.B., GORDON, R.M. and FINE, J. 1930. Metazoan immunity: A report on recent investigations. Ann. Trop. Med. Parasitol. **24**, 5 - 54.

BOHLAND, K. 1894. Uber die Eiweisszersetzung bei der Anchylostomiasis. Münch. Med. Wochschr. **41**, 901 - 904.

BOUNDOUY, T. 1910. Chimie biologique du *Sclerostomum equinum.* Thèse, Paris.

von BRAND, T. 1929. Stoffbestand und Stoffwechsel von *Moniezia expansa.* Verh. Deutsch. Zool. Ges. 1929, 64 - 66.

von BRAND, T. 1933 a. Untersuchungen über den Stoffbestand einiger Cestoden und den Stoffwechsel von *Moniezia expansa.* Z. vergl. Physiol. **18**, 562 - 596.

von BRAND, T. 1933 b. Studien über den Kohlehydratstoffwechsel parasitischer Protozoen. II. Der Zuckerstoffwechsel der Trypanosomen. Z. vergl. Physiol. **19**, 587 - 614.

von BRAND, T. 1934. Der Stoffwechsel von *Ascaris lumbricoides* bei Oxybiose und Anoxybiose. Z. vergl. Physiol. **21**, 220 - 235.

von BRAND, T. 1938 a. The nature of the metabolic activities of intestinal helminths in their natural habitat· Aerobiosis or anaerobiosis? Biodynamica **2**, No. 41, 1 - 13.

von BRAND, T. 1938 b. The metabolism of pathogenic trypanosomes and the carbohydrate metabolism of their hosts. Quart. Rev. Biol. **13**, 41 - 50.

von BRAND, T. 1951. Metabolism of Trypanosomidae and Bodonidae. In: Lwoff, A. (Ed.): Biochemistry and physiology of protozoa. **1**, 177 - 234. Academic Press, New York.

BRAULT, A., and LOEPER, M. 1904. Le glycogène dans le développement de quelques organismes inférieures (sporozoaires, coccidies, champignons, levures). J. Physiol. Pathol. Gén. **6**, 720 - 732.

BROWN, W.H. 1911. Malarial pigment (so-called melanin): Its nature and mode of production. J. Exptl. Med. **13**, 290 - 299.

BRUYNOGHE, R., DUBOIS, A., and BOUCKAERT, J.P. 1927. Le sucre du sang au cours des trypanosomiases expérimentales. Bull. Acad. Roy. Méd. Belgique. 5th Ser. **7**, 142 - 157.

BUTSCHLI, O. 1885. Bemerkungen über einem dem Glykogen verwandten Körper in den Gregarinen. Z. Biol. **21**, 603 - 612.

BUNGE, G. 1883. Uber das Sauerstoffbedürfnis der Darmparasiten. Z. physiol. Chemie **8**, 48 - 59.

BUNGE, G. 1889. Weitere Untersuchungen über die Atmung der Würmer, Z. physiol. Chemie **14**, 318 - 324.

CAILLEAU, R. 1935. La nutrition de *Trichomonas columbae.* Compt. rend. Soc. Biol. **119**, 853 - 856.

CAMERON, A.E. 1922. Bot anaphylaxis. J. Amer. Vet. Med. Assoc. **62**, 332 - 342.

CARBONE, T. 1891. Sulla natura chimica del pigmento malarico. Giorn. R. Accad. Med. Torino **54**, 901 - 906.

CERTES, A. 1889. Note sur les micro-organismes de la panse des ruminants. J. Microgr. **13**, 277 - 279.

CHRISTOPHERS, S.R., and FULTON, J.D. 1938. Observations on the respiratory metabolism of malaria parasites and trypanosomes. Ann. Trop. Med. Parasitol. **32**, 43 - 75.

CLEVELAND, L. R. 1925. The effects of oxygenation and starvation on the symbiosis between the termite *Termopsis* and its intestinal flagellates. Biol. Bull. **48**, 309 - 326.

COHN, H.L. 1896. Uber die Myxosporidien von *Esox lucius* und *Perca fluviatilis.* Zool. Jahrb. Abt. Anat. **9**, 227 - 272.

COOK, S.F., and SHARMAN, F.E. 1930. The effect of acids and bases on the respiration of tapeworms. Physiol. Zool. **3**, 145 - 163.

DAVEY, D.G. 1938. The respiration of nematodes of the alimentary tract. J. Exptl. Biol. **15**, 217 - 224.

DEEGAN, T., and MAEGRAITH, B.G. 1956 a. Studies on the nature of malarial pigment (haemozoin). I. The pigment of the simian species, *Plasmodium knowlesi* and *P. cynomolgi.* Ann. Trop. Med. Parasitol. **50**, 194 - 211.

DEEGAN, T., and MAEGRAITH, B.G. 1956 b. Studies on the nature of malarial pigment (haemozoin). II. The pigment of the human species, *Plasmodium falciparum* and *P. malariae.* Ann. Trop. Med. Parasitol. **50**, 212 - 222.

DE KOCK, G. 1919. Notes on the intoxication by *Gastrophilus* larvae. 5th and 6th Rep. Dir. Vet. Res. Dept. Agric. Union S. Africa. Onderstepoort. pp. 649 - 694.

DINULESCU, G. 1932. Recherches sur la biologie des gastrophiles. Anatomie, physiologie, cycle évolutif. Ann. Sci. Nat. Zool. Ser. 10. **15**, 1 - 183.

DONOMAE, I. 1927. Uber das Blutlipoid bei Ankylostomiasisanämien, nebst einem Anhang über Blutzucker, das Serumeiweiss und die Senkungsgeschwindigkeit der roten Blutzellen. Jap. J. Med. Sci. Part VIII. **1**, 385 - 412.

DOYERE (no initial) 1840. Vers intestinaux; acephalocystes. Extr. Procès-Verb. Séances Soc. Philomatique Paris. Year 1840: 14 (Abstract publ. 1841).

DUBOIS, A. 1926. Le phénomène de Kurt Schern dans les trypanosomiases. Compt. rend. Soc. Biol. **95**, 1130 - 1133.

EPPS, W., WEINER, H., and BUEDING, E. 1950. Production of steam volatile acids by bacteria-free *Ascaris lumbricoides.* J. Infect. Diseas. **87**, 149 - 151.

FAIRBAIRN, D. 1970. Biochemical adaptation and loss of genetic capacity in helminth parasites. Biol. Rev. **45**, 29 - 72.

FAURE-FREMIET, E. 1913. Le cycle germinatif chez l'*Ascaris megalocephala.* Arch. Anat. Micr. **15**, 1 - 136.

von FENYVESSY, B., and REINER, L. 1924. Untersuchungen über den respiratorischen Stoffwechsel der Trypanosomen. Z. Hyg. Infektionskrankh. **102**, 109 - 119.

von FENYVESSY, B., and REINER, L. 1928. Atmung und Glykolyse der Trypanosomen II. Bioch. Z. **202**, 75 - 80.

FISCHER, A. 1924. Uber den Kohlehydratstoffwechsel von *Ascaris megalocephala.* Bioch. Z. **144**, 224 - 228.

FLURY, F. 1912. Zur Chemie und Toxikologie der Ascariden. Arch. exptl. Pathol. Pharmakol. **67**, 275 - 392.

FLURY, F. 1913. Beiträge zur Chemie und Toxikologie der Trichinen. Arch. exptl. Pathol. Pharmakol. **73**, 164 - 213.

FLURY, F. and GROLL, H. 1913. Stoffwechseluntersuchungen an trichinösen Tieren. Arch. exptl. Pathol. Pharmakol. **73**, 214 - 232.

FLURY, F., and LEEB, F. 1926. Zur Chemie und Toxikologie der Distomen (Leberegel). Klin. Wochsch. **5**, 2054 - 2055.

FOSTER, M. 1865. On the existence of glycogen in the tissues of certain entozoa. Proc. Roy. Soc. London **14**, 543 - 546.

FOSTER, W.D. 1965. A history of parasitology. Livingstone. Edinburgh.

FUCHS, B. 1922. Uber eine Trichinenepidemie in Erlangen. Münch. Med. Wochsch. **69**, 1336 - 1338.

GHOSH, B.N. and NATH, M.C. 1934. The chemical composition of malaria pigment (haemozoin). Rec. Malaria Survey India **4**, 321 - 325.

GHOSH, B.N. and SINTON, J.A. 1934. Studies of malarial pigment (haemozoin). Part II. The reactions of haemozoin to tests for iron. Rec. Malaria Survey India **4**, 43 - 59.

GIARD, A. 1886. De l'influence de certains parasites Rhizocéphales sur les caractères sexuels extérieures de leur hôte. Compt. rend. Acad. Sci. **103**, 84 - 86.

GOLDSCHMIDT, R. 1931. Die sexuellen Zwischenstufen. Springer, Berlin.

GRAFTER, F. 1952. Gustav von Bunge, Naturforscher und Menschenfreund, Schweiz. Ver. abstin. Lehrer und Lehrerinnen.

GRUBE, G. 1850. Die Familien der Anneliden. Arch. Naturg. **16**, 249 - 364.

GULLIVER, G. 1840. Notes on the ova of the *Distoma hepaticum*, and on certain corpuscles obtained from the genera *Cysticercus.* Proc. Zool. Soc. Soc. London Pt. **8**, 30 - 31.

GULLIVER, G. 1841. Observations on the structure of the entozoa belonging to the genus *Cysticercus.* Medico-Chirurg. Transact. **24**, 1 - 11.

HARNISCH, O. 1932. Untersuchungen über den Gaswechsel von *Fasciola hepatica.* Z. vergl. Physiol. **17**, 365 - 386.

HARNISCH, O. 1933. Untersuchungen zur Kennzeichnung des Sauerstoffverbrauchs von *Triaenophorus nodulosus* (Cest.) und *Ascaris lumbricoides* (Nematod.). Z. vergl. Physiol. **19**, 310 - 348.

HARTNOLL, R.G. 1967. The effects of sacculinid parasites on two Jamaican crabs. J. Linn. Soc. (Zool.) **46**, 275 - 295.

HEINTZ, W. 1850. Untersuchung des flüssigen Inhalts der Echinococcenbälge (Hydatidenbälge) einer Frau. Jen. Ann. Physiol. Med. **1**, 180 - 191.

HOEPPLI, R. 1959. Parasites and parasitic infections in early medicine and science. Univ. Malaya Press, Singapore.

HOEPPLI, R. 1969. Parasitic diseases in Africa and the Western hemisphere. Early documentation and transmission by the slave trade. Acta Tropica. Suppl. 10. Verlag f. Recht u. Gesellschaft, Basel.

KEILIN, D. 1925. On cytochrome, a respiratory pigment common to animals, yeasts and higher plants. Proc. Roy. Soc. London Ser. B. **98,** 312 - 339.

KEILIN, D. 1966. The history of cell respiration and cytochrome. Cambridge Univ. Press, Cambridge.

von KEMNITZ, G.A. 1912. Die Morphologie des Stoffwechsels bei *Ascaris lumbricoides.* Ein Beitrag zur physiologisch-chemischen Morphologie der Zelle. Arch. Zellforsch. **7,** 463 - 603.

von KEMNITZ, G.A. 1914. Untersuchungen über Stoffbestand und Stoffwechsel der Larven von *Gastrophilus equi.* Verh. Deutsche Zool. Ges. Year 1914: 294 - 307.

von KEMNITZ, G.A. 1916. Untersuchungen über den Stoffbestand und Stoffwechsel der Larven von *Gastrophilus equi* (Clark), nebst Bemerkungen über den Stoffbestand der Larven von *Chironomus* (spec.?) L. (Physiologischer Teil). Z. Biol. **67,** 129 - 244.

KESTNER, O., and PLAUT, R. 1924. Physiologie des Stoffwechsels. In: Winterstein, H. (Herausg.): Handbuch der vergleichenden Physiologie **2,** Zweite Hälfte: 901 - 1112.

KRIJGSMAN, B.J. 1933. Biologische Untersuchungen über das System Wirtstier-Parasit. III-V. Z. Parasitenk. **6,** 1 - 22 and 438 - 477.

KRUGER, F. 1936. Untersuchungen zur Kenntnis des aeroben und anaeroben Stoffwechsels des Schweinespulwurmes *(Ascaris suilla).* Zool. Jahrb. Abt. Allg. Zool. **57,** 1 - 56.

KRUGER, F. 1937. Bestimmungen über den aeroben und anaeroben Stoffumsatz beim Schweinespulwurm mit einem neuen Respirationsapparat. Z. vergl. Physiol. **24,** 687 - 719.

KRUMMACHER, O. 1918. Untersuchungen über die Wärmeentwicklung der Spulwürmer. Beiträge zur Erforschung des Lebens ohne Sauerstoff. Z. Biol. **69,** 293 - 321.

KUCHENMEISTER (no initial) 1851. Kleinere helminthologische Mitteilungen. Arch. Physiol. Heilkde. **10,** 333 - 337.

KUENEN, W.A., and SWELLENGREBEL, N.H. 1913. Die Entamöben des Menschen und ihre praktische Bedeutung. Centralbl. Bakteriol. I. Abt. Orig. **71,** 378 - 410.

LANCISI, J.M. 1717. De noxiis paludum effluvis eorumque remediis. Rome (not seen).

LASSAIGNE, M. 1843. Sur le tissue tégumentaire des insectes de différents ordres. Compt. rend. Acad. Sci. **16,** 1087 - 1089.

LAVERAN, C.L.A. and MESNIL, F. 1899. De la sarcocystine, toxine des sarcosporidies. Compt. rend. Soc. Biol. **51,** 311 - 314.

LEUCKART, R. 1863. Die menschlichen Parasiten und die von ihnen herrührenden Krankheiten. Vol. 1. C.F. Winter, Leipzig.

LEVY, R. 1923. Sur la toxicité des tissues de la Sacculine *(Sacculina carcini)* vis-à-vis du crabe *(Carcinus maenas)* et sur les recherches de réactions d'immunité chez ces derniers. Bull. Soc. Zool. France **48**, 291 - 294.

LEVY, R. 1924. Sur la constatation de différences d'ordre physicochimique entre le serum des crabes sacculinés et celui des crabes normaux. Bull. Soc. Zool. France **49**, 333 - 336.

LIPSCHUTZ, A. 1924. The internal secretions of the sex glands. Williams a. Wilkins, Baltimore.

LLOYD, W.E.B. 1968. A hundred years of medicine. 2nd. ed. Duckworth, London.

LUCKE, A. 1860. Die Hüllen der Echinococcen und die Echinococcen-Flüssigkeit. Arch. Pathol. Anat. Physiol. **19**, 189 - 196.

LWOFF, A. 1934. Die Bedeutung des Blutfarbstoffes für die parasitischen Flagellaten. Zentralbl. Bakteriol. I. Abt. Orig. **130**, 498 - 518.

LWOFF, M. 1933. Recherches sur la nutrition des trypanosomides. Ann. Inst. Pasteur **51**, 55 - 116.

MAJOR, R.H. 1954. A history of medicine. Thomas, Springfield.

MANSON-BAHR, P.H. and ALCOCK, A. 1927. The life and work of Sir Patrick Manson. Wood a. Co., New York.

MARKOWICZ, W. and BOCK, D. 1931. Uber Kreatin- und Kreatininausscheidung bei der Trichinose. Z. ges. exptl. Med. **79**, 301 - 310.

MARTINI, E. 1957. Bernhard Nocht. Ein Lebensbild. Dingwort u. Sohn, Hamburg-Altona.

MAUPAS, E.F. 1886. Sur les granules amylacés du cytosome des grégarines. Compt. rend. Acad. Sci. **102**, 120 - 123.

McCOY, O.R. 1930. The influence of temperature, hydrogenion concentration, and oxygen tension on the development of the eggs and larvae of the dog hookworm, *Ancylostoma caninum*. Amer. J. Hyg. **11**, 413 - 448.

MECKEL, H. 1847. Uber schwarzes Pigment in der Milz und dem Blute einer Geisteskranken. Allg. Z. Psych. psych-gerichtl. Med. **4**, 198 - 226.

MECKEL, H. 1850. Körniger Farbstoff in der Milz und im Blute bei Wechselfieberkranken. Deutsche Klinik Year 1850, pp. 551 - 552.

MOSS, G.D. 1970. The excretory metabolism of the endoparasitic digenean *Fasciola hepatica* and its relationship to its respiratory metabolism. Parasitology **60**, 1 - 19.

NUMAN, A. 1933. Waarnemingen omtrent de horzelmaskers, welke in de maag van het paard huisvesten. N. Verhand. I. Kl. K. Nederl. Inst. **4**, 139 - 281 (not seen).

NUMAN, A. 1838. Uber die Bremsenlarven, welche sich im Magen des Pferdes aufhalten. Magazin ges. Thierheilkde. **4**, 1 - 140. (Tansl. of Numan 1833, with some additional notes by Hertwig).

OESTERLIN, M. 1937. Die von oxybiotisch gehaltenen Ascariden ausgeschiedenen Fettsäuren. Z. vergl. Physiol. **25**, 88 - 91.

ORTNER-SCHONBACH, P. 1913. Zur Morphologie des Glykogens bei Trematoden und Cestoden. Arch. Zellforsch. **11**, 413 - 449.

PACKARD, F.R. 1963. History of medicine in the United States. Hafner, New York.

PADOA, G. 1909. Il ricambio materiale nell' anchilostomenia. Il Ramazzini, Firenze **3**, 485 - 498.

PANZER, T. 1911. Beitrag zur Biochemie der Protozoen. Z. physiol. Chemie **73**, 109 - 127.

PANZER, T. 1913. Beitrag zur Biochemie der Protozoen. II. Mitteilung. Z. physiol. Chemie **86**, 33 - 42.

PFEIFFER, L. 1891. Die Protozoen als Krankheitserreger, sowie der Zellen und Zellenkernparasitismus derselben bei nicht-bakteriellen Infektionskrankheiten des Menschen. 2nd. ed. Fischer, Jena.

PINTNER, T. 1922. Die vermutliche Bedeutung der Helminthenwanderungen. Sitzungsber. Akad. Wiss. Wien. Math.-Naturw. Kl. Abt. I. **131**, 129 - 138.

PLANER, J. 1854. Uber das Vorkommen von Pigment im Blute. Z. Ges. Aerzte Wien, 10 Jahrg. **1**, 127 - 139 and 280 - 298.

RAKE, B. 1894. A note on the percentage of iron in the liver in ankylostomiasis. J. Pathol. Bacteriol. **3**, 107 - 109.

REGENDANZ, P. 1929. Erwiderung auf den Artikel von Schern: Zur Trypanosomen-arbeit von Regendanz und Tropp, Bd. III, S. 139 dieser Zeitschrift. Centralbl. Bakteriol. I. Abt. Orig. **112**, 319 - 320.

REGENDANZ, P. 1930. Der Zuckerverbrauch der Trypanosomen (nach Versuchen in vitro bei 37°C) und seine Bedeutung für die Pathologie der Trypanosomeninfek-tionen. Zentralbl. Bakteriol. I. Abt. Orig. **118**, 175 - 186.

REGENDANZ, P. 1931. Zur glykopriven Intoxikation bei den Trypanosomiasis usw. Erwiderung auf die vorstehenden Ausführungen des Prof. Schern. Zentralbl. Bakteriol. I. Abt. Orig. **119**, 303.

REGENDANZ, P. and TROPP, C. 1927. Das Verhalten des Blutzuckers und des Leberglykogens bei mit Trypanosomen infizierten Ratten. Arch. Schiffs- Tropenk. Hyg. **31**, 376 - 385.

REINHARD, E.G. and von BRAND, T. 1944. The fat content of *Pagurus* parasitized by *Peltogaster* and its relation to theories of sacculinization. Physiol. Zool. **17**, 31 - 41.

RIEVEL, H.W.L. and BEHRENS (no initial) 1903. Beiträge zur Kenntnis der Sarcosporidien und deren Enzyme. Centralbl. Bakteriol. I. Abt. Orig. **35**, 341 - 352.

ROBSON, G.C. 1911. The effect of *Sacculina* upon the fat metabolism of its host. Quart. J. Microscp. Sci. **57**, 267 - 278.

ROSS, R. 1910. The prevention of malaria. Dutton, New York.

ROSS, R. 1923. Memoirs. With a full account of the great malaria problem and its solution. Murray, London.

ROUBAUD, E., and PERARD, C. 1924. Etudes sur l'hypoderme ou varron des boeufs; les extraits d'oestres et l'immunisation. Bull. Soc. Pathol. Exot. **17**, 259 - 272.

SAZ, H.J. and LESCURE, O.L. 1969. The functions of phosphoenolpyruvate carboxykinase and malic enzyme in the anaerobic formation of succinate by *Ascaris lumbricoides*. Comp. Biochem. Physiol. **30**, 49 - 60.

SCHERN, K. 1911. Uber die Wirkung von Serum und Leberextrakten auf Trypanosomen. Arb. K. Gesundheitsamt Berlin **38**, 338 - 367.

SCHERN, K. 1925. Uber Trypanosomen. I.-VI. Mitteilung. Centralbl. Bakteriol. I. Abt. Orig. **96**, 356 - 365 and 440 - 454.

SCHERN, K. 1929. Zur Trypanosomenarbeit von Regendanz und Tropp. Centralbl. Bakteriol. I. Abt. Orig. **111**, 139 - 143.

SCHERN, K. 1931. Zur glykopriven Intoxikation bei der Trypanosomiasis etc. Erwiderung an Regendanz. Zentralbl. Bakteriol. I. Abt. Orig. **119**, 297 - 302.

SCHERN, K. and ARTAGAVEYTIA-ALLENDE, R. 1936. Die glykoprive Therapie der experimentellen Trypanosomeninfektion mit Anticoman. Z. Immunitätsforsch. exptl. Therapie **89**, 484 - 487.

SCHIMMELPFENNIG, G. 1903. Uber *Ascaris megalocephala*. Beiträge zur Biologie und physiologischen Chemie derselben. Arch. wiss. prakt. Tierheilkde. **29**, 332 - 376.

SCHOPFER, W.H. 1932. Recherches physico-chimiques sur le milieu intérieur de quelques parasites. Rev. Suisse Zool. **39**, 59 - 194.

SCHRIDDE, H. 1921. Blutbildende Organe. In: Aschoff, L. (Ed.): Pathologische Anatomie. 5th ed. Vol. **2**, 108 - 163.

SCHULTE, H. 1917. Versuche über Stoffwechselvorgänge bei *Ascaris lumbricoides*. Pflüger's Archiv **166**, 1 - 44.

SCHWAB, K.L. 1858. Die Ostraciden, Bremsen der Pferde, Rinder und Schafe. Als Manuskript für Freunde der Naturgeschichte gedruckt. München. (Quoted in von Kemnitz, 1916).

SEYDERHELM, K.R. and SEYDERHELM, R. 1914. Die Ursache der perniziösen Anämie der Pferde. Ein Beitrag zum Problem des ultravisiblen Virus. Arch. exptl. Pathol. Pharmakol. **76**, 149 - 201.

SINGER, C. and UNDERWOOD, E.A. 1962. A short history of medicine. Oxford Univ. Press, Oxford.

SINTON, J.A. and GHOSH, B.N. 1934 a. Studies of malarial pigment (haemozoin). Part I. Investigation of the action of solvents on haemozoin and the spectroscopical appearances observed in the solutions. Rec. Malaria Survey India **4**, 15 - 42.

SINTON, J.A. and GHOSH, B.N. 1934 b. Studies of malarial pigment (haemozoin). Part III. Further researches into the action of solvents, and the results of observations on the action of oxidizing and reducing agents, on optical properties, and on crystallisation. Rec. Malaria Survey India **4**, 205 - 221.

SLATER, W.K. 1925. The nature of the metabolic processes in *Ascaris lumbricoides*. Biochem. J. **19**, 604 - 610.

SLATER, W.K. 1928. Anaerobic life in animals. Biol. Rev. **3**, 303 - 328.

SMITH, G. 1911. Studies in the experimental analysis of sex. Part 7. Sexual changes in the blood and liver of *Carcinus maenas*. Quart. J. Microscop. Sci. **57**, 251 - 265.

SMITH, G. 1913. Studies in the experimental analysis of sex. Part 10. The effect of

Sacculina upon the storage of fat and glycogen, and on the formation of pigment by its host. Quart. J. Microscop. Sci. **59**, 267 - 295.

STEPANOW-GRIGORIEW, J. and HOEPPLI, R. 1926. Uber Beziehungen zwischen Glykogengehalt parasitischer Nematodenlarven und ihrer Wanderung im Wirtskörper. Arch. Schiffs. Trop. Hyg. **30**, 577 - 585.

STOLL, M. 1797. Ratio Medendi **1**, 196 (not seen).

THELOHAN, P. 1894. Nouvelles recherches sur les coccidies. Arch. Zool. exptl. gén. Ser. 3, **2**, 541 - 573.

TRAGER, W. 1932. A cellulase from the symbiotic intestinal flagellates of termites and of the roach *Cryptocercus punctulatus.* Biochem. J. **26**, 1762 - 1771.

TUBANGUI, M.A. and YUTUC, L.M. 1931. The resistance and the blood sugar of animals infected with *Trypanosoma evansi.* Philippine J. Sci. **45**, 93 - 107.

VAN ES, L. and SCHALK, A.F. 1918. Sur la nature anaphylactique de l'intoxication parasitaire. Ann. Inst. Pasteur **32**, 310 - 362.

VANEY, C. 1902. Contributions à l'étude des larves et des metamorphoses des diptères. Ann. Univ. Lyon N.S.I. Sci. Med. Fasc. 9, 171 pp. (see pp 123 - 124).

VAUQUELIN, M. 1792. Observations chimiques et physiologiques sur la respiration des insectes et des vers. Ann. de Chimie **12**, 273 - 291.

VIRCHOW, R. 1849. Zur pathologischen Physiologie des Blutes. 4. Farblose, pigmentierte und geschwärzte nicht spezifische Zellen im Blut. Arch. pathol. Anat. Physiol. Klin. Med. **2**, 587 - 598.

WEINECK, E. 1934. Die Celluloseverdauung bei den Ciliaten des Wiederkäuermagens. Arch. Protistenkde. **82**, 169 - 202.

WEINLAND, E. 1901 a. Uber den Glykogengehalt einiger parasitischer Würmer. Z. Biol. **41**, 69 - 74.

WEINLAND, E. 1901 b. Uber Kohlenhydratzersetzung ohne Sauerstoffaufnahme bei *Ascaris,* einen tierischen Gärungsprozess. Z. Biol. **42**, 55 - 90.

WEINLAND, E. 1902 a. Uber ausgepresste Extrakte von *Ascaris lumbricoides* und ihre Wirkung. Z. Biol. **43**, 86 - 111.

WEINLAND, E. 1902 b. Uber Antifermente. Z. Biol. **44**, 1 - 15.

WEINLAND, E. 1903. Uber die von *Ascaris lumbricoides* ausgeschiedenen Fettsäuren. Z. Biol. **45**, 113 - 116.

WEINLAND, E. 1904. Uber die Zersetzung stickstoffhaltiger Substanz bei *Ascaris.* Z. Biol. **45**, 517 - 531.

WEINLAND, E. 1910. Der Stoffwechsel der Wirbellosen. In: Oppenheimer, C. (Ed.): Handbuch der Biochemie des Menschen und der Tiere **4**, Part 2: 446 - 528.

WEINLAND, E. 1913. Uber einige Aufgaben und Fragen der vergleichenden Physiologie. Sitzungsber. Physik.-Med. Sozietät Erlangen **45**, 137 - 153.

WEINLAND, E. 1925. Uber den Gehalt an einigen Stoffen beim Igel im Winterschlaf. Biochem. Z. **160**, 66 - 74.

WEINLAND, E. and von BRAND, T. 1926. Beobachtungen an *Fasciola hepatica*

(Stoffwechsel und Lebensweise). Z. vergl. Physiol. **4,** 212 - 285.

WEINLAND, E. and RITTER, A. 1902. Uber die Bildung von Glykogen aus Kohlehydraten bei *Ascaris.* Z. Biol. **43,** 490 - 502.

WESTPHAL, A. 1934. Studien über Ophryoscoleciden in der Kultur. Z. Parasitenkde. **7,** 71 - 117.

YORKE, W., ADAMS, A.R.D. and MURGATROYD, F. 1929. Studies in chemotherapy I. A method for maintaining pathogenic trypanosomes alive in vitro at 37°C for 24 hours. Ann. Trop. Med. Parasitol. **23,** 501 - 518.

2

BIOCHEMICAL EFFECTS OF ANTISCHISTOSOMAL DRUGS

Ernest Bueding*

Department of Pathobiology
School of Hygiene and Public Health
and
Department of Pharmacology and Experimental Therapeutics
School of Medicine
The Johns Hopkins University
Baltimore, Maryland

More than a century ago Claude Bernard recognized the value of pharmacologically active substances in the study of physiological and biochemical mechanisms. He considered such compounds "as instruments far more selective than our mechanical means, well suited for dissecting one by one the properties of the elements of the living organisms. In this manner it is possible to dissociate and analyze attentively the mechanism of death and thus indirectly to learn about the physiological mechanisms of life" . Using this approach in his classical experiments wih curare, Claude Bernard developed the concept of the neuromuscular junction; shortly thereafter, he uncovered the role of hemoglobin in the transport of oxygen by studying the mode of the toxic action of carbon monoxide. Numerous other examples can be cited for the subsequent contributions of drugs to the elucidation of physiological and biochemical mechanisms; mention may be made of the use of cholinergic and adrenergic agents in investigations of the autonomic nervous system or of the analysis of the mode of action of sulfonamides, various antibiotics and other chemotherapeutic agents which advanced our knowledge about bacterial biosynthetic reactions and about their similarities with, and differences from, biochemical mechanisms in the mammalian host.

In the following an attempt will be made to illustrate that antischistosomal drugs have provided information about the physiology and biochemistry of schistosomes whose functional integrity is interfered with by these compounds. Like the adult stages of other helminths, those of

* The investigations of the author were supported by grants from the National Institutes of Health (U.S. Public Health Service) and The Rockefeller Foundation.

schistosomes are rather refractory to inhibitors of growth and multiplication, but they have proved vulnerable to interference with reactions providing for the generation of metabolic energy or with mechanisms concerned with muscular activity and coordination.

Trivalent organic antimonials, the oldest group of compounds used as chemotherapeutic agents in schistosomiasis, have a selective action on one enzyme of the parasite. The activity of schistosome phosphofructokinase (PFK) is inhibited by low concentrations of antimonials. This inhibition is competitive with, and reversed by, one of the substrates, fructose-6-phosphate. The isofunctional enzyme of the host is 70 to 80 times less susceptible to inhibition by these compounds[2-4]. These effects have demonstrated the principle that enzymes catalyzing the same chemical reaction in the parasite and the host are not necessarily identical with each other and that their activities can be interfered with in a selective manner. Further analysis of the relationship between the inhibition of PFK activity and the chemotherapeutic action of antimonials in schistosomiasis has revealed that the rate of glycolysis is limited by the rate of the PFK reaction[3] and that glycolysis is the major, if not exclusive, source of metabolic energy for *S. mansoni.* In contrast to many other helminths, schistosomes are homolactic fermenters and this, in turn, is explained by the high rate of pyruvic kinase activity in relation to the PEP carboxykinase activity[5]. The dependence of *S. mansoni* on glycolysis is not affected by the presence or absence of oxidative metabolism, because the rate of glucose utilization and lactic acid formation is the same under aerobic and anaerobic conditions; in other words, *S. mansoni* does not exhibit a Pasteur effect[6]. Therefore, glycolysis is the major source of metabolic energy for *S. mansoni.* Since, in this parasite, the rate of glycolysis is limited by the rate of the PFK reaction, inhibition of the activity of this enzyme by antimonials could reduce the glycolytic rate to such an extent that this becomes incompatible with the functional integrity of the worm.

Inhibition of PFK activity by trivalent antimonials has been demonstrated not only **in vitro,** but also **in vivo.** After the administration of subcurative doses of potassium antimony tartrate or of stibophen to mice infected with *S. mansoni*, the substrate of the PFK reaction, fructose-6-phosphate, accumulates in the parasite while the concentration of the product, fructose-1,6-diphosphate, is decreased, indicating an inhibition of PFK activity[3]. These changes in the levels of hexosephosphate esters coincide with the "hepatic shift" and are reversed when the worms have recovered from the effects of these drugs and have relocated themselves in the mesenteric veins[4]. The close association of reversible chemothera-

26

peutic effects with reversible biochemical changes provides further eviden-ce for the causal relationship between inhibition of schistosome PFK activity and the chemotherapeutic action of trivalent organic antimonials; this, in turn, is based on the differential susceptibilities of an iso-functional enzyme of the host and the parasite to a given chemical agent. Therefore, it should be possible to use such quantitative differences in chemotherapeutic studies. A more recently reported example in support of this principle is the selective inhibition of dihydrofolate reductase activity of *P. berghei* by pyrimethamine. In this instance the iso-functional enzyme of the host is 2000 times less sensitive to inhibition by this antimalarial agent[7].

The selective action of trivalent organic antimonials on schistosome PFK differs from that of other known antischistosomal drugs, none of which have demonstrable effects on this enzyme of the parasite. In the case of the salts of tris (p-aminophenyl) carbonium salts (TAC, p-rosaniline)[8], the hepatic shift of the worms observed following the administration of this compound is preceded by a loss of the coordinated movements, and subsequently by a paralysis, of two muscular organs of schistosomes, the acetabulum and the oral sucker. These changes are reversed almost immediately by exposure **in vitro** of the worms to certain cholinergic blocking agents, e.g., atropine or mecamylamine. These effects suggests that p-rosaniline might produce in these two organs an accumulation of endogenous acetylcholine, which in turn could be caused by an inhibition of acetylcholinesterase activity. This indeed has been observed[9], and in this manner studies of the physiological and biochemical effects of p-rosaniline have contributed to a recognition of the role of acetylcholine in *S. mansoni,* a topic which will be reviewed in more detail in a subsequent paper (chapter 6).

Up to the present it has not been possible to relate biochemical changes observed in schistosomes following the administration of the thioxantho-ne derivatives lucanthone and hycanthone to the host, with the mode of action of these drugs. Furthermore, the mechanism remains to be elucidated which brings about the development of hycanthone resistance in the progeny of worms which had been exposed to a single dose of this compound[10]. This type of drug resistance has remained stable for 4 subsequent generations; it is maternally inherited and may be related to the property of hycanthone as a frameshift mutagen in bacteria[11].

Unlike antimonials, and like p-rosaniline, the antischistosomal drug niridazole has relatively slow onset of action; a hepatic shift of the worms is observed only 2 to 3 days after a chemotherapeutically effective dosage schedule of niridazole has been initiated. If biochemical changes in the worms brought about by niridazole are to be related to the mode of action of this drug, they should precede the functional damage reflected in the loss of attachment of the worms to the mesenteric veins of the host. Prior to, or even after the hepatic shift produced by niridazole, no inhibition of PFK or AChE activities is detectable in the worms. Therefore, the biochemical actions of this drug differ from those of trivalent antimonials and of p-rosaniline. On the other hand, the niridazole-induced hepatic shift is preceded by a dose-dependent glycogen depletion of the worms[12]. However, this is not associated with an inhibition of the activities of the two forms of UDPG-glycogen synthetase of the worms. Therefore, the possibility of an increased rate of glycogenolysis has been explored. Following the administration of niridazole to the host, the inactivation of glycogen phosphorylase activity, catalyzed by phosphorylase phosphatase, is reduced. Again, this effect is dose-dependent and it coincides with, or precedes, the glycogen depletion of the worms[12]. This, in turn, can be accounted for by an increased phosphorylase activity, due to a reduced rate of inactivation of this enzyme. These biochemical effects, which are reversible following the administration of subcurative doses of niridazole, raise the question whether loss of glycogen is related to the mode of action of this drug. Besides being a source of metabolic energy, glycogen may serve another function in the parasite. Since this polysaccharide interacts with proteins and possibly other tissue constituents, the latter may be protected from the action of enzymes present in the cytoplasm when bound to glycogen. The disappearance of glycogen produced by niridazole would result in the release and subsequent degradation by cytoplasmic enzymes of one or several constituents essential for the functional integrity of the parasite.

Niridazole belongs to a large number of nitroheterocyclic compounds which have antibacterial and antiprotozoal activities. However, extremely few of them are endowed with antischistosomal activity. Therefore, the structural features conferring antischistosomal activity to a nitroheterocyclic compound must be far more selective and specific than the structural requirements for antibacterial and antiprotozoal properties. It is of interest that, while the three-dimensional configurations of most antibacterial and antiprotozoal nitroheterocyclic compounds differ from that of niridazole, the latter has conformational similarities with a nitrovinylfuran, trans-5-amino-3-(2-[5-nitro-2 furyl]-vinyl)-1,2,4-oxadiazole (SQ 18,506).

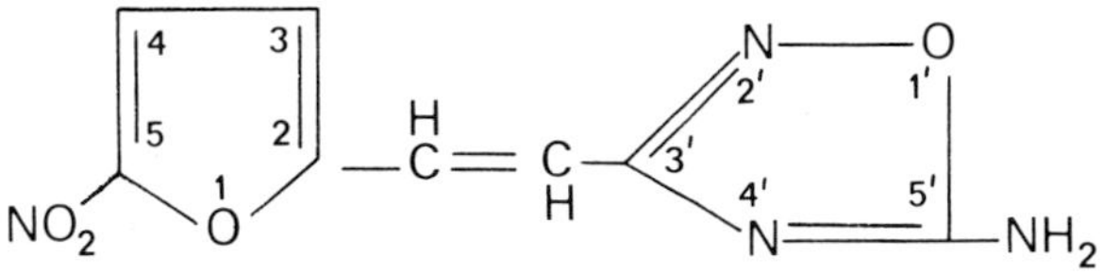

SQ 18,506

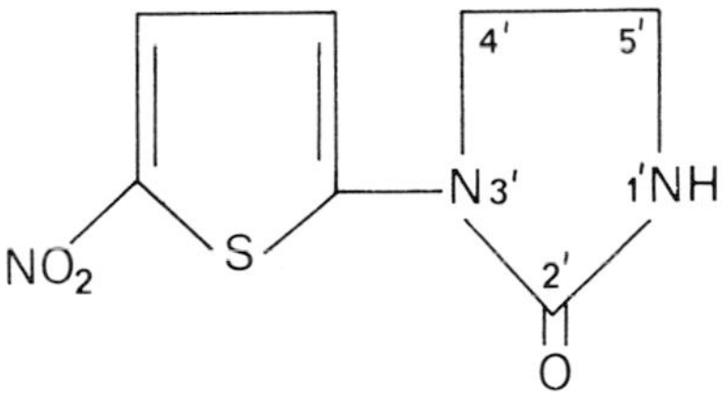

Niridazole

If, in space-filling models, the nitro groups of these two compounds are superimposed, the 3' nitrogen of niridazole and the vinyl bridge of the furan derivative are superimposed also. In addition, the 1'-nitrogen of niridazole and the 4'-nitrogen of the furan derivative are coincident in space. In an attempt to determine whether these conformational characteristics have a bearing on antischistosomal activity, it has been found that administration of this compound to mice infected with *S. mansoni* produces the same effects on the parasite as niridazole. In addition, the time course is identical also. At first, there is a reduction in the activity of glycogen phosphorylase phosphatase and a reduction in the glycogen stores. This is followed by a hepatic shift and, depending on the dosage schedule, either by a recovery of the worms or their elimination [13].

The availability of an isomer of this nitrofuran derivative in which the positions of the 4'-nitrogen and of the oxygen in the oxadiazole ring are reversed has provided an opportunity to determine the specificity of the structural requirements for antischistosomal activity. The isomer virtually has no activity; also, its administration does not result in a marked glycogen depletion or reduction in phosphorylase phosphatase activity, indicating again a close association between the biochemical effects and the chemotherapeutic action of this group of compounds [13]. In view of this highly selective relationship, an attempt is being made with C. H. Robinson, P. Hulbert, and D. Henry to delineate the structural characteristics conferring this type of antischistosomal activity. While these studies are still in progress, the results obtained so far can be summarized as follows.

1. Replacement of the nitrofuran or the nitrothiazole by a nitrophenyl or nitropyridine ring results in a loss of activity.

2. Nitrothiophenes are less active than the corresponding nitrofuran analogs.

3. If the nitro group is replaced by less electronegative methyl-, carboxy-, or carboxymethyl groups, activity is lost.

4. The vinyl side chain is essential for activity, since analogs with a saturated ($-CH_2-CH_2-$), acetylenic ($-C{\equiv}C-$), hydrazo ($-NH-NH-$), azo ($-N{=}N-$), or amide ($-\underset{\underset{O}{\|}}{C}-NH-$) bridge are inactive.

5. Nitrofurans with heterocyclic rings attached to the vinyl bridge and containing a nitrogen in a position similar to that in SQ 18,506 (position 4′) have antischistosomal activity; again, this is preceded by a reduction in phosphorylase phosphatase activity and of the glycogen levels of the worms. Analogs lacking a nitrogen in this position of the ring are inactive.

6. While the spatial relationship of the 4′-nitrogen to the nitro group of SQ 18,506 is critical inconferring antischistosomal activity, this nitrogen does not need to be part of a substituent ring. For example,

N-isopropyl-5-nitro-2-furyl-acrylamide

N-isopropyl-5- nitro-2- furylacrylamide has antischistosomal activity [14]. Compared with SQ 18,506, this compound is considerably more toxic, but its biochemical effects on the worms are similar because its chemotherapeutic action is preceded by a reduction of phosphorylase phosphatase activity and a glycogen depletion of the worms.

The effect of niridazole in reducing glycogen phosphorylase phosphatase activity is not entirely selective for *S. mansoni* because administration of subcurative and curative doses of this drug brings about similar changes in the skeletal muscle of the host[12]. This effect can account for the glycogen depletion observed in the muscle of rhesus monkeys to which niridazole has been administered[15]. By contrast, the antischistosomal nitrovinylfuran SQ 18,506 administered at dosage schedules exceeding several fold those producing parasitological cures does not produce any change in the rate of glycogen phosphorylase inactivation of skeletal muscle of mice[16]. Therefore, structural alternations of antischistosomal nitroheterocyclic compounds can increase the selectivity of inhibitory effects on the parasite, provided certain structural and conformational characteristics can be maintained.

It is concluded that studies of the effects of compounds active against schistosomes not only can contribute to a better understanding of the biochemistry and physiology of the parasite, but also can suggest approaches for the rational design of antischistosomal agents.

References

1. BERNARD, C. La science expérimentale. Ballière, Paris, (1878).

2. MANSOUR, T.E. and BUEDING, E., Brit. J. Pharmacol., **9**, 459, (1954).

3. BUEDING, E. and MANSOUR, J.M., Brit. J. Pharmacol., **12**, 159, (1957).

4. BUEDING, E. and FISHER, J., Biochem. Pharmacol., **15**, 1197, (1966).

5. BUEDING, E. and SAZ, H.J., Comp. Biochem. Physiol., **24**, 511, (1968).

6. BUEDING, E., J. Gen. Physiol., **33**, 475, (1950).

7. FERONE, J., BURCHALL, J. and HITCHINGS, G.H., Molec. Pharmacol., **5**, 49, (1969).

8. THOMPSON, P.E., MEISENHELDER, J.E. and NAJARIAN, H., Am. J. Trop. Med. and Hyg., **11**, 31, (1962).

9. BUEDING, E., SCHILLER, E.L. and BOURGEOIS, J.G., Am. J. Trop. Med. and Hyg., **16**, 500, **(1967).**

10. ROGERS, S.H. and BUEDING, E., Science, **172**, 1057, (1971).

11. HARTMAN, P.E., LEVINE, K., HARTMAN, Z. and BERGER, H., Science, **172**, 1058, (1971).

12. BUEDING, E. and FISHER, J., Molec. Pharmacol., **6**, 532, (1970).

13. ROBINSON, C.H., BUEDING, E. and FISHER, J., Molec. Pharmacol., **6**, 604, (1970).

14. LEI, H., CHING, M., HSU, H., CHANG, K., CHENG, M.-C., LU, Y.-L., CHANG, M., YEN, T., T'ANG, P.S. and TING, S., Chin. Med. J., **82**, 90, (1963).

15. BUEDING, E., ERICKSON, D.G., SCHEIBEL, L.W., FISHER, J. and KEY, J.C., Am. J. Trop. Med. and Hyg., **19,** 459, (1970).

16. BUEDING, E., NAQUIRA, C., BOUWMAN, S. and ROSE, G., J. Pharmacol. Exp. Ther., In Press, (1971).

3

COMPARATIVE BIOCHEMISTRY OF CARBOHYDRATES IN NEMATODES AND CESTODES[1]

Howard J. Saz
Department of Biology
University of Notre Dame
Notre Dame, Indiana 46556
U.S.A.

All parasitic helminths examined to date are different biochemically from the tissues of their mammalian hosts. These differences are most readily demonstrable in their carbohydrate or energy metabolisms. This becomes more apparent when it is realized that in spite of the fact that all helminths examined are capable of assimilating oxygen under appropriate conditions, none of them are capable of the complete oxidation of substrates to carbon dioxide and water. All of them examined accumulate organic end products, indicating a limited or incomplete terminal respiratory pathway. Although this concept is by no means new, parasitologists and biochemists alike only recently have come to generally accept it. In so doing, we can much better appreciate other concepts pertaining to possible chemotherapy and the modes of action of anthelmintics.

In spite of the fact that all parasitic helminths investigated are capable of consuming oxygen, in many this gas does not appear to be required for the energy metabolism. Similarly, the aerobic or anaerobic nature of the energy metablism of a given worm is not necessarily a function of its physiological environment. For example, adult *Schistosoma mansoni* lives in the blood stream where an ample supply of oxygen is present. In spite of this abundance of oxygen, Bueding[1] has shown that *Schistosoma mansoni* is a homolactate fermentor. That is, all of the glucose dissimilated even aerobically can be accounted for as lactate, the only product formed. Lactated is formed **via** the anaerobic glycolytic sequence. Thus,

1. Recent studies from the laboratory of the author which are discussed in this review were supported in part by the N.I.H., U.S. Public Health Service grants AI 09483 and TOI-AI 00400.

all of the worm's energy seems to be derived anaerobically, despite the presence of oxygen in its habitat. Furthermore schistosomes isolated from animals which have been treated with a cyanine dye show a 90 % inhibition of oxygen uptake[2]. Nevertheless, neither survival of the schistosomes in the animals nor their ability to produce viable eggs in the animals is affected by the dye treatment.

On the other hand, the small nematode, *Nippostrongylus brasiliensis* resides in the intestine, but is nevertheless aerobic, requiring the presence of air for survival. In this connection, the larger intestinal nematodes and cestodes generally appear to possess a predominantly anaerobic type of carbohydrate metabolism while the smaller ones can be either aerobic or anaerobic in their reactions. Von Brand[3] previously pointed out that the larger survace area to weight ratio of the smaller worms would allow for a more rapid diffusion of oxygen into the tissues. However, reports by Bueding **et al.**[4] and Van den Bossche **et al.**[5] indicate, that the dog whipworm, *Trichuris vulpis* and the small rat pinworm, *Syphacia muris,* respectively, may both possess predominantly anaerobic energy metabolisms. Therefore, it is not possible to determine oxygen requirements of worms on the basis of their physiological habitat.

Except possibly for *H. diminuta,* it can not be said at this time, that any single helminth is a complete anaerobe. Many adult worms can survive for extended periods of time under anaerobic conditions, and have no apparent requirement for oxygen in their carbohydrate or energy metabolisms. It is difficult to rule out the possibility, however, that small quantities of oxygen may be required for biosynthetic sequences within the organisms. For example, recent findings indicate that *Ascaris* may employ an oxygen requiring system for the synthesis of hydroxyproline which, in turn, is a constituent of nematode cuticles. In addition, the possible requirement for oxygen for the biosynthesis of components necessary for later stages in the life cycle can not be dismissed. Therefore, until an organism can be cultured completely around the life cycle, as are bacteria, it will be essentially impossible to distinguish between microaerophillic and anaerobic requirements.

The situation is much simpler to determine the requirements of those parasites which are obligate aerobes. Even in this case, however, relatively few parasitic helminths have been proven to be aerobic in the adult stage, although evidence has been presented that a number of nematode larvae are aerobic, such as the Eustrongyle[6], Ascaris[7] and Trichinella[8] larvae.

The findings of Bueding[9] have established that the adult filarial worm, *Litomosoides carinii* is highly dependent upon oxygen for motility and survival. This inhabitant of the pleural cavity of cotton rats or, more recently, of gerbils, metabolizes carbohydrates as illustrated in Figure 1. Under aerobic conditions in the presence of glucose, a mixture of lactate, acetate and CO_2 is formed. In addition glycogen synthesis from glucose takes place in the presence of air. On the other hand, placing the worms under nitrogen leads to a shift in metabolism with the concomitant formation of considerably less acetate and more lactate. A Pasteur effect was also demonstrated, and it was reported that glycogen could be broken down anaerobically but could not be synthesized unless air was present. All of these findings indicate the obligate aerobic character of *Litomosoides* adults.

Of particular interest is the fact that *L. carinii* has been employed as a screening organism for antifilarial compounds. One such group of compounds, the cyanine dyes, were found to be particularly effective filaricides when tested against *L. carinii*. The cyanine dyes, of which dithiazinine is a member, acted presumably by virtue of their inhibitory effect upon oxygen uptake by the worm[9]. The parasite reacted to these drugs much as it did to anaerobiosis. Unfortunately, those species of filarial worms which infest humans were not noticeably affected by cyanine dyes; possibly because these invaders of humans may be more nearly anaerobic.

How then does this aerobic parasite, *Litomosoides carinii,* obtain its energy? Two likely possibilities are illustrated in Figure 2. Either some amounts of the acetate formed are being oxidized completely to CO_2 and water **via** the tricarboxylic acid cycle with the concomitant production of considerable ATP; or, little if any acetate is further oxidized, all of the aerobic energy generated during the oxidative decarboxylation of pyruvate to acetate. A third possibility, not shown in Figure 2 would be that acetate were arising independent of glucose, for example, form lipids.

These possiblities could be distinguished with the aid of isotopically labeled glucose. For example, as shown in Figure 3, the mechanism of the Embden Meyerhof glycolytic scheme results in the splitting of glucose between the 3 and 4 carbon atoms, each of which becomes the carboxyl carbons of pyruvate. When pyruvate is further oxidatively decarboxylated to acetate and carbon dioxide, the CO_2 arises only from the carboxyl carbon of pyruvate which, in turn, originated from the 3 and 4 carbons of glucose. Therefore, if *Litomosoides* metabolism stopped at the acetate level, then, glucose-1-C^{14} or glucose-6-C^{14} should give rise to radioactive acetate, but non-radioactive CO_2. If, on the other hand, glucose were being completely oxidized to CO_2, then all species of labeled glucose should give rise to radioactive CO_2.

Miss Emma Jen in our laboratory has performed some of these experiments and has obtained preliminary results which would indicate that the tricarboxylic acid cycle is of doubtful or at best, minor significance to the physiology of this parasite. Her findings, then, indicate that very little, if any, CO_2 arises from the complete oxidation of glucose. It would appear rather that essentially all of the aerobic energy obtained by this parasite is associated either with the single step conversion of pyruvate to acetate and CO_2 or with the oxidation of other substrates to acetate. Since this organism is very dependent upon aerobiosis, it seems likely that the very existance of this parasite is dependent upon this acetate forming system.

If then *Litomosoides carinii* is a model of other aerobic helminths, the physiological significance of a tricarboxylic acid cycle mechanism for terminal respiration should be reevaluated in all of the aerobic forms. The TCA cycle may not serve the same function as in mammalian tissues. Merely demonstrating the presence of some of the enzymatic reactions associated with this pathway is insufficient. It must be remembered that reactions of the TCA cycle also serve as synthetic pathways for amino acids in all cells. Therefore, the physiological significance as an energy yielding pathway in the parasite should be evaluated independently.

Another nematode which is commonly used as a screen for antinematodal agents is the intestinal parasite of the rat, *Nippostrongylus brasiliensis.* Roberts and Fairbairn[10] reported that the adult form of this parasite survived considerably longer in the presence of air as compared to an anaerobic environment. These authors suggested, however, that the parasite might be microaerophillic or even anaerobic in its energy metabolism. Unfortunately, glucose utilization by *N. brasiliensis* under **in vitro** conditions is extremely low and did not allow for the characterization of the products formed. Recently, Daniel Saz **et al.** [11] determined the disappearance of carbohydrate and the appearance of some of the products of the endogenous metabolism of these parasites. It can be seen from the results recorded in Table 1 that the end products formed aerobically are quantitatively quite different from those found under anaerobic incubations. Lactate appears to be a major product either aerobically or anaerobically. Considerably more lactate is formed, however, in the absence of air. Similarly, although succinate is present in the adult worms on isolation, there is little change in succinate levels after aerobic incubation. Anaerobically, on the other hand, considerably more succinate is formed. Most significant is the fact that carbohydrate disappears more rapidly in a given time period under anaerobic conditions than in the presence of air. In experiment 2, for example, two flasks were incubated simultaneously; one in air, the other under N_2. After 3.5 hrs., 133.8 mμ moles more glucose equivalents disappeared anaerobically than aerobically. This phenomenon is known as the Pasteur effect, and signifies a competition for cofactors between aerobic and anaerobic pathways resulting in more rapid utilization anaerobically because the comptetion for the aerobic cofactors is removed. These data indicate again, the aerobic nature of this adult nematode. Unfortunately, succinate and lactate can not account for all of the carbohydrate disappearance in these experiments. Therefore, the significance of an oxygen requirement in this metabolism still remains to be determined. The question may be raised at this point, however, as to the efficacy of this organism as a test or screen for antinematodal compounds. The definite aerobic nature of *N. brasiliensis* immediately sets it apart somewhat from a large number of other intestinal helminths which are either anaerobic or less obviously aerobic in their requirements. The major point in favor of employing this helminth as a screening organism is the relative ease with which it can be maintained through its life cycle in the laboratory.

In recent years, a great deal of attention has been paid to the anaerobic metabolism of many worms. *Ascaris lumbricoides* appears to have served as a model system for a number of other helminths which derive energy from an anaerobic pathway of reactions involving the fumarate reductase system resulting in the formation of succinate. Scheibel and Saz[12] and Saz and Bueding[13] have listed some of the parasitic helminths which are known to accumulate significant quantities of succinate or products presumably derived from succinate. Representative examples are shown in Table 2. Whether or not all of these helminths form succinate by the same mechanism remains to be determined. Evidence has accumulated, however, to indicate that many of them are quite similar to *Ascaris* in this respect. It should be pointed out that this list is far from complete. Nematodes which fall into this category include *Ascaris, Heterakis gallinae, Trichuris vulpis, Trichinella spiralis* (larvae), *Syphacia muris* (the rat pinworm) and *Dictyocaulus viviparis* (the cattle lungworm). Some cestodes include *Hymenolepis diminuta, Moniezia expansa, Taenia taeniaformis* (adults and larvae) and *Echinococcus granulosus* (cysts). The trematode, *Fasciola hepatica* and the acanthocephalan, *Moniliformis dubius* also form succinate or products presumed to be derived from succinate.

Employing C^{14}-labeled substrates, the anaerobic pathway for succinate formation in *Ascaris* muscle was investigated[14]. As shown in Figure 4, findings indicated that glucose was disimilated **via** glycolysis to a three carbon moiety presumed to be pyruvate. Carbon dioxide was then fixed into the three carbon compound followed by reduction of the product to succinate. Acetate and propionate, arising from pyruvate and succinate respectively, then would serve as precursors for the volatile fatty acids which are major *Ascaris* fermentation products. It was suggested further, that the succinate dehydrogenase in this tissue acted physiologically in a manner opposite to that of mammalian tissues, that is, as a "fumarate reductase", and served to reoxidize the reduced DPN formed during glycolysis. Subsequently, the succinate dehydrogenase from *Ascaris* muscle was partially purified by Kmetec and Bueding [15]. These authors demonstrated that it did indeed behave as a "fumarate reductase" system. Studies of this nature, in addition to the report of Seidman and Entner [16], made it clear that *Ascaris* muscle relied on this backward pathway for much of its energy supply, since ATP generation was found to be associated with the electron transport system coupled to the "fumarate reductase" reaction.

More detailed investigations of the **Ascaris** muscle system have demonstrated that this parasite has adapted itself remarkably well to its environment. Although mitochondria are readily discernable in *Ascaris* tissues, physiologically they function quite differently from the corresponding organelles of mammalian tissues. *Ascaris* mitochondria function anaerobically, while mammalian mitochondria require oxygen in order to generate ATP **via** the electron transport system [17].

Figure 5 illustrates schematically the pathway by which mammalian cells dissimilate carbohydrates. Glycolytic enzymes catalyze the splitting of each glucose molecule into two C_3 units each of which gives rise to phosphoenolpyruvate (PEP). PEP, in turn, donates its high energy phosphate to ADP to form pyruvate and ATP, the reaction being catalyzed by the important glycolytic enzyme, pyruvate kinase. Under conditions of anaerobiosis, such as in rapidly contracting muscle, mammalian tissues can then reduce pyruvate to lactate catalyzed by lactate dehydrogenase, with the concomitant oxidation of DPNH formed during glycolysis. Normally, however, in the presence of air pyruvate enters the mitochondrion where it is completely oxidized to CO_2 and water by the reactions of the tricarboxylic acid cycle with the release of energy in the form of ATP. These reactions are coupled through the cytochrome electron transport system to oxygen.

Ascaris differs quite remarkably from mammalian tissues in these reactions. First, cytochrome oxidase activity could not be detected in this helminth by a number of investigators [18, 19, 20]. Second, oxygen is toxic to some of the enzyme systems by virtue of the hydrogen peroxide formed. Third, the tricarboxylic acid does not function as an cycle energy pathway in *Ascaris* mitochondria which operate anaerobically [21]. Figure 6 illustrates some of the major differences which occur in the soluble, or cytoplasmic portion of *Ascaris* muscle cells in comparison to the cells of the host. Both cells dissimilate carbohydrates to PEP by similar glycolytic enzymes. Mammalian cells then catalyze the cytoplasmic formation of pyruvate by means of the enzyme, pyruvate kinase, reaction 1 in the figure. Bueding and Saz [22] reported that *Ascaris* can not form pyruvate from PEP in the cytoplasm, since pyruvate kinase activity is only barely detectable. Therefore, PEP becomes the compound at which the metabolism of the host and parasite cells become divergent. Instead of forming pyruvate at this point, *Ascaris* cells fix CO_2 into PEP to form oxalacetate (reaction 2 in Figure 6). This reaction is catalyzed by the enzyme PEP carboxykinase [23]. It is of interest that mammalian tissues are thought to employ PEP carboxykinase primarily for the reverse reaction, that is, for the decarboxylation of oxalacetate to form PEP. By such a scheme, glyconeogenesis from succinate can be explained in mammalian cells. It has been demonstrated in *Ascaris* tissues, however, that 1. the high levels of endogenous PEP, 2. the K_m of *Ascaris* PEP carboxykinase and 3. the high malate dehydrogenase activity would all favor reaction in the direction proposed.

One major question remained unanswered by the proposed scheme for succinate formation in *Ascaris*. Acetate is an end product of *Ascaris* fermentation. In addition, alpha-methylbutyrate is formed as a major end product and acetate has been shown to be one of the precursors for this branched chain acid. How then does this nematode form acetate in the absence of a pyruvate kinase and thus in the absence of cytoplasmically formed pyruvate? Recent findings have helped elucidate an answer to this question. Findings [24] are in accord with the scheme depicted in Figure 7, which is our current concept of *Ascaris* metabolism.

As suggested above, the glycolytic enzymes *of Ascaris* are present in the soluble portion of the cell and function similarly to those of the host tissues up to the point of PEP accumulation. PEP can not be dephosphorylated directly to pyruvate by *Ascaris*. Instead, CO_2 is fixed into PEP to form oxalacetate, as catalyzed by the helminth PEP carboxykinase. This dicarboxylic keto acid is rapidly reduced by glycolytically formed DPNH, the reaction being catalyzed by what appears to be the most active of all *Ascaris* enzymes presently described, malate dehydrogenase. This reaction, like the lactate dehydrogenase reaction in mammalian tissues, serves to regenerate cytoplasmic DPN so that glycolysis may continue.

Malate thus formed in the cytoplasm, now crosses over into the

39

mitochondrion and becomes the mitochondrial substrate. Malate, within the mitochondrion, must then undergo a dismutation reaction, since normally direct oxidations do not occur in this anaerobic organelle. Intramitochondrial reducing power, in the form of DPNH, is obtained by the oxidative decarboxylation of malate to pyruvate and CO_2, thereby giving rise to pyruvate in the absence of pyruvate kinase. This reaction is catalyzed by the mitochondrial malic enzyme which was shown to be present in *Ascaris* in 1957[25], but whose function was not understood until 1969[24]. DPNH formed from this reaction then serves to reduce a corresponding amount of malate to succinate **via** fumarate and the fumarate reductase reaction with the concomitant formation of ATP. Pyruvate may then serve as a precursor of acetate, but again within the mitochondrion. This reaction might result in additional mitochondrial reducing power which could conceivable enter into the reductive formation of volatile fatty acids which are end products of *Ascaris* fermentations. The site of fatty acid formation is, however, still unknown.

It becomes apparent then, that the metabolism of this parasite has been modified considerably to conform with its environment. In addition, it is becoming apparent that a number of other parasitic helminths have modified their metabolisms in a similar manner.

Probably the best example of an anerobic helminth currently available is the cestode, *Hymenolepis diminuta.* Scheibel and co-workers[12, 26] found 1. no evidence of cytochrome oxidase activity 2. no significant quantity of isotopic CO_2 was formed from glucose-1-C^{14} or glucose-6-C^{14} indicating the absence of both the tricarboxylic acid cycle and the pentose shunt mechanisms; 3. mitochondria from the cestode incorporated P^{32} into ATP by means of an anaerobic, electron transport associated, exchange reaction; and most important, 4. the adult stage has been cultured anaerobically. Oxygen is quite inhibitory to the cultivation procedures [27, 28]. Circumstantial evidence, at least, indicates that similarities exist between *H. diminuta* and *Ascaris* metabolisms.

In conclusion, it must be recognized that although biochemical similarities do exist between parasites and their hosts, more detailed investigations show a number of striking differences, particularly in the energy metabolisms. In most instances, where the modes of action of anthelmintics have been examined, their major effects appear to be associated with the energy yielding patways or the neurochemical processes of the parasites. The "rational approach to chemotherapy" has taught us a great deal concerning parasite biochemistry. We are now, however, at a crossroads awaiting some new concepts pertaining to the design of chemotherapeutic agents. Certainly in the case of *Ascaris,* we have learned of a number of enzyme systems which should be vulnerable to chemotherapeutic attack without corresponding toxicity to the host. Unfortunately, however, the problem arises as to how does one design a specific inhibitor or a given enzyme? The answer to this question may have to wait until we learn a great deal more concerning the configurations of the active sites of such enzyme systems. Perhaps a more realistic approach at present would be a compromise between the rational and the empirical approaches, whereby studying modes of actions of existing drugs might lead to rational structural modifications to reduce toxicities or enhance required inhibitions. Such an approach recently has met with some degree of success in the laboratory of Dr. Bueding. One fact appears certain. Detailed biochemical studies of parasites and the modes of action of anthelmintics must be continued if there is to be any rational approach to chemotherapy of parasite infections in the future.

References

1. BUEDING, E., J. Gen. Physiol., **33,** 475 (1950).
2. BUEDING, E., PETERS, L., KOLETSKY, S. and MOORE, D.V., Brit. J. Pharmacol., **8,** 15 (1953).
3. BRAND, T. von, and ALLING, D.W., Comp. Biochem. Physiol., **5,** 141 (1962).
4. BUEDING, E., KMETEC, E., SWARTZWELDER, C., ABADIE, S. and SAZ, H.J., Biochem. Pharmacol., **5,** 311 (1960).
5. VAN DEN BOSSCHE, H., SCHAPER, J. and BORGERS, M., Comp. Biochem. Physiol., **38B,** 43 (1971).
6. BRAND, T. von, J. PARASITOL., **24,** 445 (1938).
7. SAZ, H.J., LESCURE, O. and BUEDING, E., J. Parasitol., **54,** 457 (1968).
8. BRAND, T. von, WEINSTEIN, P.P., MEHLMAN, B. and WEINBACH, E.C., Exptl. Parasitol., **1,** 245 (1952).
9. BUEDING, E., J. Exptl. Med., **89,** 107 (1949).
10. ROBERTS, L.S. and FAIRBAIRN, J. Parasitol., **51,** 129 (1965).
11. SAZ, D.K., BONNER, T.P., KARLIN, M. and SAZ, H.J., In preparation.
12. SCHEIBEL, L.W. and SAZ, H.J., Comp. Biochem. Physiol., **18,** 151 (1966).
13. SAZ, H.J. and BUEDING, E., Pharmacol. Rev., **18,** 871 (1966).
14. SAZ, H.J. and VIDRINE, A., Jr., J. Biol. Chem., **234,** 2001 (1959).
15. KMETEC, E. and BUEDING, E., J. Biol. Chem., **236,** 584 (1961).
16. SEIDMAN, I. and ENTNER, N., J. Biol. Chem., **236,** 915 (1961).
17. SAZ, H.J., Am. Zoologist, **11,** 125 (1971).
18. BUEDING, E. and CHARMS, B., J. Biol. Chem. **196,** 615 (1952).
19. KATSUME, T. and OBO, F., Acta Medica Univ. Kagoshima, **4,** 56 (1962).
20. CHANCE, B. and PARSONS, D.F., Science, **142,** 1176 (1963).
21. SAZ, H.J., in Chemical Zoology, Vol. III, p. 329 (ed. Florkin, M. and Scheer, B.T.). Academic Press, N.Y. 1969.
22. BUEDING, E. and SAZ, H.J., Comp. Biochem. Physiol., **24,** 511 (1968).

23. SAZ, H.J. and LESCURE, O.L., Comp. Biochem. Physiol., **22,** 15 (1967).
24. SAZ, H.J. and LESCURE, O.L., Comp. Biochem. Physiol., **30,** 49 (1969).
25. SAZ, H.J., and HUBBARD, J.A., J. Biol. Chem., **225,** 921 (1957).
26. SCHEIBEL, L.W., SAZ, H.J. and BUEDING, E., J. Biol. Chem., **243,** 2229 (1968).
27. BERNTZEN, A.K., J. Parasitol., **47,** 351 (1961).
28. SCHILLER, E.L., J. Parasitol., **51,** 516 (1965).

Table 1: Aerobic vs. Anaerobic Carbohydrate Utilization by *Nippostrongylus brasiliensis* *

Expt.	Conditions	Carbohydrate		Lactate		Succinate	
		Found	Change	Found	Change	Found	Change
1	0 Time	461.6		48.5		53.7	
	Air - 3.5 Hrs.	365.7	- 95.9	80.0	+ 31.5	54.9	+ 1.2
2	Air - 3.5 Hrs.	304.5		65.4		62.3	
	N₂ - 3.5 Hrs.	170.7	-133.8	259.7	+194.3	84.8	+22.5

* All figures represent mμ moles/mg worm protein.

Table 2: Some helminths reported to form succinate or products presumed to be derived from succinate.

Nematodes

Ascaris lumbricoides
Heterakis gallinae
Trichuris vulpis
Trichinella spiralis (larvae)
Syphacia muris (?)
Dictyocaulus viviparus (?)

Cestodes

Hymenolepis diminuta
Moniezia expansa
Echinococcus granulosus (cysts)
Taenia taeniaformis (adults and larvae)

Trematodes

Fasciola hepatica

Acanthocephala

Moniliformis dubius

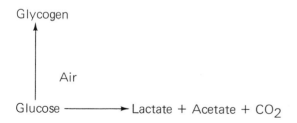

Fig. 1.: Overall Metabolism of Adult
        Litomosoides carinii

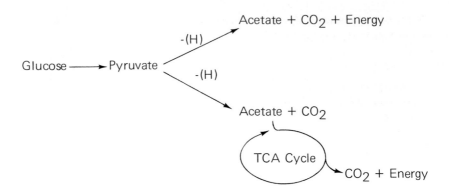

Fig. 2: Possible Origins of Respiratory CO_2 in *Litomosoides carinii*.

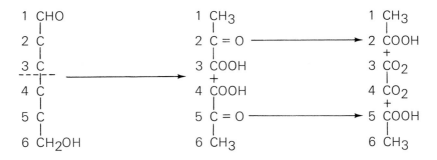

Fig. 3: Conversion of Glucose to CO_2 and Acetate

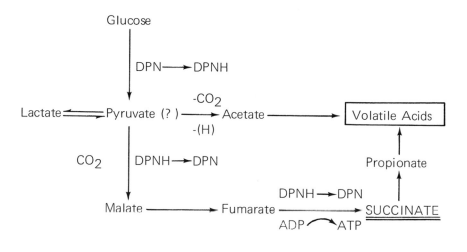

Fig. 4: Proposed pathway for the formation of succinate and volatile acids in Ascaris muscle.

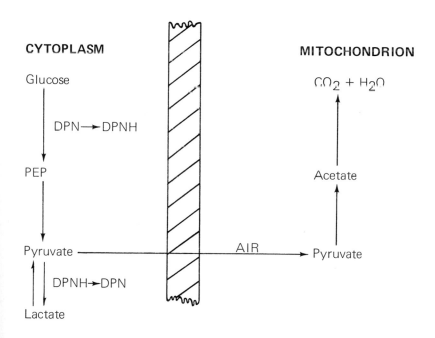

Fig. 5: Carbohydrate dissimulation in mammalian tissues.

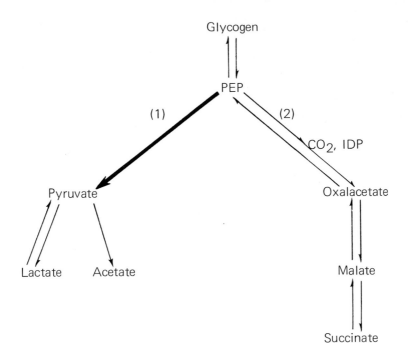

Fig. 6: Comparison of metabolic reactions of mammalian and Ascaris tissues. (1) = Pyruvate kinase. (2) = Phosphoenolpyruvate carboxykinase.

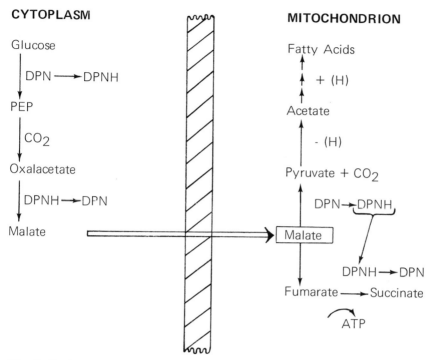

Fig. 7: Pathway of carbohydrate dissimilation in *Ascaris* muscle.

4

THE UTILIZATION OF CARBON DIOXIDE BY MONIEZIA EXPANSA: ASPECTS OF METABOLIC REGULATION.

C. Bryant

Department of Zoology, Australian National University,
Canberra, Australia.

Introduction

In recent years, considerable attention has been focussed on the ability of parasitic helminths to utilize exogenous sources of CO_2 for metabolic purposes. Among the earliest reports was that of Fairbairn (1954), who showed that *Heterakis gallinae* rapidly incorporated $^{14}CO_2$ into propionic and succinic acids. Later, Saz and Hubbard (1957), and Saz and Vidrine (1959) investigated this phenomenon in *Ascaris,* and concluded that it was mediated by a "malic enzyme". These demonstrations, as well as stimulating interest in parasite biochemistry, also resulted in the appraisal of other invertebrates from the point of view of CO_2 fixation and there have been published numerous accounts of the role the process plays in the metabolism of organisms as diverse as the oyster and the terrestrial planaria (Hammen and Wilbur, 1959; Hammen and Lum, 1962, 1964; Awapara and Campbell, 1964; Bryant and Janssens, 1969). The role of CO_2 fixation in the urea cycle has also been examined (Campbell and Lee, 1963; Campbell, 1965; Janssens and Bryant, 1969; Bryant and Janssens, 1969).

It therefore seems that, far from being exceptions, parasitic helminths conform to a general pattern amongst invertebrates, and that the pattern of CO_2 incorporation is worth studying, not only in the specific context of parasitism but also in a much wider, biological context. Saz (1971) has recently drawn attention to the general significance of studies of parasite biochemistry.

Considerably more is known about CO_2 incorporation in intestinal helminths than in other organisms. Thus, it is now generally accepted that there are in parasites important pathways involving phosphoenolpyruvate carboxykinase and/or malic enzyme, the products of which are oxaloacetic, malic, fumaric and succinic acids (Agosin and Repetto, 1963; Graff, 1965; Prescott and Campbell, 1965; Scheibel and Saz, 1966; Saz and Lescure, 1967; Prichard and Schofield, 1968 a; Ward **et al.,** 1968 a, 1969; Davey and Bryant, 1969; Horvath and Fisher, 1971). In addition, Scheibel and Saz (1966) have shown that there is probably a stoichiometrical relationship, in *Hymenolepsis diminuta* at least, between CO_2 incorporated and succinate produced.

Moniezia expansa is a large cestode inhabiting the small intestine of the sheep. It was early shown to be an enthusiastic excretor of succinate (von Brand, 1933), and much later to possess a system of electron transfer in which NADH, the electron donor, was linked to fumarate, the electron acceptor (Cheah and Bryant, 1966; Cheah, 1968). To date little attention has been paid to the relationship of this pathway to the pathway of CO_2 fixation in *M. expansa,* although Davey and Bryant (1969) showed that homogenates of the worm were capably utilising $^{14}CO_2$, and that the utilisation was enhanced by the addition of pyruvate and ATP.

Biochemical studies on a large tapeworm like *M. expansa* bring with them their own problems. Amongst these is the fact that all stages of development, from embryonic growth to the production of fertile ova, are represented in one individual. In the following experiments, the scolex end of the organism was used to ensure uniformity and to confine the work to the persistently growing material. The paper explores more fully the route by which $^{14}CO_2$ is incorporated into soluble intermediates in this material, and investigates its significance in the intact animal. A scheme for metabolic regulation is proposed on the basis of these findings, but it must be borne in mind that, even if the hypothesis is found to have validity, it may not be applicable to more mature sections of the helminth.

Materials and Methods

Moniezia expansa was obtained from freshly slaughtered sheep at Queanbeyan Abattoir, N.S.W., Australia, on the morning of the day upon which the experiment was to be performed. The worms were removed immediately from the intestines of the sheep and stored in ice cold Krebs'-Ringer Phosphate solution (KRP) comprising 100 parts NaCl, 4 parts KCl, 1 part KH_2PO_4, 1 part $MgSO_4.7H_2O$, all in concentrations of 0.154 M, 3 parts 0.11 M $CaCl_2$, and 21 parts 0.1 M sodium phosphate buffer, pH 7.4.

Thirty minutes collecting usually sufficed to produce twenty to thirty worms, which were taken to the laboratory, wiped clean of mucus and intestinal contents, whashed several times in fresh KRP, blotted and weighed. Only the anterior, actively strobilating, portion of the worm was used, which weighed about 1 gram. After separation from the remainder of the tape, the anterior portions remained active at 37° in KRP for at least 6 hours.

Acetyl coenzyme A, guanosine diphosphate (GDP), inosine diphosphate (IDP), adenosine di- and tri-phosphates (ADP, ATP), sodium pyruvate, sodium phosphoenolpyruvate (PEP), nicotinamide adenine dinucleotide, and its reduced form (NAD, NADH), nicotinamide adenine dinucleotide phosphate and its reduced form (NADP, NADPH), and fructose 1,6 diphosphate were obtained from the Sigma Chemical Co., St. Louis, U.S.A. Enzymes were purchased from Boehringer Mannheim GmbH., Germany. $NaH^{14}CO_3$, specific activity 51 mCi/mM, was obtained from the Radiochemical Centre, Amersham, Bucks., U.K. Where indicated in the tables and figures, the specific activity was reduced with $NaH^{12}CO_3$. All other reagents were analytical grade and glass distilled water was used throughout.

For enzyme determinations, for the determinations of the distribution of radiocarbon in homogenates, and for the determination of protein by the methods of Lowry et al, (1951) and Layne (1957), a known weight of worm was homogenised in an all-glass Dounce-type hand homogeniser, in an equal weight of 0.25 M glycylglycine buffer, pH 7.4. Where necessary, the homogenate was diluted further in glycylglycine buffer, or a supernatant fraction was prepared by subjecting it to centrifugation at 30,000 g for 30 minutes. In those experiments involving intact worms, the animals were incubated in five times their own weight of KRP to which appropriate additions had been made. For further details, see tables and figures.

Phosphoenolpyruvate carboxykinase (E.C. 4.1.1.32), pyruvate carboxylase (E.C. 6.4.1.1.), malic enzyme activity (E.C. 1.1.1.38-40), and pyruvate kinase (E.C. 2.7.1.40) were measured as indicated in the tables and figures. It was found that lactic dehydrogenase (E.C. 1.1.1.27) activity in homogenates, as measured by the method of Bergmeyer **et al.** (1963), was negligible. Therefore, as there was little pyruvate depletion, the direct methods of observing CO_2 fixation, described in the tables and figures, were employed.

Most of the experiments were carried out under air and under nitrogen. In each case the incubation vessels were serum vials fitted with self-sealing rubber caps through which injections could be made. The vials were gassed with the appropriate gas phase through 18 gauge hypodermic needles for one minute, after which the radioactive sodium bicarbonate was added by syringe, followed by the reagent to initiate the reaction.

Results

The activities of three enzymes possibly concerned with the incorporation of carbon dioxide into organic acids in *Moniezia expansa* were measured. The enzymes are phosphoenolpyruvate carboxykinase, pyruvate carboxylase and malic enzyme, and are responsible for the synthesis of oxaloacetic and malic acids according to the following reactions:

$$\text{Phosphoenolpyruvate} + HCO_3^- + IDP \overset{Mn^{++}}{\rightleftharpoons} \text{oxaloacetate} + ITP$$
$$\text{phosphoenolpyruvate}$$
$$\text{carboxykinase}$$

$$\overset{Mg^{++}}{\underset{\text{Acetyl coenzyme A}}{}}$$
$$\text{Pyruvate} + ATP + HCO_3^- \rightleftharpoons \text{oxaloacetate} + ADP + P_i$$
$$\text{pyruvate}$$
$$\text{carboxylase}$$

$$\overset{Mn^{++} \text{ or } Mg^{++}}{}$$
$$\text{Pyruvate} + HCO_3^- + NADPH \rightleftharpoons \text{Malate} + NADP$$
$$\text{malic enzyme}$$

In addition, the activity of pyruvic kinase, which catalyses the following interconversion, was investigated.

$$\overset{Mg^{++}}{}$$
$$\text{Phosphoenolpyruvate} + ADP \rightleftharpoons \text{pyruvate} + ATP$$
$$\text{pyruvic kinase}$$

Figure 1 shows clearly that, under the conditions of the enzyme assays, incorporation of CO_2 took place. The activities of pyruvate kinase, pyruvate carboxylase and malic enzyme were linear up to a concentration of protein of at least 1.5 mg/ml. The activity of phosphoenolpyruvate carboxykinase was much greater, and reached an asymptote at a concentration of 0.3 mg protein/ml. Figures 2 and 3 demonstrate that the rates of reaction for all four enzymes were constant for at least 15 minutes.

The requirements for optimum activity of phosphoenolpyruvate carboxykinase are given in table 1. Optimum activity was observed in the presence of PEP, manganous ions and GDP. GDP and IDP could be interchanged without a significant loss of activity, but although magnesium ions caused a slight increase in activity compared with systems in which there were no divalent metal ions, their effect was much less than that of simlar concentrations of manganese.

Table 2 shows the optimum requirements for "malic enzyme" activity in homogenates from *M. expansa.* The addition of NADPH to the endogenous system had no effect on the incorporation of CO_2; independently, however, pyruvate and manganous ions doubled the endogenous rate. The greatest activity was observed when both pyruvate and manganous ions were included in the reaction mixture in the absence of pyridine nucleotide; the rate of incorporation of CO_2 was elevated to four times that of the endogenous system. Magnesium ions were not able to substitute for manganous ions, and further additions of NADH, NADPH or ATP had no effect.

Pyruvate carboxylase activity was found in homogenates of *M. expansa.* Acetyl coenzyme A, ATP and magnesium ions were necessary for maximum activity (table 3).

The pyruvic kinase assay is linked to lactic dehydrogenase, and involves following optically, at 340 mμ, the formation of NAD from added NADH and pyruvate generated by pyruvate kinase. In the absence of lactic dehydrogenase, no oxidation of NADH occurred. Assays for lactic dehydrogenase in homogenates of *M. expansa* also proved negative.

Pyruvic kinase activity was observed when PEP, magnesium ions and ADP were present in the reaction medium. Addition of fructose 1,6 diphosphate more than doubled the rate of reaction (table 4, figure 3). The presence of calcium ions in equimolar concentrations with magnesium caused a 50 % inhibition.

Table 5 shows that over 85 % of the activity of all four enzymes was located in the supernatant fraction. The enzymes were assayed under nitrogen as well as under air. There were no significant differences in the activities observed in either gas phase.

Once fixed, the subsequent fate of carbon from CO_2 is illustrated in table 6. Homogenates of *M. expansa* incubated with PEP and cofactors necessary for maximum phosphoenolpyruvate carboxykinase activity incorporated radiocarbon from $H^{14}CO_3^-$ predominantly into succinic acid.

Less than 10 % appeared in lactic acid. The major differences between incubation in air and under nitrogen were that only in the former were small amounts of radiocarbon found in α-ketoglutaric, citric and glutamic acids; and also that, in air, the total amount of radiocarbon fixed was 50 % greater.

In table 7 the effects of including metabolites in the incubation mixtures are listed. The stimulatory effect of PEP on homogenate CO_2 uptake is clearly demonstrated, but it is more pronounced in nitrogen than in air because the uptake in the absence of PEP was much greater in air. The addition of pyruvate produced effects similar to that of PEP. Fumarate was remarkable in inhibiting endogenous uptake in air, and reducing aerobic incorporation to the anaerobic level in the presence of PEP.

The fixation of CO_2 by intact anterior portions of the tapeworm was also studied. Table 8 shows the effects on uptake of including a range of substances in the KRP in which worms were maintained for three hours. Addition of NaCl, PEP, pyruvate, propionate, acetate or glucose did not bring about a change in the amount of CO_2 taken up when compared with the control. In all these cases, however, the uptake of CO_2 in nitrogen was significantly higher (P = 0.01) than the corresponding incubation in air. Addition of fumarate stimulated uptake in the aerobic system; under nitrogen the stimulation was much less marked.

The distributions of radiocarbon in soluble intermediates within the whole worm, and amongst those excreted into the incubation medium are given in table 9. Glucose did not cause an increased uptake of CO_2 in either air or nitrogen. However, in air, about 75 % of the total fixed CO_2 was found in the medium whether glucose was included or not. Under nitrogen, between 75 % and 80 % was retained by the worms in the absence of exogenous glucose. When glucose was present, the majority of radiocarbon appeared in the medium.

In air, the worm extract contained a high proportion (66 %) of succinic acid and a low proportion of lactic acid (23 %). Incubation of the animals in glucose reversed this to 27 % and 66 %, respectively. These relative proportions were largely reflected in the distribution of radiocarbon in succinate and lactate in the media, although a rather higher proportion of the latter was excreted. The next most important metabolite was malate, which dropped from 4.2 % to 1.5 % when glucose was included in the medium, and was excreted only in its absence. Other metabolites detected were aspartate (worm extract only) fumarate, alanine, α-ketoglutarate, citrate, glutamate and glycine and serine.

Under nitrogen, the endogenous succinate/lactate ratio was reversed in the presence of glucose. When glucose was present, worm extracts contained 63 % succinate and 30 % lactate. When it was omitted, the proportions were 35 % and 53 %. This distribution was generally reflected in the incubation media, but in the absence of glucose, the proportion of succinate excreted decreased by 23 %, and that of lactate increased by 33 %. There was little difference between the distributions in the medium and in the worm when glucose was present. Other metabolites included aspartate, malate, fumarate and alanine; and glycine plus serine, α-keto-glutarate, citrate and glutamate when glucose was omitted. Less malate and more fumarate was produced under nitrogen than under air. Larger total amounts of CO_2 were incorporated in each case under nitrogen.

Of particular interest in these experiments is the behaviour of the succinate and lactate pools. Figure 4 shows that, in air, considerably more succinate than lactate was produced when glucose was absent. As the glucose concentration was increased, the relative proportion of lactate increased until it was the major end product. In nitrogen, the opposite was the case.

The pattern of excretion is shown in figures 5 a and b. They show the ratios of succinate or lactate remaining in the animals to the amounts of succinate or lactate excreted into the medium. In air, these ratios were independent of glucose concentration, and the proportion excreted was high compared with that retained by *M. expansa.* In nitrogen, much more succinate and lactate were retained by the worms, but as glucose concentration increased, the proportion excreted increased until, at 25 mM glucose, it did not differ significantly from that obsered in air.

Discussion

The results presented in figures 1 to 3 and in tables 1 to 4 show unequivocally that phosphoenolpyruvate carboxykinase is the most active of the three CO_2-fixing enzymes which were studied in *Moniezia expansa.* Its requirements are similar to those of vertebrate enzymes in that manganous ions, and IDP or GDP are essential for optimal activity (Bandurski and Lipmann, 1956). It was difficult to distinguish any differences between the effects of the nucleotides: of the two, GDP was slightly more active. These findings are in complete accord with those of Ward **et al.** (1969) in *Trichinella spiralis;* and in general agreement with those obtained by Kurahashi **et al.** (1957) with chicken liver, by Agosin and Repetto (1965) with *Echinococcus granulosus,* by Prescott and Campbell (1965) with *Hymenolepis diminuta,* by Saz and Lescure (1967) with *Ascaris lumbricoides* muscle, by Ward **et al.** (1968) with *Haemonchus contortus,* and by Prichard & Schofield (1968) with *Fasciola hepatica*, although in the latter group, IDP was somewhat more effective than GDP.

The results in table 2 suggest that "malic enzyme" is absent from *M. expansa*. Although pyruvate, NADPH, and Mn^{++} together stimulate CO_2 uptake by a factor of 3, this can be accounted for by summing the individual stimulatory effects of pyruvate and manganese, which are probably exerted on phosphoenolpyruvate carboxykinase and pyruvate carboxylase. NADPH (and NADH) have no effect on subsequent addition. It is not impossible that there is sufficient of these reduced cofactors already present in the homogenate. Even in this event the activity of "malic enzyme" compared with that of phosphoenolpyruvate carboxykinase is negligible. On the other hand, the evidence for some pyruvate carboxylase activity is good (table 3), but again, it is relatively unimportant compared with phosphoenol pyruvate carboxykinase (table 10).

Pyruvate kinase was found to be present in *M. expansa.* Its activity was less that a tenth of phosphoenolpyruvate carboxykinase, but in the presence of fructose 1,6 diphosphate there was an allosteric activation which more than doubled its rate of reaction. A similar activation has been demonstrated in rat liver pyruvate kinase (Bailey **et al.,** 1968), although muscle pyruvate kinase is not influenced by fructose 1,6 diphosphate. The observation for *M. expansa,* and the relatively low activities of this enzyme in this and other helminths suggest that it could well be rate limiting in the pathway from glucose to lactate, and hence exert an important regulatory function in parasitic worms.

After fixation, the distribution of radiocarbon in the soluble intermediates of homogenates of *M. expansa* (table 6) was consistent with the pathway of oxaloacetate → malate → fumarate → succinate which has been reported in a whole range of parasitic helminths (Scheibel and Saz, 1966). Some aspartic acid was formed, presumably from oxaloacetic acid by transamination. In air, small amounts of intermediates associated with tricarboxylic acid cycle activity were detected. This observation is consistent with previous results (Davey and Bryant, 1969). Most interestingly, radiocarbon was detected in alanine and lactate. There are two important implications from this observation. Firstly, radiocarbon must have been incorporated into pyruvate: the only way this could have been achieved is by randomising the distribution of radiocarbon after its fixation into oxaloacetic acid. The reactions to malate and fumarate are reversible, and the latter is a symmetrical molecule. Thus, the radiocarbon in pyruvate must have initially passed into the fumarate pool and returned to the malate or oxaloacetate pools, causing these acids to be labelled in the 1 **or** 4 position. Decarboxylation to yield labbelled pyruvate must then have ensued. Secondly, the appearance of radiocarbon in lactate implies the presence of lactic dehydrogenase at very low activity, as the enzyme was not detectable by routine spectrophotometric methods. Lactic dehydrogenase assumes a greater significance in subsequent experiments.

Finally, there was little difference in the overall **distribution** of $^{14}CO_2$ in either air or nitrogen, but in contrast to the intact worm experiments, the total amount of CO_2 fixed in air was 50 % greater than that in nitrogen.

In homogenates, over extended periods of incubation, pyruvate was almost as effective as PEP in enhancing CO_2 fixation (table 7). This probably indicated an endogenous recruitment of PEP in the experiments in which pyruvate was the substrate, and an exhaustion of nucleotide requirements for phosphoenolpyruvate carboxykinase activity when PEP was the substrate. Studies in this laboratory (Furlonger, 1971) of absolute metabolic pool size have shown that, in *M. expansa* there are considerable resources of 3- and 2- phosphoglyceric acid present. Addition of fumarate had an inhibitory effect in air on both endogenous and PEP-stimulated uptake of CO_2 by homogenates of *M. expansa.* Cheah (1967) has postulated that a major electron transfer pathway in *M. expansa* involves b and o type cytochromes capable of transferring electrons either to oxygen or to fumarate. Thus, in air, fumarate ought to compete with oxygen for the terminal oxidase. The inhibitory effect of fumarate on CO_2 uptake in air in the homogenates in the present experiments can perhaps be ascribed to this competition. Fumarate addition effectively converts the system to an anaerobic one. In corroboration, there was little effect of fumarate on the CO_2 uptake of the homogenate under nitrogen.

When intact portions of *M. expansa* were incubated with a whole range of substances under air and nitrogen, in all cases CO_2 incorporation was greater under anaerobic conditions (table 8). NaCl was included in this experiment to act as a control for the osmotic effect of making the additions. As CO_2 uptake in the presence of additional NaCl did not differ significantly from that observed when it was omitted, osmotic considerations can be ignored. Neither PEP, pyruvate, propionate, acetate or glucose exerted any significant effect on CO_2 uptake by intact worms in either the aerobic or anaerobic systems. The lack of effect of glucose is surprising, but presumably over the short period of the experiment the tapeworms were able to mobilise sufficient reserves to keep their metabolic pathways saturated. Addition of fumarate increased the amount of CO_2 taken up in air almost to the level of uptake observed in nitrogen. Bryant and Morseth (1968) have shown that another cestode, *Echinococcus granulosus,* is capable of utilising fumarate added to its maintenace medium, so that in the present instance it can once again be postulated that fumarate penetrated the worm and competed with oxygen to produce an effectively anaerobic situation. The stimulation of CO_2 uptake due to fumarate in air is of the order of 4 times that of the control; in nitrogen it is a little over 2. The effect of fumarate can be ascribed to the presence of a larger fumarate pool competing successfully with oxygen for oxidase sites in the aerobic system. In the anaerobic case, the enlarged pool presumably enabled the more rapid turnover of reduced cofactors.

Although the addition of 5mM glucose had no effect on the total incorporation of carbon dioxide in intact worms under either air or nitrogen, this and higher concentrations caused marked modifications to the distribution of radiocarbon from $H^{14}CO_3^-$ amongst the soluble intermediates. (tabel 9; figure 4). Under nitrogen, when no glucose was present, considerably more lactic acid than succinic acid was produced. As the concentration of glucose was increased in the medium these proportions reversed, until at a concentration of 25 mM glucose 60-70 % of the radiocarbon was present in succinic acid. Under air, exactly the opposite situation applied. The levels of radiocarbon in the metabolic pools accurately reflect changes in absolute pool size (Furlonger, 1971), and these results conform to those of von Brand **et al.** (1968) with adult and larval *Taenia taeniaeformis.* These workers reported succinate/lactate ratios of 3.3 and 3.0 under anaerobic conditions, and 0.8 and 0.4 under aerobic conditions in the presence of 11 mM glucose. Figure 4 shows that, in *M. expansa,* the ratio is dependent on both glucose concentration and on gas phase.

Figures 5 a and b illustrate one further point about the behaviour of the succinate and lactate pools. Under aerobic conditions, the proportions of succinate and lactate excreted over the duration of the experiment are high even at zero glucose concentration. Under nitrogen much of the succinate and lactate is retained by the worm in the absence of glucose, and as glucose concentration increases, proportionately more excretion of the two acids occurs. The conclusion is, therefore, that excretion is an energy dependent process; that with oxygen available there is also more energy available in the absence of an exogenous supply of glucose than under anaerobic conditions. In the latter case, the presence of an external source of glucose is necessary for effective excretion. At concentrations of 25 mM glucose, the difference between the aerobic and anaerobic systems is removed.

To summarise the conclusions from the preceding discussion, it is possible to surmise that there are at least two alternative pathways for glucose oxidation in *Moniezia expansa,* one leading to the formation of succinate, the other to lactate. The absence of lactic dehydrogenase in homogenates is therefore artefactual.

The switch from one pathway to the other is brought about by varying the environmental conditions. Two such conditions would appear to be the presence or absence of glucose in the external medium, and presence or absence of oxygen. Addition of fumarate, by competing with oxygen for the terminal oxidase, effectively mimics the anaerobic situation. Pyruvate kinase can perhaps be implicated in the switch from one pathway to the other because of its low activity, perhaps rate-limiting, and susceptibility to allosteric activation by fructose 1,6 diphosphate. The availability of energy under anaerobic conditions is less than under aerobic conditions. These conclusions lead to an hypothesis for metabolic regulation in *M. expansa,* the model for which is illustrated in figure 6. It incorporates the suggestion of Bueding and Saz (1968) that the two enzymes, pyruvate kinase and phosphoenolpyruvate carboxykinase compete for substrate.

Under aerobic conditions, oxygen is the terminal acceptor in electron transfer, and there is thus an increased availability of ATP, which leads to an increased production of fructose 1,6 diphosphate from glucose. Furlonger (1971) has measured ATP and fructose 1,6 diphosphate levels under aerobic and anaerobic conditions and found that they are substantially increased in the former situation. Elevated levels of fructose 1,6 diphosphate lead to increased levels of phosphoenolpyruvate, presumably by the normal reactions of glycolysis. The elevated levels of fructose 1,6 diphosphate also result in the activation of pyruvate kinase which competes successfully with phosphoenol pyruvate carboxykinase for PEP, resulting in increased production of lactate. (A criticism which can be applied here is that the activity of pyruvate kinase is only one fifth of that of phosphoenolpyruvate carboxykinase. However, it is not impossible that, like lactic dehydrogenase, it suffers some loss of activity during homogenisation. Work is in progress to attempt to resolve this point). At the same time, some succinate production occurs, and to account for the appearance of radiocarbon in lactate in these experiments, it is suggested that the reversible reactions between oxaloacetate and fumarate lead to random distribution of radiocarbon in malate and oxaloacetate. Decarboxylation of either of these two acids would lead to the appearance of radiocarbon in pyruvate and hence, in lactate.

Under anaerobic conditions, electron transfer is achieved by way of the fumarate/succinate interconversion. By performing the rather facile sum it is clear that NADH oxidation could, by analogy with mammalian systems, yield 3 molecules of ATP. Likewise the conversion of fumarate to succinate could require an amount of energy equivalent to 2 molecules of ATP. The net yield, if the appropriate enzymes and carriers exist, and there is no evidence for this, is therefore 1 molecule of ATP. In the aerobic system, however, Cheah (1971) has shown that for succinate or α-glycerophosphate oxidation, a P/O ratio of 2 is likely. From this it is clear that the aerobic system is energetically more favourable but that other environmental conditions are limiting.

Thus, in the anaerobic situation, depressed levels of ATP are reflected in depressed levels of production of fructose 1,6 diphosphate and the allosteric activation of pyruvate kinase is removed. Phosphoenolpyruvate carboxykinase competes more successfully for PEP and the result is an increased incorporation of CO_2 into oxaloacetate and an increased yield of succinate.

Under aerobic conditions, the participation of the enzymes of the tricarboxylic acid cycle in general metabolism is more probable, as the brake due to the accentuation of the phosphoenolpyruvate carboxykinase system is applied less heavily. Aerobically, the helminth has several potentially functional ATP-producing systems at its disposal, of which the most important are presumably the pyruvate kinase and oxidative phosphorylation system. Anaerobically, it has a high dependence on the phosphoenolpyruvate carboxykinase system. A recent study by Scheibel **et al.** (1968) illustrates this well in another cestode, *Hymenolepis diminuta*. In addition, the fumarate/succinate transformation is a general feature of parasites (for a review see Bryant, 1970), and the relatively low level of pyruvate kinase also appears to be a common phenomenon so it is possible that the model outlined here has a wider application.

Saz (1970, 1971) has, in two recent reviews, high-lighted the differences between aerobic and anaerobic systems in invertebrates, and has drawn attention to the low activity of pyruvate kinase in many parasitic helminths. It would be interesting to know whether the pyruvate kinases of these animals are susceptible to activation by fructose 1,6 diphosphate, especially in cases where the enzyme has not been detected.

Acknowledgements

I wish to express my gratitude to Miss L. Shaw for expert technical assistance, and to Dr. F. Bygrave and other members of the Department of Biochemistry, A.N.U., for much fruitful discussion. The shortcomings of the present paper are in no way their responsibility. I would also like to acknowledge the generous grant towards the cost of this work made by The Rural Credits Fund of the Reserve Bank of Australia.

References

ADAM, H., In "Methods of Enzymatic Analysis" (edited by Bergmeyer, H-U.) p 573 (1963). Academic Press, N.Y.

AGOSIN, M. and REPETTO, Y., Comp. Biochem. Physiol. **8,** 245 (1963).

AGOSIN, M. and REPETTO, Y., Comp. Biochem. Physiol. **14,** 299 (1965).

AWAPARA, J. and CAMPBELL, J.W., Comp. Biochem. Physiol. **11,** 231 (1964).

BAILEY, E., STIRPE, F., and TAYLOR, C.B., Biochem. J. **108,** 427 (1968).

BANDURSKI, R.S. and LIPMANN, F., J. biol. Chem. **219,** 741 (1956).

BERGMEYER, H-U., BERNT, E. and HESS, B., In "Methods of Enzymatic Analysis" (edited by Bergmeyer, H-U.) p 736 (1963), Academic Press, N.Y.

BRYANT, C., Adv. in Parasitol. **8,** 139 (1970), Academic Press, New York.

BRYANT, C. and JANSSENS, P.A. Comp. Biochem. Physiol. **30,** 841 (1969).

BRYANT, C. and MORSETH, D.J., Comp. Biochem. Physiol. **25,** 541 (1968).

BUEDING, E. and SAZ, H.J., Comp. Biochem. Physiol. **24,** 511 (1968).

CAMPBELL, J.W., Nature, Lond. **208,** 1299 (1965).

CAMPBELL, J.W. and LEE, T.W., Comp. Biochem. Physiol. **8,** 29 (1963).

CHEAH, K.S., Comp. Biochem. Physiol. **23,** 277 (1967).

CHEAH, K.S., Biochim. biophys. Acta, **153,** 718 (1968).

CHEAH, K.S. (1971). This volume, chapter 33.

CHEAH, K.S. and BRYANT, C., Comp. Biochem. Physiol. **19,** 197 (1966).

DAVEY, R.A. and BRYANT, C. Comp. Biochem. Physiol. **31,** 503 (1969).

FAIRBAIRN, D. Expl. Parasit. **3,** 52 (1954).

FURLONGER, C.A. (1971). M.Sc. thesis, Australian National University. In preparation.

GRAFF, D.J., J. Parasit. **51,** 72 (1965).

HAMMEN, C.S. and LUM, S.C., J. biol. Chem., **237,** 2419 (1962).

HAMMEN, C.S. and LUM, S.C., Nature, Lond. **201,** 414 (1964).

HAMMEN, C.S. and WILBUR, K.M., J. biol. Chem. **234,** 1268 (1959).

HORVATH, K. and FISHER, F.M. Jr., J. Parasitol. **57,** 440 (1971).

JANSSENS, P.A. and BRYANT, C., Comp. Biochem. Physiol. **30,** 261 (1969).

KURAHASHI, K., PENNINGTON, R.J. and UTTER, M.F., J. Biol. Chem. **226,** 1059 (1957).

LANE, M.D., CHANG, H.C. and MILLER, R.S. In Methods in Enzymology (edited by Lowenstein, J.M.) Vol. 13, p. 270, (1969). Academic Press, N.Y.

LAYNE, E., In Methods in Enzymology (edited by Colowick, S.P. & Kaplan, N.O.) Vol. 3, p. 447. (1957). Academic Press, N.Y.

LOWRY, O.H., ROSEBROUGH, N.J., FARR, A.L. and RANDALL, R.J., J. biol. Chem. **193,** 265 (1951).

PRESCOTT, L.M. and CAMPBELL, J.W., Comp. Biochem. Physiol. **14** 491 (1965).

PRICHARD, R.K. and SCHOFIELD, P.J., Comp. Biochem. Physiol. **24,** 773 (1968 a).

PRICHARD, R.K. and SCHOFIELD, P.J., Comp. Biochem. Physiol. **24,** 697 (1968 b).

SAZ, H.J., J. Parasitol. **56,** 634 (1970).

SAZ, H.J., Am. Zoologist, **11,** 125 (1971).

SAZ, H.J. and HUBBARD, J.A., J. biol. Chem. **225,** 921 (1957).

SAZ, H.J. and LESCURE, O.L., Comp. Biochem. Physiol. **22,** 15 (1967).

SAZ, H.J. and VIDRINE, A., J. biol. Chem. **234,** 2001 (1959).

SCHEIBEL, L.W. and SAZ, H.J., Comp. Biochem. Physiol. **18,** 151 (1966).

SCHEIBEL, L.W., SAZ, H.J. and BUEDING, E., J. biol. Chem. **243,** 2229 (1968).
SMITH, M.J.H. and MOSES, V., Biochem. J. **76,** 579 (1960).
VON BRAND, T. (1933), Z. vergl. Physiol. **18,** 562 (1933).
VON BRAND, T., CHURCHWELL, F. and ECKERT, J., Expl. Parasit. **23,** 309 (1968).
WARD, C.W., CASTRO, G.A. and FAIRBAIRN, D., J. Parasitol. **55,** 67 (1969).
WARD, C.W., SCHOFIELD, P.J. and JOHNSTONE, I.L., Comp. Biochem. Physiol. **26,** 537 (1968 a).
WARD, C.W., SCHOFIELD, P.J. and JOHNSTON, I.L., Comp. Biochem. Physiol. **24,** 643 (1968 b).

Table 1: Requirements for Phosphoenolpyruvate Carboxykinase Activity in *Moniezia expansa.*

The enzyme was assayed routinely at 30° in a total volume of 1.0 ml, containing 100.0 μmoles imidazole buffer (Cl^-), pH 6.6; 49.0 μmoles $KHCO_3$; 1.0 μmoles $NaH^{14}CO_3$, diluted to give a specific activity of 1.0 m Ci/mM; 1.25 μmoles IDP; 1.0 μmoles $MnCl_2$; 2.0 μmoles reduced glutathione; 2.5 μmoles NADH; 5 units malic dehydrogenase; 0.5 ml homogenate of *M. expansa* prepared in four times its own weight of 0.25 M glycyl glycine buffer, pH 7.4. The reaction was started by the addition of 1.25 μmoles PEP. The subsequent procedure was that of Lane **et al.** (1969). The reaction mixture was varied according to the entry in the column headed "system". Where present, 1.0 μmoles $MgCl_2$ was substituted for $MnCl_2$; 1.25 μmoles GDP for IDP. The total volume of the reaction micture was not varied. Results are expressed in mμmoles CO_2 incorporated/mg protein/min.

System	Activity	No. of Estimations	Standard deviation
Without PEP, IDP, Mn^{++}	0.73	10	± 0.67
+PEP	1.27	5	± 0.13
+PEP, IDP	1.81	5	± 0.04
+PEP, Mn^{++}	6.43	5	± 0.41
+PEP, Mn^{++}, IDP	72.76	10	± 6.97
+PEP, Mg^{++}, IDP	11.32	5	± 4.02
+PEP, Mn^{++}, GDP	85.76	5	± 3.42

Table 2: "Malic Enzyme" activity in *Moniezia expansa*

The enzyme was assayed routinely at $30°$ in a total volume of 1.0 ml, containing 80.0 μmoles triethanolamine HCl buffer, pH 7.4; 10.0 μmoles sodium pyruvate; 1.2 μmoles $MnCl_2$; 1.5 μmoles NADPH, 49.0 μmoles $KHCO_3$; 1.0 μmole NaH $^{14}CO_3$, specific activity 1.0 m Ci/mM. The reaction was started by the addition of 0.5 ml homogenate from *M. expansa* prepared in its own weight of 0.25 M glycylglycine buffer, pH 7.4. The determination of acid stable activity after incubation was as described by Lane **et al.** (1969). The reaction mixture was varied according to the entry in the column headed "system". Where present 1.2 μ moles $MgCl_2$ was subsituted for $MnCl_2$; 1.5 μmoles NADH for NADPH. In one instance 10 μmoles ATP was present.

Results are expressed as mμmoles CO_2 incorporated/mg protein/ min.

System	Activity	No. of Estimations	Standard deviation
Without pyruvate, Mn^{++}, NADPH	2.75	10	± 0.27
+ pyruvate	5.76	5	± 0.87
+ Mn^{++}	6.57	5	± 0.94
+ NADPH	2.08	5	± 0.20
+ pyruvate, NADPH	4.62	5	± 0.13
+ pyruvate, NADPH, Mn^{++}	9.25	5	± 0.40
+ pyruvate, Mn^{++}	10.38	5	± 0.54
+ pyruvate, NADH, Mn^{++}	9.78	5	± 0.80
+ pyruvate, NADH, Mg^{++}	5.49	5	± 0.20
+ pyruvate, NADPH, Mn^{++}, ATP	9.46	5	± 0.51

Table 3: Requirements for Pyruvate Carboxylase Activity in *Moniezia expansa*.

The enzyme was assayed routinely at 30°, in a total volume of 1.0 ml, containing 100.0 μmoles triethanolamine buffer, pH 7.4; 10.0 μmoles sodium pyruvate; 1.0 μmoles ATP; 49.0 μmoles $KHCO_3$; 1.0 μmoles NaH $^{14}CO_3$ specific activity 1.0 m Ci/mM; 5.0 μmoles $MgCl_2$; 1.0 μmole acetyl coenzyme A. The reaction was started by the addition of 0.5 ml of a 1:1 homogenate of *M. expansa* in 0.25 M glycylglycine buffer, pH 7.4. The procedure after incubation was that of Lane **et al.** (1969). The reaction mixture was varied according to the entry in the column headed "system". The total volume of reaction mixture was not varied.

Results are expressed as mμmoles CO_2 incorporated/mg/min.

System	Activity	No. of Estimations	Standard deviation
Without pyruvate, Acetyl coenzyme A, Mg^{++}, ATP	1.41	6	$\pm$0.40
+ pyruvate	3.28	6	$\pm$0.47
+ pyruvate, acetyl coenzyme A.	3.35	6	$\pm$0.27
+ pyruvate, acetyl coenzyme A, Mg^{++}	4.56	6	$\pm$0.41
+ pyruvate, acetyl coenzyme A, Mg^{++}, ATP	9.38	6	$\pm$0.87

Table 4: Requirements for Pyruvate Kinase Activity in *Moniezia expansa.*

The enzyme was assayed routinely at 30°, in a total volume of 2.5 ml according to the method described by Adam (1963), in a spectrophotometer at 340 mμ. The reaction mixture contained 150.0 μmoles triethanolamine buffer, pH 7.5; 225.0 μmoles KCl; 0.75 μmoles ADP; 0.15 μmoles NADH; 0.6 μmoles PEP; 10.0 μmoles MgCl$_2$; 10 μg lactic dehydrogenase. The reaction was initiated by the addition of 100 μl of the supernatant fraction abtained after centrifuging a 1:1 homogenate of *M. expansa* in 0.25 M glycylglycine buffer, pH 7.4 at 20,000 g for 30 min.

The reaction mixture was varied according to the entry in the column headed "system". Where present, fructose 1, 6 diphosphate was at a concentration of 1.0 mM; CaCl$_2$ was in equimolar concentrations with MgCl$_2$.

Results expressed as mμmoles pyruvate formed/mg protein/min.

System	Activity	No. of Estimations	Standard deviation
Without PEP, Mg^{++}, ADP	0	6	—
+ PEP	0	6	—
+ PEP, Mg^{++}	0	6	—
+ PEP, Mg^{++}, ADP	6.57	6	$\pm$0.47
+ PEP, Mg^{++}, ADP, fructose 1,6 diphosphate	15.68	6	$\pm$1.01
+ PEP, Mg^{++}, ADP, fructose 1,6 diphosphate, Ca^{++}	7.84	6	$\pm$0.402

Table 5: Distribution of Enzyme Activities in Soluble and Insoluble Fractions from *Moniezia expansa.*

Activities were determined as in tables 1 to 4. The soluble fraction was prepared by centrifuging homogenates of **M. expansa** in 0.25 M glycylglycine buffer, pH 7.4 at 20,000 g for 30 minutes. The insoluble fraction from this treatment was reconstituted in glycylglycine buffer and used in the assays.

Results are expressed as percentage of total activity.

Enzyme	Soluble Fraction	Insoluble Fraction
Phosphoenolpyruvate carboxykinase	85.5	14.5
"Malic enzyme"	90.9	9.1
Pyruvate kinase	100.0	—
Pyruvate carboxylase	89.6	10.4

Table 6: The Distribution of Radiocarbon from $H^{14}CO_3^-$ amongst the Soluble Intermediates in Homogenates of *Moniezia expansa*.

The reaction mixture was identical with that described in Table 1, except that volumes and quantities of reactions added were multiplied by 6, NADH and malic dehydrogenase were omitted, the $NaH^{14}CO_3$ was undiluted, and the homogenate was prepared in a concentration of one part *M. expansa* and one part glycylglycine buffer. The reactions were carried out under air or high purity nitrogen after preliminary gassing. After one hour's incubation at 37° aliquots were taken, extracted and the soluble intermediates investigated by chromatography and autoradiography by the methods of Smith and Moses (1960) and Davey and Bryant (1969).
Results are expressed as a percentage of total incorporation.

Soluble Intermediate	Air	Nitrogen
Aspartic acid	0.2	0.2
Malic acid	0.5	0.3
Fumaric acid	0.4	0.3
Succinic acid	88.4	91.8
Lactic acid	7.5	6.3
Alanine	1.0	1.1
α-ketoglutaric acid	0.1	0
citric acid	1.4	0
glutamic acid	0.5	0
Total CO_2 fixation (μmoles/mg protein)	15.1	11.6

Table 7: The Effects of Various Additions on the Total Incorporation of CO_2 by Homogenates of *Moniezia expansa*.

Methods as described in Table 6. 30.0 μmoles of the appropriate addition were present in the incubation medium.

Results are expressed as μmoles CO_2 fixed/mg protein.

Addition	Air	Nitrogen
None	7.3 ± 0.5	2.5 ± 0.3
PEP	15.1 ± 0.9	11.6 ± 0.7
Pyruvic acid	14.5 ± 1.1	10.5 ± 0.6
Fumaric acid	3.8 ± 0.4	2.2 ± 0.3
PEP + fumaric acid	10.8 ± 0.6	11.3 ± 0.8

Table 8: The Effect, on Total Incorporation of CO_2 by Intact *Moniezia expansa*, of Various Inclusions in the Maintenance Medium.

The anterior gram of *M. expansa* was incubated, under air or high-purity nitrogen, with 5.0 ml KRP containing 30 μmoles of the additives indicated below, and 1.0 mmoles $NaH^{14}CO_3$, diluted to give a specific activity of 1.0 μCi/mM. At the end of 3 hours incubation at 37°, the worms were homogenised in their own medium and the acid stable radioactivity determined by the method of Lane **et al.** (1969). Results are expressed as total μmoles CO_2 fixed/mg. protein, and are the means of 10 estimations.

System	Air		Nitrogen	
	Mean	S.D.	Mean	S.D.
No additions	8.02	±4.36	20.18	±3.19
+ NaCl	8.10	±7.40	25.79	±6.15
+PEP	7.95	±2.80	22.51	±2.10
+ Pyruvate	6.62	±4.52	17.52	±2.03
+ Propionate	7.71	±0.94	24.15	±5.38
+ Acetate	5.45	±1.32	20.57	±6.86
+ Glucose	8.10	±1.09	17.53	±4.05
+ Fumarate	33.11	±3.12	46.51	±7.43

Table 9: The Distribution of Radiocarbon from $H^{14}CO_3^-$ amongst the Soluble Intermediates In and Excreted by Intact *Moniezia expansa*.

The anterior gram of *M. expansa* was incubated, under air or high-purity nitrogen, with 5.0 ml KRP containing either no additions or 150 μmoles glucose, and 1.0 mmole $NaH^{14}CO_3$, diluted to a specific activity of 50 μCi/mM. At the end of 3 hours incubation at 37°, the worms were separated from the medium, washed rapidly three times in KRP, and homogenised in 4 ml ethanol. The soluble intermediates in the worms and in the media were analysed separately as described in Table 6.
Results are expressed as the percentage distribution of radiocarbon after 3 hours.

Soluble Intermediate	Air			
	medium		worm extract	
	No additions	25mM+ glucose	No additions	25mM+ glucose
Aspartic acid	0	0	1.0	0.5
Malic acid	4.9	0	4.2	1.5
Fumaric acid	1.3	0.4	1.5	1.0
Succinic acid	53.9	19.0	66.1	27.2
Lactic acid	38.4	79.9	23.2	66.1
Alanine	1.0	0.6	2.1	2.6
α ketoglutaric acid	0	0	0.2	0
citric acid	0.3	0.1	0.3	0.3
glutamic acid	0	0	0.4	0
glycine + serine	0.2	0	1.0	0.8
Total CO_2 fixation, (μmoles/mg protein)	6.72	7.29	2.49	2.07

continued

Soluble Intermediate	Nitrogen			
	medium		worm extract	
	No additions	25mM$^+$ glucose	No additions	25mM$^+$ glucose
Aspartic acid	0	0	0.4	0.6
Malic acid	0.1	0.2	1.2	0.8
Fumaric acid	0.5	1.0	4.3	4.7
Succinic acid	11.7	71.3	34.9	63.2
Lactic acid	86.1	27.2	53.0	29.9
Alanine	1.0	0.3	3.8	0.8
α-ketoglutaric acid	0	0	0.2	0
citric acid	0.6	0	0.1	0
glutamic acid	0	0	0.1	0
glycine + serine	0	0	2.0	0
Total CO_2 fixation (μmoles/mg protein)	4.22	13.36	12.49	3.19

Table 10: The Relative Activities of the Enzymes Metabolising PEP and Pyruvate in *Moniezia expansa*.

Enzyme	Relative Activity
Phosphoenolpyruvate carboxykinase	12.8
"malic enzyme"	1.6
Pyruvate carboxylase	1.4
Pyruvate kinase	1.0
Pyruvate kinase + fructose 1,6 diphosphate	2.5

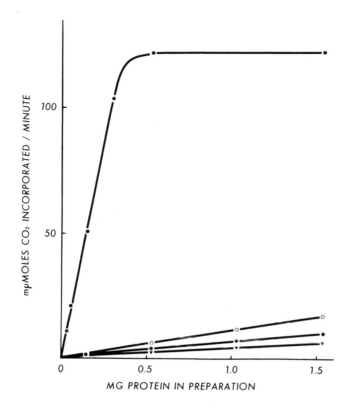

Fig. 1.

The Activity of Enzymes Metabolising PEP and Pyruvate in Preparations from *Moniezia expansa.*

Uptake of CO_2 as a function of the concentration of protein in the preparation. ● — ●, phosphoenolpyruvate carboxykinase; ◆ — ◆, pyruvate carboxylase; ▼ — ▼, "malic enzyme".

○ — ○, pyruvate kinase activity plotted on the same scale, but with mμmoles pyruvate formed/min as the ordinate. The enzymes were assayed as described in tables 1 to 4.

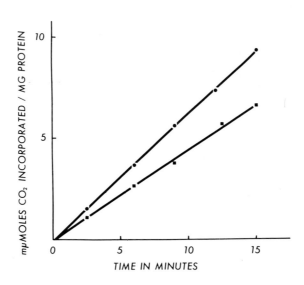

Fig. 2.
The Activity of Enzymes Metabolising Pyruvate in Preparations from
Moniezia expansa.

Uptake of CO_2 as a function of time. ●—●, pyruvate carboxylase; ■—■, "malic enzyme". The enzymes were essayed as described in table 2 and 3.

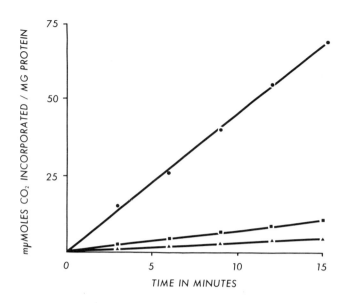

Fig. 3.
The Activity of Enzymes Metabolising PEP in Preparations from *Moniezia expansa.*

The figure shows phosphoenolpyruvate carboxykinase activity ($\bullet - \bullet$), expressed as mμmoles CO_2 incorporated/mg of protein, as a function of time. Pyruvate kinase activity is plotted on the same scale, but with mμmoles pyruvate formed/min as the ordinate. $\blacktriangle - \blacktriangle$ pyruvate kinase; $\blacksquare - \blacksquare$ pyruvate kinase + 1 mM fructose 1,6 diphosphate. The enzymes were assayed as described in tables 1 and 4.

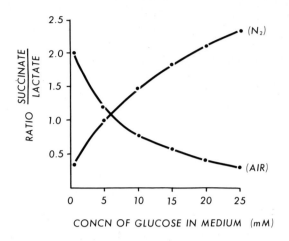

Fig. 4.

Succinate and Lactate Production by Intact *Moniezia expansa.*

The figure shows the canges in the ratio $\frac{\text{total succinate produced}}{\text{total lactate produced}}$ with increasing concentrations of glucose in nitrogen (N_2) and in air. Experimental conditions as described in table 9.

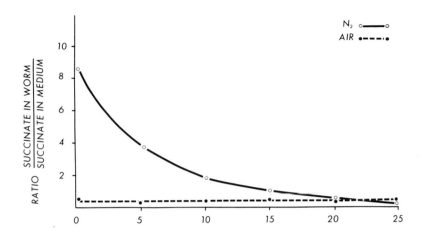

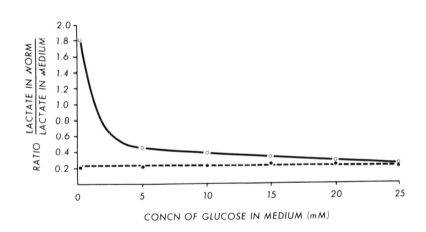

CONCN OF GLUCOSE IN MEDIUM (mM)

Fig. 5.

Succinate and Lactate Production by Intact *Moniezia expansa.*

The figures show the changes in the ratio $\frac{\text{succinate (or lactate) retained}}{\text{succinate (or lactate) excreted}}$ $\frac{\text{by the worm}}{\text{into the medium}}$ with increasing concentrations of glucose, in nitrogen and in air. Figure 5a (above), succinate; figure 5b (below), lactate. ○—○, in nitrogen; ●---●, in air.

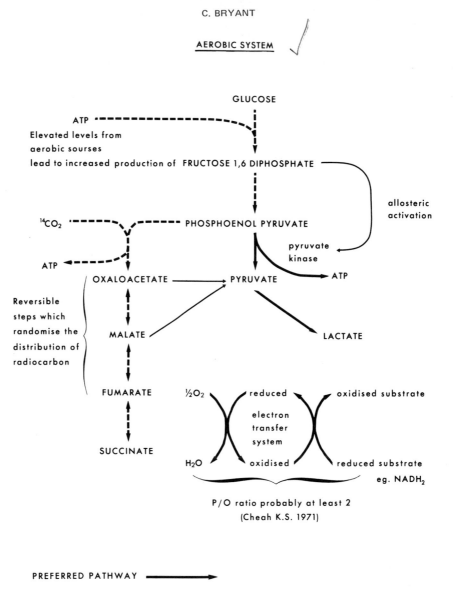

Fig. 6.

Regulation of Succinate and Lactate Production in *Moniezia expansa*: **an Hypothesis.**

Under aerobic conditions (left) with oxygen as the terminal electron acceptor, elevated levels of ATP lead to increased production of FDP,

ANAEROBIC SYSTEM

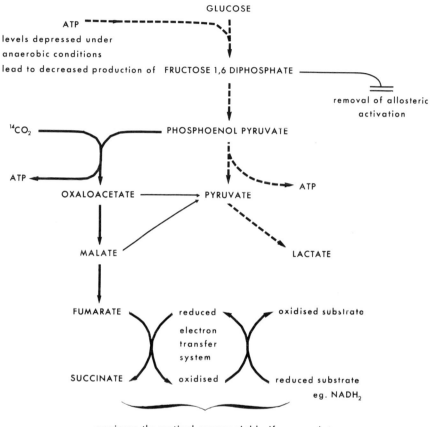

GLUCOSE

ATP

levels depressed under
anaerobic conditions

lead to decreased production of FRUCTOSE 1,6 DIPHOSPHATE

removal of allosteric
activation

$^{14}CO_2$ PHOSPHOENOL PYRUVATE

ATP   ATP

OXALOACETATE   PYRUVATE

MALATE   LACTATE

FUMARATE   reduced   oxidised substrate

electron
transfer
system

SUCCINATE   oxidised   reduced substrate
eg. $NADH_2$

maximum theoretical energy yield , if appropriate
enzyme systems exist, is 1 molecule ATP

PREFERRED PATHWAY

which activates pyruvate kinase. Lactate production is favoured. Under anaerobic conditions (right), fumarate is the terminal electron acceptor, resulting in decreased ATP and FDP production, thus depressing pyruvate kinase activity. Succinate production is favoured.

5

HORMONES, THE ENVIRONMENT, AND DEVELOPMENT IN NEMATODES

K.G. Davey

Institute of Parasitology, Macdonald
Campus of McGill University, Box 231,
Macdonald College, P.Que., Canada.

The existence of parasitology as a universally recognized and distinct discipline suggests that there is a common thread which links the diverse groups of animals recognized as parasitic. While a symposium on the biochemistry of parasites might suggest that parasitic organisms share biochemical mechanisms uncommon in free-living groups, a much less subtle characteristic of parasites is worth exploring.

Parasites are, after all, called parasites because they live in or on other animals, and parasitology is the study of animals which live in a particular sort of environment. But do the various parasitic environments have a common characteristic rather different from other environments?

At first sight the answer is no. There is no obvious similarity between, for example, the hind gut of insects and the blood system of mammals - certainly their biochemical attributes are very different - yet both act as environments for a variety of parasites. These parasitic environments do, however, share an important general characteristic - that of discontinuity. The environment in which a parasite finds itself is discontinuous in SPACE. Thus in *Ascaris,* for example, one pig intestine is very much like another, yet the total environment available to the nematodes consists of a large number of identical units, each separated spatially from the others by virtue of being in different pigs. Similarly, because pigs are not immortal, the environment of an *Ascaris* is also sharply discontinuous in TIME.

Parasite populations are, therefore, faced with the task of getting from one unit of their environment to another. In order to do so they may have to pass through a radically different sort of environment, either as a free-living or dormant stage or in a vector or intermediate host. Natural selection would predict a high degree of developmental and physiological plasticity among parasitic organisms. Indeed, it may be fruitful to regard the ability to exhibit polymorphic development in the sense in which Wigglesworth (1954, 1961) has developed the concept for insects as a prerequisite for successful parasitism. The trematodes, with their succession of different forms, each suited to the particular environment in which it finds itself, and each developing by a renewal of embryonic growth among undifferentiated and genetically identical cells, form an excellent example of polymorphic development.

Polymorphic development is also well illustrated in other parasitic groups, including, as we shall see in a moment, the nematodes. It is important first of all, however, to emphasize the close link which exists between the environment and the different developmental and physiological events which occur during the life of a parasitic organism. Thus, to use the obvious example of a trematode, it is important that the undifferentiated mass of cells in each of the larval stages gives rise to the appropriate larval form. Since the forms arise by a kind of polyembryony, the genetic constitution of each of the embryonic cells must be identical and it is the environment in which the cells develop that, in the last analysis, determines the direction of development. This is not to say that there is necessarily a direct link between the environment and the developing cells. In many cases the progression of developmental stages of flukes is invariable and regarded as a characteristic of the species. On the other hand, it is well known, for example, that only one generation of rediae is produced in *Fasciola hepatica* during the cooler months, whereas a second generation of daughter rediae may be interposed before the production of cercariae in the summer. Apparently the environment of the host is in some way communicated to the parasite, directing the development of the embryonic cells into one or the other developmental pathway. A more spectacular example of the plasticity of the develomental stages of trematodes is described by Sewell (1922) in his monumental work on the cercaria of India. He describes the production of miracidia by sporocysts of his "Cercaria Indica XV". Apparently sporocysts in this trematode are able to give rise to either miracidia or to daughter sporocysts, and daughter sporocysts can contain either cercaria or miracidia, depending upon unspecified environmental conditions.

It is clear, then, that parasites possess a remarkable developmental plasticity which can be compared to the developmental polymorphism which occurs in insects. Because the environment is discontinuous in space and time, developmental events are likely to be closely associated with the abrupt environmental changes which occur when the parasite moves from one unit of its environment to another.

Let us now look more closely at the nematodes, which are not, at first sight, very promising material for such studies. In the first place development is held to be highly determinate and in addition, there are supposedly no cell divisions, except for the reproductive system, during the postembryonic life of the nematode. The result is that nematodes are deceptively simple organisms as far as structure is concerned. There are rather few tissue types, and the number of cells is small. Even in *Ascaris,* Goldschmidt (1908, 1909, 1910) was able to identify and number every cell in the central nervous system.

Yet nematodes, particularly parasitic forms, are able to execute physiological and developmental maneuvers of a complexity that we normally associate with higher organisms. Moreover, these maneuvers are often associated with environmental changes. One has only to think of *Strongyloides stercoralis* with its complex polymorphic development, at least part of which appears to depend on environmental triggers, to appreciate what is possible for these organisms.

If there is a link between the environment and developmental and physiological events in nematodes, what is its nature? How are the environmental stimuli mediated? Given the relative lack of sophistication of the integrative aspects of the nematode central nervous system, and the apparent scarcity of peripheral connections, it is unlikely that there is much in the way of direct nervous control of developmental events. Another means of linking the environment to various developmental events is by endocrines, and since the study of endocrinology becomes increasingly the study of neurosecretion as one proceeds down the evolutionary scale, it is hardly surprising that nematodes have proved to contain nerve cells which exhibit the staining properties of neurosecretory cells. Gersch and Scheffel (1958) described a cell in each of the lateral ganglia of *Ascaris* which stained with both of the classical stains for neurosecretion, paraldehyde-fuchsin and chrome-haemotoxylin. Ishikawa (1961) has also described a neurosecretory cell in *Ascaris.* We have found up to 20 nerve cells in *Ascaris* which stain with paraldehyde-fuchsin, including some of the primary sense cells on the lips (Davey 1964). This latter observation suggests that stimulation of the sense cells may lead directly to release of hormones without further integration, providing the possibility of a direct link between the environment and as yet undetermined target organs.

Affinity for dyes is perhaps not the best indicator of neurosecretory activity. Rogers (1968) has described axones containing neurosecretory granules in *Haemonchus contortus.* In addition, we find abundant evidence of neurosecretion in electron micrographs of the ventral nerve cord of *Phocanema,* even when there is no staining of nerve cells by paraldehyde-fuchsin (Fig. 1).

Ascaris, for a variety of reasons, does not form suitable material for studying possible endocrine relationships. Another ascarid, *Phocanema (Porracaecum, Terranova) decipiens* presents some opportunities for manipulation in the laboratory, and we have exploited it in a series of studies on the control of development in this nematode. It occurs as a fourth stage larva encysted in the muscles of cod-fish and reaches maturity in the intestines of seals. The earlier stages in its life history are uncertain. The worms can be recovered from heavily infected fillets in the laboratory and maintained in 0.9 % NaCl at 5°C in good conditions for a month or more. If the larval worms are put into a complex culture medium (Townsley et al. 1963) at 37°C, development is initiated, and the worms lay down a new adult cuticle and shed the old larval cuticle. We have examined this process in the light microscope (Davey 1965; Kan and Davey 1967, 1968). The details of this study are not important to the present discussion, but it is of interest to point out that the adult cuticle is distinctly different from the larval cuticle. A prominent layer of fibres appears in the larval cuticle, but is absent from the adult cuticle. There is thus evidence in this species of polymorphic development in that the hypodermis is able to produce at least two different sorts of cuticle at different developmental stages, each of which occurs in a different environment. Similarly, the larval cuticle of *Ascaris* (Thust 1966) is very different from the adult cuticle (Watson 1965), and the cuticle of the third stage of *Nippostrongylus brasiliensis* is distinct from that of the adult (Lee 1965, 1966 b). While generalizations are not yet possible, it is interesting to note that the cuticle of the various stages of free-living nematodes such as *Turbatrix aceti* (Lee 1966 a) or *Panagrellus silusae* (Samoiloff and Pasternak 1968) are remarkably similar in structure. Other examples of differences in structure during the life cycle of nematodes could be brought forward, but cuticular structure is an obvious and satisfactory example. This is not to suggest that there is now necessarily a direct control emanating from the environment to produce a particular cuticular pattern: this appears not to be the case. On the other hand we shall see in a moment that environmental stimuli may determine when the various cuticles become functional.

There are neurosecretory cells in the ventral ganglion of *Phocanema* which undergo a cycle of staining with paraldehyde-fuchsin. This cycle is correlated with molting (Davey 1966). Now the process of molting in nematodes is made up of at least two steps. The first involves the formation of the new cuticle and the second is the process by which the old cuticle is shed. In some cases, notably the infective stages of trichostrongyles, there two events are widely separated in time. In *Phocanema,* ecdysis is closely associated with cuticle formation; indeed, the old cuticle may be shed before the new cuticle is fully differentiated. Since there is wide variation among individual worms in the timing of these events, it is not possible to correlate the cycle of neurosecretion with either of these events on observational grounds alone.

However, a simple experimental approach demonstrates that the neurosecretory cells of the anterior end play no part in cuticle formation. If the worms are ligatured so as to separate completely the anterior and posterior halves, both portions will produce adult cuticles which are normal in structure (Davey 1966). Not only does this demonstrate that the formation of the cuticle is not under a control localised in either the anterior or posterior of the worm, but it also reveals that there is no localised control of the **sort** of cuticle - larval or adult - which is formed. Ecdysis is a process about which rather more is known. Much of our knowledge comes from a study of the infective third stage larva of trichostrongyles, which remains ensheathed within the cuticle of the second stage until it receives the appropriate stimuli from the host. The analysis of these stimuli, and of some of the subsequent developmental events, has been carried out largely by W.P. Rogers and R.I. Sommerville in Australia. Their experiments have shown that exsheathment occurs when the infective larvae are exposed to a combination of stimuli provided by the host or by appropriate **in vitro** conditions (Rogers 1960). These stimuli result in the appearance in the space between the two cuticles of a complex mixture of enzymes, the moulting fluid, which digests the old cuticle and allows the worms to free themselves (Rogers and Sommerville 1963). They have further shown (Rogers and Sommerville 1960) that a small area just posterior to the excretory pore is essential to the release of moulting fluid. More recently, Rogers (1963, 1965) has shown that one active principle in the exsheathing fluid is the enzyme leucine aminopeptidase, and that neurosecretory vesicles occur in the ventral nerve of infective larvae.

Work at the Institute of Parasitology has relied heavily on the work of Rogers and Sommerville, and has extended their results to include *Phocanema.* We believe that the neurosecretory cells of the ventral ganglion in *Phocanema* act as a link between an environmental stimulus and the process of ecdysis. In summary, the evidence is as follows:

1. Worms reared in the complete medium ecdyse in a characteristic way

between 3 and 6 days after being put into culture (Davey and Kan 1968).

2. Worms reared in 0.9 % saline produce a completely normal adult cuticle, but fail to ecdyse (Davey and Kan 1967, 1968).

3. The neurosecretory cells of the ventral ganglion are inactive in worms reared in saline (Davey and Kan 1967, 1968).

4. The excretory glands of worms cultured in the complete medium show a cycle of secretion of protein culminating in the release, just prior to ecdysis, of material into the space between the two cuticles. This cycle of secretion is absent in worms cultured in saline (Davey and Kan 1968).

5. The enzyme leucine aminopeptidase, which is known to be the principal component of exsheathing fluid in trichostrongyles (Rogers 1965), can be demonstrated histochemically to increase in activity in the excretory gland during the first three days in culture, after which there is a decrease in activity. This cycle is absent from worms cultured in 0.9 % NaCl (Davey and Kan 1967, 1968).

6. Isolated excretory glands will exhibit an increase in leucine aminopeptidase activity when exposed to extracts of the heads of worms which obtain active neurosecretory cells. They fail to do so when exposed to extracts of heads in which the neurosecretory cells are inactive (Davey and Kan 1967, 1968).

The implications of these results seem clear enough. The complete medium contains elements which stimulate the neurosecretory cells, probably via the cephalic sense organs, to release a hormone. This hormone then brings about an increase in activity of leucine aminopeptidase. The hormone may also stimulate the excretory cell to release its mixture of enzymes via the excretory duct into the space between the two cuticles, resulting in the weakening of the old cuticle so that it is ruptured by the adult worm. On the other hand, the saline lacks the elements necessary to stimulate the neurosecretory cells, the enzyme is not activated, and ecdysis does not occur.

Thus, the neurosecretory cells and the excretory gland constitute two links in the chain between the environmental stimulus and a developmental event, ecdysis. Because of the large size of the excretory cell, and the ease with which it can be handled **in vitro,** this system presents unparalleled opportunities for examining the mode of action of the hormone.

R.I. Sommerville and I have recently had an opportunity to collaborate on this aspect of the work, and what follows is a very brief summary of some of our observations. There is sufficient enzyme activity in a single gland to perform a colorimetric assay for aminopeptidase activity using homogenates (Green et al. 1955). Although the histochemical assay which we have used continues to demonstrate that the enzyme activity increases upon exposure to head extracts, this proved not to be the case when the colorimetric assay was performed on homogenates. We have been unable to demonstrate an increase in activity of the enzyme in glands which have been exposed to head extracts, when the assay involved homogenizing the gland. Furthermore, preliminary experiments have demonstrated that there is no increase in incorporation of labelled leucine or tyrosine in glands exposed to the head extract. These experiments suggest that the effect of the hormone may be to activate or free the enzyme which is held in an inactive or bound form. Disrupting the cell apparently also frees the bound enzyme.

In the electron microscope, the most obvious organelle in the gland consists of a membrane bound granule of uniform electron density (Fig. 2). In excretory glands which have been exposed to the hormone, these granules exhibit changes in their electron density, and the membranes surrounding them tend not to survive the fixation process (Fig. 3). This evidence suggests that the hormone acts by altering the permeability of the membranes surrounding the granule, but we have no concrete evidence yet which suggests that the granules contain an aminopeptidase. For the purposes of the present paper, however, we are less interested in the mode of action of the hormone than we are in the existence of this relatively simple link between the environmental stimulus and the developmental event. We know nothing about the nature of the stimulus, present in the medium, but absent from saline, which stimulates the neurosecretory cells to produce and release their hormone. Indeed, there is no reason to believe that the factors present in the **in vitro** system are the same as the conditions which obtain **in vivo** in the intestines of seals. Given the involvement of the nervous system, one can envisage a variety of different environmental stimuli impinging on the neuroendocrine system to yield the same developmental event in a variety of species. Thus, the environmental stimuli that evoke exsheathment in *Haemonchus* are presumably different from those that bring about ecdysis in *Phoca-nema,* but in *Haemonchus* it is likely that the neurosecretory system is also the mediator between stimulus and exsheathment (Rogers 1968). The possibility that the cephalic sense organs in *Ascaris* may be neurosecretory enhances the potential functional complexity of an anatomically simple system. Thus, the cephalic sense organs may respond to some stimuli by secreting as independent effectors and to others by conducting information to the ganglia of the central nervous system.

The possibility exists that neurosecretory cells may control other important developmental events. For example, neurosecretion disappears from the ganglia of *Phocanema* after ecdysis, but reappears again within two weeks when development of the reproductive system is proceeding apace. We have not yet investigated a possible functional link between these two events, but on **a priori** grounds alone, such a link would scarcely be surprising.

The fact that a developmental event is controlled by the environment acting **via** the endocrine system contains the kernel of a valuable lesson for those who work with nematodes in **in vitro** culture. This lesson is well illustrated by recent work on the effect of compounds with insect juvenile hormone activity on development in nematodes.

Meerovitch (1965) first described an effect of farnesol, which is a mimic of insect juvenile hormone, on *Trichinella spiralis* **in vitro** and later (Shanta and Meerovitch) extended these results to include the more potent farnesyl methyl ether (FME). Treatment with FME at concentrations as low as 10^{-7}M produced a delay in development of the worms. Johnson and Viglierchio (1970) produced a variety of bizarre developmental abnormalities in the sugar beet nematode, *Heterodera schactii* as a result of exposure to high concentrations of FME and farnesydiethylamine.

Our own work (Davey 1971) has shown that the effects of FME and of a synthetic juvenile hormone depend on the timing and conditions of their application. If the materials are present from the beginning in the normal culture medium, ecdysis does not occur. If, on the other hand, the worms are reared in 0.9 % NaCl (where they are normally destined not to ecdyse) and the materials are added 2 1/2 days after the beginning of culture, when the new cuticle is well formed, a normal ecdysis occurs in many of the worms. We have shown that the effect of FME or juvenile hormone is to stimulate the neurosecretory system. When the hormone is added to the saline culture, in which neurosecretory cells are normally not observed, the neurosecretory cells of the ventral ganglia stain prominently, and the excretory gland secretes a material with leucine aminopeptidase activity. The inhibition of ecdysis when the compounds are present in the normal culture medium from the beginning is interpreted as a result of exhaustion of the system due to its premature activation. The ability of rather high doses of FME to stimulate the neurosecretory system is not limited to nematodes: we have demonstrated a similar effect in *Rhodnius prolixus* (Pratt and Davey 1971). On the other hand, we have been unable to bring about exsheathment in *Haemonchus contortus* with FME or juvenile hormone.

But these compounds have other, less obvious effects in *Phocanema* (Davey 1971), including some disruption of the outer layer of the new cuticle, and of the attachment of the new cuticle to the hypodermis, as well as an apparent prolongation of the period of lipid mobilisation which normally accompanies molting (Kan and Davey 1968). None of these phenomena is known to be associated with neurosecretion in the normal worm, and it is thus not clear whether the insect juvenile hormone mimetics are also having a direct effect on developmental processes.

However, it is clear that worms treated with FME or juvenile hormone exhibit abnormally large amounts of neurosecretion in the ventral ganglion and that neurosecretory axons within the ventral nerve cord become much more apparent (Davey 1971). These compounds are apparently not selective in their action but appear to act generally on the whole neurosecretory apparatus of the animal. Whether they act directly on the neurosecretory cells or more generally by disrupting the normal pattern of sensory input remains to be seen. The fact that the compounds can depolarise membranes is surely of interest in this regard (Baumann 1968).

While juvenile hormone and its mimics can interfere with development in some nematodes, it is probably unwise to ascribe any functional significance to these phenomena. Even if, as Shanta and Meerovitch (1970) have suggested, substances with juvenile hormone activity should be found in nematodes, great caution should be exercised. Substances with juvenile hormone activity can be extracted from a wide variety of plant and animal sources (Williams, Moorhead and Pulis 1959; Schneiderman and Gilbert 1958; Highnam and Hill (1969). It would scarcely be surprising if nematodes were found to contain similar material. It is important to remember that relatively high concentrations are required to yield effects in nematodes, and that in the only case in which we have an inkling of their mode of action, the compounds act through the neuroendocrine system of the worm.

Thus I view these compounds as simply another stressful factor in the **in vitro** environment. I have already suggested elsewhere that in insects (Davey 1963) and nematodes (Davey 1964), massive and unaccustomed sensory input brings about, *inter alia,* a release of endocrines. Perhaps some of our failures with **in vitro** culture of parasitic organisms may not be due to the absence of nutritional factors but to the presence of materials or conditions which impose a stress on the organism. It may be just as important to make the worms comfortable as it is to offer them nutritious food.

References

BAUMAN, G. Journal of Insect Physiology **14,** 1459 (1968).
DAVEY, K.G. Journal of Insect Physiology **9,** 375 (1963).
DAVEY, K.G. Canadian Journal of Zoology **42,** 731 (1964).
DAVEY, K.G. Canadian Journal of Zoology **43,** 997 (1965).
DAVEY, K.G. American Zoologist **6,** 243 (1966).
DAVEY, K.G. International Journal for Parasitology **1,** 61 (1971).
DAVEY, K.G. and KAN, S.P. Nature **214,** 737 (1967).
DAVEY, K.G. and KAN, S.P. Canadian Journal of Zoology **46,** 893 (1968).
GERSCH, M. and SCHEFFEL, H. Naturwissenschaften **45,** 345 (1958).
GOLDSCHMIDT, R.B. Zeitschrift für Wissenschaftliche Zoologie **90,** 73 (1908).
GOLDSCHMIDT, R.B. Zeitschrift für Wissenschaftliche Zoologie **92,** 306 (1909).
GOLDSCHMITDT, R.B. Festschrift für 60 Geburtstag Richard Hertwigs München **2,** 253 (1910).
GREEN, M.N., TSOU SWAN-CHUNG, BRESSLER, R. and SELIGMAN, A.M. Archives of Biochemistry and Biophysics **57,** 458 (1955).
HIGHNAM, K.C. and HILL, L.C. 1969. The comparative endocrinology of the invertebrates. 270 pp. Edward Arnold, London.
ISHIKAWA, M. 1961. Kiseichagaku Zasshi. **10,** 1 (1961).
JOHNSON, R.N. and VIGLIERCHIO, D.R. Experimental Parasitology **27,** 301 (1970).
KAN, S.P. and DAVEY, K.G. Canadian Journal of Zoology **46,** 235 (1967).
KAN, S.P. and DAVEY, K.G. Canadian Journal of Zoology **46,** 723 (1968).
LEE, D.L. Parasitology **55,** 173 (1965).
LEE, D.L. **In** Advances in Parasitology. (**Ed.** B. Dawes) **4,** 187 (1966 a).
LEE, D.L. Parasitology **56,** 127 (1966 b).
MEEROVITCH, E. Canadian Journal of Zoology **43,** 81 (1965).
PRATT, G.E. and DAVEY, K.G. Journal of Experimental Biology (1971). In the press.
ROGERS, W.P. Proceedings of the Royal Society B. **152,** 367 (1960).
ROGERS, W.P. Annals of the New York Academy of Science. **113,** 208 (1963).
ROGERS, W.P. Comparative Biochemistry and Physiology. **14,** 311 (1965).
ROGERS, W.P. Parasitology **58,** 657 (1968).
ROGERS, W.P. and SOMMERVILLE, R.I. Parasitology **50,** 329 (1960).
ROGERS, W.P. and SOMMERVILLE, R.I. **In** B.E. Dawes (ed.), Advances in Parasitology **1,** 109 (1963). Academic Press, London and New York.
SAMOILOFF, M.R. and PASTERNAK, J. Canadian Journal of Zoology **46,** 1019 (1968).
SEWELL, R.B. Seymour. Cercariae Indicae. Indian Journal of Medical Research **10** Supplement 310 pp. (1922).
SCHNEIDERMAN, H.A. and GILBERT, L.I. Biological Bulletin **115,** 530 (1958).
SHANTA, C.S. and MEEROVITCH, E. Canadian Journal of Zoology **48,** 617 (1970).
THUST, R. Zool Amz. **177,** 411 (1966).
TOWNSLEY, P.M., Wight, H.G., SCOTT, M.A. and HUGHES, M.L. Journal of the Fisheries Research Board of Canada **20,** 743 (1963).
WATSON, B.D. The Quarterly Journal of Microscopical Science **106,** 83 (1965).
WIGGLESWORTH, V.B. The Physiology of Insect Metamorphosis. Cambridge University Press (1954), Cambridge.

WIGGLESWORTH, V.B. Insect polymorphism - A tentative synthesis. In Insect Polymorphism. ed. J.S. Kennedy Symp. 1. Royal Entomological Society London (1961).
WILLIAMS, C.M., MOORHEAD, L.V. and PULIS, J.V. Nature **183,** 405 (1959).

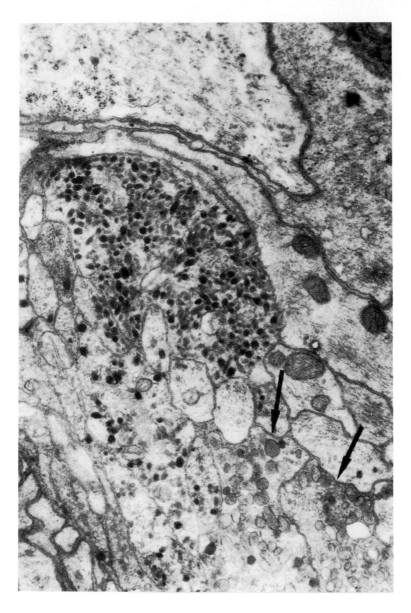

Fig. 1.
Section of the ventral nerve cord of a fourth stage larva of *Phocanema decipiens* to show neurosecretion material in the axones. Two types of membrane-bounded granules are apparent. Many axones contain small (about 500A) dense granules, while other axones (arrow) contain larger (about 1200A) less dense granules.   x 28,000

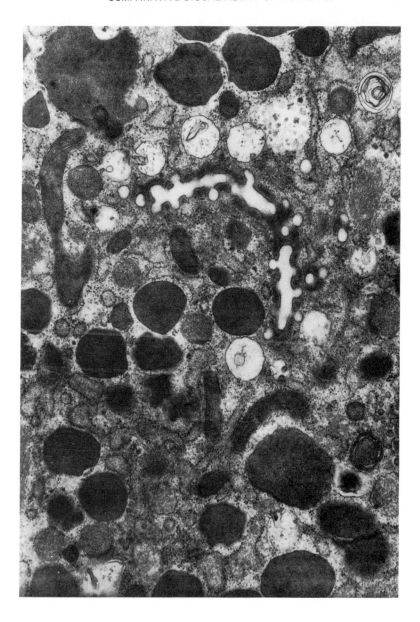

Fig. 2.
Section of a portion of excretory gland from a worm incubated in 0.9 %
NaCl. The large granules are of uniform density.   x 41,000

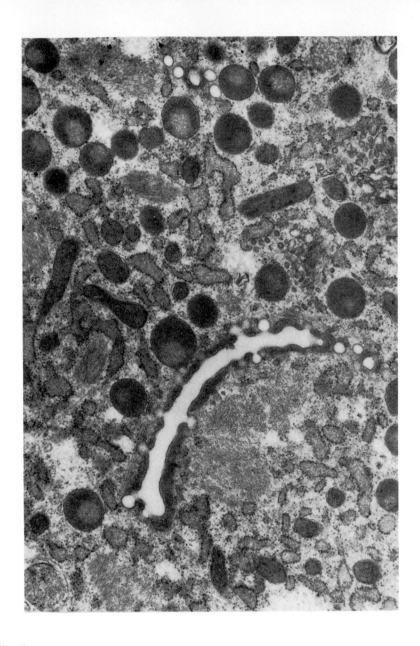

Fig. 3.
A section similar to Figure 2 from a worm incubated with saline
containing FME in order to activate the neurosecretory system. The
granules are no longer of uniform density, and the membranes surrounding them are wrinkled. x 28,000

6

NEUROTRANSMITTERS IN TREMATODES

Ernest Bueding and James Bennett*
Department of Pathobiology
School of Hygiene and Public Health
and
Department of Pharmacology and Experimental Therapeutics
School of Medicine
The Johns Hopkins University
Baltimore, Maryland

The first observations indicating that acetylcholine (ACh) may be of functional significance in parasites were made by Bülbring et al. [1]. They demonstrated the presence of ACh in *Trypanosoma rhodesiense* and its absence in *Plasmodium gallinaceum.* These findings suggested a role of ACh in mediating the motility of flagellate protozoa. Subsequently, ACh and enzymes catalyzing its hydrolysis and its synthesis have been found in two trematodes, *Schistosoma mansoni* and *Fasciola hepatica* [2-4] . In *S. mansoni* the concentration of ACh (5μg per Gm wet weight) is of the same order of magnitude as that of the gray matter of mammalian brain cortex. Furthermore, in the head region of *S. mansoni* AChE and choline acetylase (ChAc) activities are 2 and 4 to 5 times higher, respectively, than in the remainder of the body of the parasite and are of the same order of magnitude as in mammalian brain cortex [5] . The association of AChE with the nervous system of this trematode has been demonstrated by the histochemical localization of this enzyme in the central ganglia of three stages of the life cycle of *S. mansoni,* the miracidia [6] , the cercariae [7] , and the adult worm, as well as in the lateral nerve trunks and the nerves supplying the musculature of adult *S. mansoni, S. hematobium* and *S. japonicum* [8]. Pharmacological evidence indicates that ACh is an inhibitory neuromuscular transmitter in *S. mansoni, F. hepatica,* and possibly in other trematodes. For example, an increase in the level of endogenous ACh produced by exposure of the worms to AChE inhibitors results in a depression of muscular activity which is

* The investigations of the author were supported by grants from the National Institutes of Health (U.S. Public Health Service) and The Rockefeller Foundation

indistinguishable from that induced by carbachol and other cholinomimetic agents, **e.g.,** arecoline [3, 4]. Furthermore, some cholinergic blocking agents, such as atropine and mecamylamine stimulate the motor activity of *S. mansoni* [4]. In this connection mention should be made of the paralysis of the acetabulum and of the oral sucker of *S. mansoni* produced by the administration of the antischistosomal drug p-rosaniline which is associated with a histochemically demonstrable inhibition of AChE in the two muscular organs of the parasite. This paralysis also is relieved by atropine and mecamylamine [8]. The effects of these cholinergic blocking agents can be explained by a block of an interaction of the worm's cholinergic receptors with endogenous ACh. In this manner ACh cannot exert its inhibitory action on the motor activity of the worm.

While the cholinergic receptors of *S. mansoni* and of *F. hepatica* exhibit some similarities with those of their hosts, their responses also suggest differences from any of the three pharmacologically defined cholinoceptive mammalian receptors. For example, in contrast to cholinergically innervated effector organs of vertebrates, the motor activity of schistosomes is not affected by pilocarpine, muscarine, nicotine, or methacholine. In addition, the motor activity of schistosomes is stimulated by only a few ganglionic blocking agents (mecamylamine, pempidine) but it is unresponsive to quaternary ganglionic blocking agents (tetramethylammonium, hexamethonium, pentolinium, chlorisondamine) and to myoneural blocking agents (d-tubocurarine, decamethonium, succinylcholine) [4]. Differences between the isofunctional enzymes catalyzing the degradation and synthesis of ACh in the parasite and the host have been observed also. For example, the KM and optimal substrate concentration of ACh are 10 times higher for schistosome AChE [2], and ChAc of mammalian brain is far more sensitive to the inhibitory effects of some naphthylvinylpyrimidines than is schistosome ChAc [5]. In view of these differences, there is, at least theoretically, a possibility for designing cholinomimetic agents and inhibitors of AChE and ChAc which are selective for the parasite. However, it is unlikely that the reported antischistosomal activity of the organic phosphorus compound metrifonate can be ascribed to a selective inhibitory effect on the activity of AChE of *S. hematobium.* AChE of mouse brain is only three times less sensitive to the inhibitory effects of metrifonate than the isofunctional enzyme of *S. hematobium.* Furthermore, the inhibitory potencies of metrifonate on the acetylcholinesterases of *S. hematobium* and *S. mansoni* are the same [9]; yet metrifonate has been reported to be effective in the treatment of infections caused by the former, but not by the latter, species of schistosomes.

The stimulation of motor activity of schistosomes induced by cholinergic blocking agents suggests the presence in these helminths of an excitatory transmitter whose action opposes that of ACh and which becomes unmasked when the cholinergic receptors are blocked. The motor activity of schistosomes is markedly enhanced by 5-hydroxytryptamine (5-HT) as well as by reserpine, tyramine, chlorimipramine and 4α-dimethyl-meta-tyramine, compounds known to release this amine or to inhibit its reuptake by its storage sites. Exploration of the possible role of 5-HT as an excitatory neurotransmitter in *S. mansoni* has revealed that acid extracts of *S. mansoni* contain a substance whose behavior is identical with that of 5-HT when the following criteria were used: (a) chromatography on silica gel (thin layer) and on a cation exchange resin (Dowex CG-50); (b) fluorescence characteristics and intensities either induced by strong acid or after reaction with ninhydrin [11], and (c) by several highly specific bioassay systems [12]. The concentration of 5-HT in both male and female adults of *S. mansoni* has been found to be 2 to 5 μg per Gm (fresh weight) [13] or 4 to 10 times higher than in mammalian brain cortex. The head portion of the worms contains approximately twice as much 5-HT as the remainder of the body [13], indicating that, in contrast to other invertebrates, the localization of this amine is not limited to the central nervous system of the worms.

In addition to 5-HT, *S. mansoni* also contains norepinephrine (NE); its concentration is 5 to 10 times lower than that of 5-HT [13]. NE has been identified in extracts of the worms by the fluorescence characteristics of its oxidation (trihydroxyindole) product preceded by ion exchange chromatography or by adsorption on, and elution from, alumina [11]. Using these criteria, the properties of the material extracted from the worms are indistinguishable from authentic samples of NE. By contrast, no - or at least only traces of - dopamine have been detected in these worms [14]. This contrasts with the relative concentrations of catecholamines in other invertebrates in which dopamine has been found to predominate and often to be the only one to be present [15, 16].

An attempt has been made to localize 5-HT and NE within the worm by the use of the histochemical fluorescence method of Falck [17]. This procedure is based on the observation that in frozen-dried tissues the reaction products of catecholamines with formaldehyde are recognizable by their green, and that of 5-HT by its yellow fluorescence. When this method is applied to *Schistosoma mansoni,* green (NE) fluorescence is found in many structures which have high AChE activity, i.e., the central ganglia, the commissure connecting them in the head region, and in the two bilateral, longitudinal nerve trunks which end anteriorly near the oral sucker, and posteriorly in the tail region of the worm. Within the nerve trunks green bulb-like structures are separated from each other by a distance of 300 to 400 μ. Throughout the worm the two nerve trunks are connected with each other by a fine network of green fibers which also extend into the acetabulum and other muscle layers [18]. When a nerve trunk had been cut prior to freeze-drying, accumulation of green fluorescent material is observed on both sides of the damaged area, suggesting the occurrence of axonal flow of NE within the nerve trunk. Yellow (5-HT) fluorescent structures are found adjacent to the NE neurons and in small granules throughout the entire worm. After the concentration of 5-HT in the parasite has been increased 5- to 10-fold by prior incubation of the worm with 5-HT or 5-hydroxytryptophan, many yellow granules are observable throughout the worms, giving the appearance of a string of yellow dots. Some penetrate the muscle layer and end just below the dorsal surface or at the base of the tubercle. In addition, on the surface of the gynecophoric canal of the male, protruding yellow fluorescent knobs, suggesting sensory nerve endings proposed by Smith **et al.** [19] , are connected by fibers to larger structures (2 to 3μ in diameter) located in the region adjacent to the muscle layer. [18] . It is noteworthy that under these conditions there is no change in the green fluorescence [11] . Therefore, in contrast to 5-HT containing storage sites, NE neuronal structures fail to take up and to store 5-HT.

Both green and yellow fluorescence disappears or is greatly reduced after the amine stores have been depleted by preincubation of the worms in a medium containing reserpine and chlorimipramine [11] . This provides support for the association of green and yellow fluorescent structures with NE and 5-HT, respectively; this is confirmed also by the identity of the spectral characteristics of the fluorescent structures of *S. mansoni* with those of these two biogenic amines as determined by microspectrofluorimetry by Van Orden [20] .

The localization of AChE, of NE, and of 5-HT in the nervous system of *S. mansoni* does not rule out the possibility - if not the likelihood - that individual nerve cells and fibers contain only a single transmitter. This is suggested also by observations that exogenous 5-HT is taken up and stored by 5-HT, but not by NE-containing structures. Possibly NE acts an intraneural transmitter and also induces the release of 5-HT from its storage sites. Besides a role as an excitatory moter transmitter, 5-HT and NE may play a role in sensory transmission in *S. mansoni* because of green fluorescent fibers terminating on the dorsal, and yellow fluorescent fibres terminating on both the dorsal and the ventral surface of the worm.

While the concentrations of 5-HT in *S. hematobium* are similar to those of *S. mansoni,* no 5-HT could be detected in two other trematodes, *Fasciola hepatica* and *Paragonimus westermanii.* These parasites, as well as *S. japonicum* and *Clonorchis sinensis,* contain significant amounts of dopamine, but NE has been detectable only in *S. japonicum* [14]. The functional significance of these qualitative and quantitative species differences among trematodes needs to be elucidated. This could contribute to a better understanding of the mechanisms involved in the control and the coordination of muscular activity in these parasites.

References

1. BULBRING, E., LOURIE, E.M. and PARDOE, Brit. J. Pharmacol., **4,** 290 (1949).
2. BUEDING, E., Brit. J. Pharmacol., **7,** 563 (1952).
3. CHANCE, M.R.A. and MANSOUR, Brit. J. Pharmacol., **8,** 134 (1953).
4. BARKER, L.R., BUEDING, E. and TIMMS, A.R., Brit. J. Pharmacol., **26,** 656 (1966).
5. GOLDBERG, A. and BUEDING, E., Unpublished observations.
6. PEPLER, W.J., J. Histochem. Cytochem., **6,** 139 (1958).
7. LEWERT, R.L. and HOPKINS, D.R., J. Parasitol., **51,** 616 (1965).
8. BUEDING, E., SCHILLER, E.L. and BOURGEOIS, J.G., Am. J. Trop. Med. and Hyg., **16,** 500 (1967).
9. LIU, C.L., ROGERS, S. and BUEDING, E., Unpublished observations.
10. DAVIS, A. and BAILEY, R.D., Bull. Wld. Health Org., **41,** 209 (1969).
11. BENNETT, J. and BUEDING, E., Unpublished observations.
12. VANE, J., Personal communication.
13. BENNETT, J., BUEDING, E., TIMMS, A.R. and ENGSTROM, R.G., Molec. Pharmacol., **5,** 542 (1969).
14. CHOU, D., BENNETT, J. and BUEDING, E., Unpublished observations.
15. COTTRELL, G.A. and LAVERACK, M.S., Ann. Rev. Pharmacol., **8,** 273 (1968).
16. WELSH, J.H. and KING, E.C., Comp. Biochem. Physiol., **36,** 683 (1970).
17. FALCK, B., Acta Physiol. Scand., **56,** Suppl. 197 (1962).
18. BENNETT, J. and BUEDING, E., Comp. Biochem. Physiol. In Press (1971).
19. SMITH, J.H., REYNOLDS, E.J. and VON LICHTENBERG, F., Am. J. Trop. Med. Hyg., **18,** 28 (1969).
20. VAN ORDEN, L., Personal communication.

7

PHARMACOLOGICAL ASPECTS OF TETRAMISOLE

J.M. Van Nueten
Department of Pharmacology
Research Laboratories
Janssen Pharmaceutica, Beerse, Belgium

Introduction

Tetramisole racemic 2,3,5,6-tetrahydro-6-phenyl-imidazo[2,1-b]thiazole hydrochloride, has been introduced as a potent drug in the treatment of nematodal infestation [1].

Biochemical studies have been performed recently [2, 3] (Van den Bossche, this book).

In the present study some pharmacological properties of tetramisole and its optical isomers are reported.

The first part of this study on its mechanism of action will describe the results obtained on isolated mammalian and avian tissues.

In the second part, results obtained on Ascaris strips and on whole Ascaris will be given.

Experiments were done with both tetramisole (dl-tetramisole), levamisole (l-tetramisole) and the dextro-isomer (d-tetramisole).

I. Mammalian and avian tissues.

A. Methods

Strips of rabbit duodenum, guinea-pig ileum, rat stomach, rabbit spleen, chicken rectal caecum and vas deferens of rat or guinea-pig were suspended in an oxygenated Tyrode or Krebs-Henseleit solution at 37° C. Durg solutions were added in single or increasing cumulative doses while the activity, i.e. contraction, relaxation or tension changes, was recorded. Transmural stimulation was applied on guinea-pig ileum by means of 2 platinum electrodes, one of which was placed inside the ileum and the other in the Tyrode solution. The tissue contracted in response to single submaximal rectangular pulses of 1msec. duration and a frequency of 6 pulses per minute [4].

The same transmural technique was applied for sympathetic stimulation of the vas deferens; however repetitive pulse trains were used [5]. Stimulation was for 3 seconds at 1 min. intervals at a frequency of 2.5, 5 and 10 pulses per sec. with a pulse duration of 1msec. at 50 mA. Peristaltic reflex was induced in the guinea-pig ileum, using a modified Trendelenburg method [6]; both muscle activity and outflow were recorded. Interaction with KCl-depolarisation was studied on the guinea-pig ileum and on the perfused central artery of the rabbit ear, as described by Van Nueten [7].

Two nerve-muscular preparations were used; the phrenic-nerve diaphragm preparation of the rat, first described by Bülbring [8] and the biventer cervicis muscle of the chick, as described by Ginsborg and Warriner [9]. This muscle consisted of 2 muscle bellies separated by a tendon enclosing the nerve supply.

The 2 preparations were suspended in an oxygenated Tyrode bath and twitch responses were induced by single-pulse stimulation of the nerve (supramaximal rectangular pulses of 0.25 msec. duration and a frequency of 6 per minute). The twitch respones were recorded by means of an isometric strain gauge transducer.

B. Results

Both 1-tetramisole and its racemic form induced contractions of the rabbit duodenum at concentrations of 0.63 mg/l (Fig. 1) or more. With dl-tetramisole, however, these contractions were rather slight and relaxation was observed with doses of 160 and 640 mg/l.

With the l-isomer a dose of 640 mg/l was necessary to produce relaxation. The d-isomer mainly produced relaxation and a decrease of spontaneous movements (10 mg/l or more).

Acetylcholine contractions were not enhanced by dl-tetramisole and its isomers, as they are by cholinesterase inhibitors.

Dose related spasmogenic effects were also observed on guinea-pig ileum with the laevo- and racemic form (Fig. 2); 50 % of the maximum response was obtained with 9 and 14 mg/l respectively. This is comparable with 0.04 mg/l of Acetylcholine in the same conditions. The intrinsic activity was 0.94 for the laevo- and 0.64 for the racemic form, comparable with a value of 1.00 for Acetylcholine. The dextro-isomer was almost completely inactive on guinea-pig ileum, as shown by its low intrinsic activity of ± 0.10 for its spasmogenic effect.

These spasmogenic responses were inhibited by atropine (0.02 mg/l), morphine (0.04 mg/l) and hexamethonium (0.63 mg/l and more) (Fig. 3). Cumulative dose-response curves of l-tetramisole and dl-tetramisole were shifted to the right and maximum contraction was depressed at the same time. The antagonism was non-competitive.

The antihistaminic compound pyrilamine failed to inhibit these dl- and l-tetramisole-induced contractions.

Contractions produced by the ganglion-stimulating agents nicotine and DMPP (dimethylphenylpiperazinium) were potentiated by dl- and l-tetramisole.

From all these results we conclude that the contractions produced in mammalian intestinal tissues were due to ganglionic stimulation. D-tetramisole-induced relaxation, which is slightly inhibited by α-blockers, may be due in part to stimulation of adrenergic receptors.

The laevo- and racemic forms of tetramisole also potentiated the contractions of the longitudinal muscle in response to single stimulation, whereas the dextro-isomer acted as an inhibitor (Fig. 4), thus resembling adrenergic drugs.

This increase in response was interpreted as a lowering of the threshold for excitation of the intramural ganglia, since it could be prevented by hexamethonium.

Further proof of the ganglion-stimulating activity was the facilitation of the peristaltic reflex of the ileum. At concentrations of 0.63 and 2.5 mg/l of l-tetramisole the reflex response was triggered by lower degrees of distension of the intesinal wall.

Cumulative dose-response curves of methacholine were not shifted to the left in the presence of the laevo- and racemic forms of tetramisole (0.63 to 10 mg/l), as they are by both reversible (i.e. neostigmine) and irreversible (i.e. organophosphorus compounds) inhibitors of cholinesterase.

Contractions or tension development of guinea-pig ileum was also produced by K-depolarisation. Dl-tetramisole and its isomers relaxed previously contracted depolarised muscle at concentrations of 2.5 mg/l and more. This effect was confirmed on isolated Taenia Coli of the guinea-pig by Godfraind (personal communication) and on isolated perfused arteries in our laboratory. It is possible that dl-tetramisole and isomers temporarily reduce the permeability of the depolarised membrane to Ca-ions, rather than by acting directly on the contractile machinery.

Indeed, they were devoid of any anticholinergic, antihistaminic or antispasmogenic activity in various tissues. Finally no adrenergic blocking effect at α- or β-sites was observed.

On the vas deferens of guinea-pig contractions induced by sympathetic stimulation, were potentiated by l-tetramisole at doses as low as to 0.63 mg/l and at higher doses by the dextro-isomer and the racemic mixture (Fig. 5).

The prevention of this effect by hexamethonium suggested that dl-, l- and d-tetramisole stimulate the sympathetic ganglia.

L-tetramisole was found to potentiate norepinephrine on the vas deferens of rat or guinea-pig, while the dextro-isomer and the racemic form either potentiated or inhibited norepinephrine, depending on the experiment and the doses.

The experiments on nerve-muscle preparations showed that the twitch response of the two preparations was affected at high concentrations (10 to 40 mg/l) of the racemic form and both of its isomers. The progressive neuromuscular inhibition observed could not be prevented by cholinesterase-inhibitors. In the chick preparation, it was accompanied by a sustained contracture (Fig. 6).

This neuromuscular inhibition presented the typical characteristics of drugs producing depolarisation of the motor endplate, such as succinylcholine.

C. Conclusions

The following conclusions may be drawn from the reported results on mammalian and avian tissues:

Tetramisole (laevo- and dl-form) produced a reversible ganglionstimulating effect on mammalian tissues at both parasympathetic and sympathetic sites, as demonstrated in various experiments and various tissues.

In this respect we are at variance with the results of Eyre[10], who failed to observe an inhibition by hexamethonium.

Further studies on smooth muscle showed that dl-tetramisole and its isomers are devoid of antispasmodic activity. In particular, no anticholinergic or adrenergic blocking activity was obsered.

They were devoid of an anticholinesterase effect on intestinal tissues and on neuromuscular preparations at concentrations up to 10 mg/l.

Earlier studies, using a biochemical method, revealed inhibition of cholinesterase activity at fairly high concentrations of dl-tetramisole[10], (Veenendaal, personal communication). This was confirmed in our laboratory for l-tetramisole, which was at least 100 times less active as a cholinesterase inhibitor than physostigmine (Van Belle, unpublished results).

The interaction with norepinephrine in the vas deferens may be due to interference with an adrenoceptive mechanism and deserves further study.

Dl-tetramisole and isomers blocked neuromuscular transmission. As a neuromuscular inhibitor they resembled the non-competitive depolarizing compounds, as observed earlier by Eyre[10].

Finally, they were observed to cause a transitory relaxation of depolarised muscular strips and peripheral arteries. The hypothesis of a temporary reduction of the permeability of the cell membrane to Calcium ions is to be worked out in further experiments.

II. Effect of tetramisole on Ascaris

A. Methods

A dorsal nerve-muscle strip, cut from the anterior region of an adult female worm[11] was suspended in a nitrogenated* Tyrode solution at 37° C with a preload of 1 g.
Tension changes were recorded isometrically.
In another series of experiments, whole adult female worms were suspended vertically in nitrogenated* Tyrode solution, containing double the amount of glucose and 100 mg of thiamine per litre.
A narrow pararubber tubing, 1 cm long, was fitted around the anterior tip i.e. the head and connected to an isotonic lever with a preload of 1 g. Movements were recorded continuously on a kymograph, without magnification.

* 95 % N_2: 5 % CO_2

B. Results

Tetramisole induced an increase in muscle tension of Ascaris strips at concentrations of 0.63 mg/l or more of the laevo-isomer, 1.25 mg/l of the racemic form and 2.5 mg/l of the dextro-isomer (Fig. 7).

Following a rapid onset, maximum effect was reached within 8 to 10 minutes and was sustained for more than 20 minutes. A similar increase in muscle tension was induced by DMPP.

The ganglion-blocking agent hexamethonium produced a weak and transitory inhibition. Anticholinergic drugs had no effect.

The whole Ascaris reacted to dl-, l- and d-tetramisole with a marked increase in muscular tone, which was followed by partial relaxation and subsequent irreversible paralysis and death (Fig. 8).

This was the usual pattern of reaction for the three forms of tetramisole, although different doses had to be used and quantitative differences were frequent.

Contracture started after 2 to 7 minutes and reached a maximum of 2 to 13 cm within 4 to 25 minutes. During this phase searching movements of the head and spontaneous movements of the body progressively disappeared.

The relaxation, observed in most treated worms, had a sudden onset in some, but followed a slower course in others.

As shown in Table 1 paralysis was observed within 45 minutes at concentrations of 2.5 mg/l or more of the laevo- and racemic form, and 40 mg/l of the dextro-isomer.

Using the paralysis as a criterion for effectiveness, the laevo-isomer was about twice as active as the racemic form and about 25 times more active than the dextro-isomer.

It was observed that, even at the highest doses, paralysis occurs much later when the worm is not submitted to preload.

At the end of the experiments, the worms were incubated in a drug free medium and observed for periods of up to 16 hours.

The paralysis was irreversible since no recovery of motility was observed after 16 hours, except in some experiments at the lowest dose levels used.

In the presence of hexamethonium, l-tetramisole produced a less pronounced contracture but paralysis was not inhibited. Therefore, the two phases might be independent.

C. Conclusions

Dl-tetramisole reduced the membrane potential of the muscle cell bellies [12]. Gaitonde and Mahajani [13] confirmed its depolarizing action. Hence dl-tetramisole and isomers seem to act by stimulation of ganglion-like structures, followed by neuromuscular inhibition of the depolarising type.

A possible explanation for the irreversible block of the contractile machinery is drug interference with metabolic energy supply. One supporting argument is the observation that paralysis occurred much later in worms lying in a petri-dish than in those submitted to traction and thus forced to use energy.

The hypothesis of a mechanism linking the paralysing effect of tetramisole with an inhibition of the energy-producing system needs to be borne out by biochemical arguments.

Summary

Tetramisole (laevo- and dl-form) showed reversible spasmogenic activity on isolated intestinal tissues. It potentiated responses of these tissues to nicotine, DMPP and transmural stimulation and facilitated the peristaltic reflex of the ileum. These effects result from ganglia-stimulation.

Dl-tetramisole and isomers were also found to potentiate responses of the ductus deferens to sympathetic stimulation and to norepinephrine.

They produced a transitory relaxation of previously contracted depolarised mammalian smooth muscle.

At higher doses, they inhibited the neuromuscular transmission and this effect was due to depolarisation of the motor endplate.

These compounds were devoid of any adrenergic blocking activity at α- or β-receptor sites and of anticholinergic, antihistaminic or antispasmogenic activity on smooth muscle.

No cholinesterase-inhibition was observed on intestinal tissues and on neuromuscular preparations.

Dl-tetramisole and isomers produced a marked contracture of anterior strips of the Ascaris and of the whole parasite.

This effect was slightly reduced by hexamethonium. On the whole parasite the contracture was followed by relaxation and subsequent irreversible paralysis, which could not be prevented by hexamethonium.

A comparative study of the racemic form and the dextro- and laevo-isomers showed that the laevo-isomer was the most active form.

A possible explanation for the block of the contractile machinery is drug interference with metabolic energy supply.

Acknowledgements

I would like to thank Mr. L. Helsen, Mr. P. De Winter and Mr. J. Van Beek for skilful assistance with the experiments.

References

1. THIENPONT, D., VANPARIJS, O.F.J., RAEYMAEKERS, A.H.M., VANDEN-BERK, J., DEMOEN, J.A., ALLEWIJN, F.T.N., MARSBOOM, R.P.H., NIEME-GEERS, C.J.E., SCHELLEKENS, K.H.L. and JANSSEN, P.A.J., Nature **209**, 1084, 1966.
2. VAN DEN BOSSCHE, H. and JANSSEN, P.A.J., Life Sci. **6**, 1781, 1967.
3. VAN DEN BOSSCHE, H. and JANSSEN, P.A.J. Biochem. Pharmacol. **18**, 35, 1969.
4. PATON, W.D.M., J. Physiol. (Lond.) **127**, 40 P, 1955.
5. BIRMINGHAM, A.T. and WILSON, A.B., Brit. J. Pharmacol. **21**, 569, 1963.
6. VAN NUETEN, J.M., Arch. int. pharmacodyn. **171**, 243, 1968.
7. VAN NUETEN, J.M., Europ. J. Pharmacol. **6**, 286, 1969.
8. BULBRING, E., Brit. J. Pharmacol. **1**, 38, 1946.
9. GINSBORG, B.L. and WARRINER, J., Brit. J. Pharmacol. **15**, 410, 1960.
10. EYRE, P., J. Pharm. Pharmac. **22**, 26, 1970.
11. BALDWIN, E. and MOYLE, V., Brit. J. Pharmacol. **4**, 145, 1949.
12. ACEVES, J., ERLY, D. and MARTINEZ-MARANON, R., Brit. J. Pharmacol. **38**, 602, 1970.
13. GAITONDE, B.B. and MAHAJANI, S.S.: 22nd Indian Pharmaceutical Congress Association Conference, 1970.

Table 1.

Effect of tetramisole on the whole *Ascaris.*

positive responses	l-			dl-			d-		
mg/l	0.63	2.5	10	2.5	10	40	10	40	160
n*	5	5	5	5	5	5	3	5	5
contracture (⟩10 mm)	4	5	5	5	5	5	1	5	5
relaxation (⟩10 mm)	3	4	5	4	4	4	0	3	3
paralysis within 45 minutes	0	3	5	2	3	5	0	2	5

* n = number of determinations

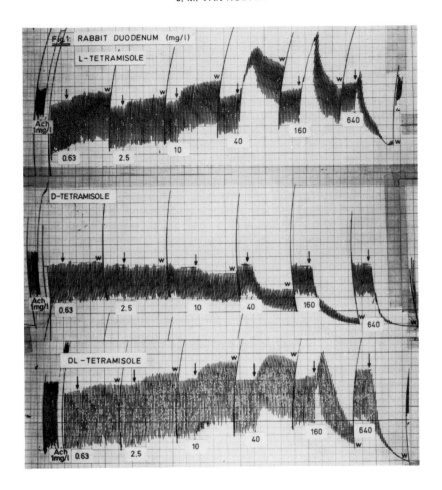

Fig. 1.

Isotonic records of response of rabbit duodenum to l-tetramisole, d-tetramisole and dl-tetramisole (mg/l). Drugs were added every 6 minutes at ↗ and removed after 2 minutes at W (washing).

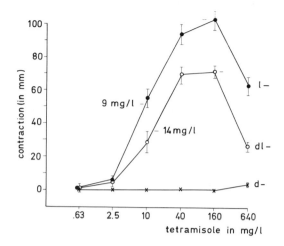

Fig. 2.
Dose-effect curves of tetramisole on guinea-pig ileum; ordinate: isotonic contraction in mm; abscissa: concentrations in mg/l. Each point represents the mean value $\pm$ S.E.M. from 6 experiments.

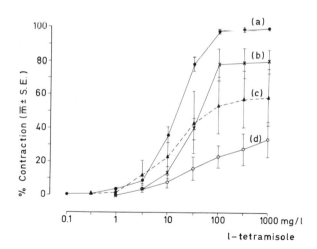

Fig. 3.
Cumulative dose-response curves of l-tetramisole on guinea-pig ileum in absence and in presence of hexamethonium 0.63, 2.5 and 10 mg/l. Mean values and S.E.M. from 20 experiments for the control curve and from 5 experiments for the other curves. Isometric recordings.

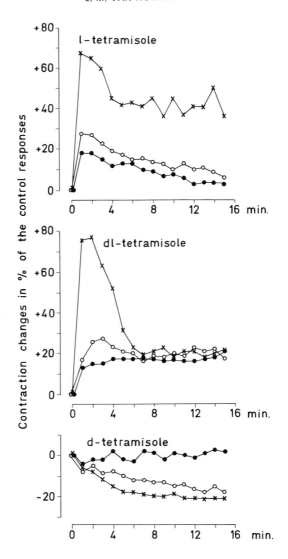

Fig. 4.
Interaction of tetramsiole with isometric contractions induced on guinea-pig ileum by single coaxial stimulation. Mean values (6 experiments) of the contraction changes in % of the contraction observed before addition of the drug at concentrations of 0.63 (●), 2.5 (○) and 10 (x) mg/l.

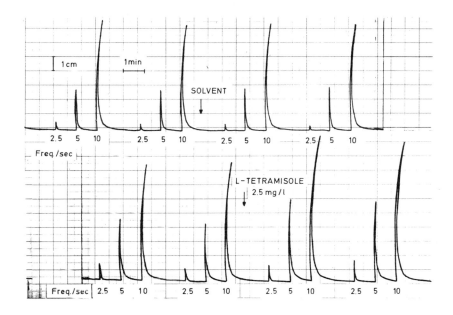

Fig. 5.
Isotonic contractions of vas deferens of the guinea-pig in response to sympathetic stimulation with repetitive pulse trains of 3 sec. at an increasing frequency of 2.5, 5 and 10 pulses per sec. Records in absence and in presence of Solvent and l-tetramisole respectively.

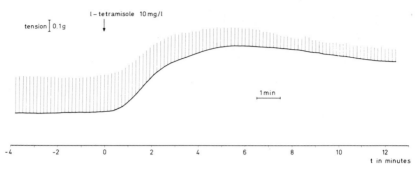

Fig. 6.
Isometric record of the effect of l-tetramisole (10 mg/l) on the response of chick biventer cervicis muscle to indirect stimulation at a frequency of 6 per min.

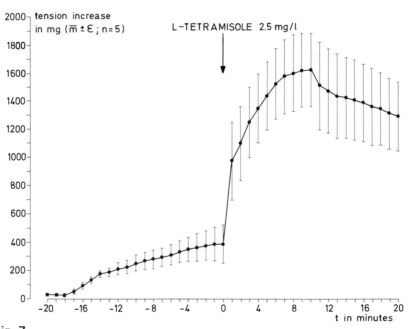

Fig. 7.
Increase by l-tetramisole (2.5 mg/l) of the muscle tension of the isolated dorsal muscle strip of Ascaris suum. Mean values of S.E.M. from 5 experiments. Isometric recordings.

114

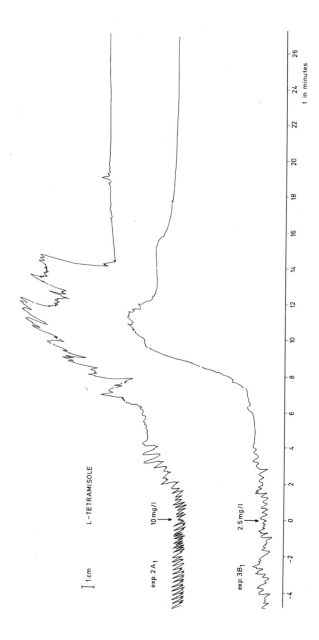

Fig. 8.

Isotonic records of movements of the whole Ascaris suum in response to l-tetramisole at concentrations of 2.5 mg/l (lower record) and 10 mg/l (upper record).

8

BIOCHEMICAL EFFECTS OF TETRAMISOLE

H. Van den Bossche
Department of Comparative Biochemistry
Janssen Pharmaceutica - Research Laboratories
Beerse - Belgium.

Tetramisole* is an antinematodal drug effective both in animals and man [1]. The laevo-isomer** has been shown in animals and man to be approximately twice as active as the racemic mixture, and depending on the parasite, several times more active than the dextro-isomer [2].

In 1967 and 1969, we forwarded the hypothesis that the anthelmintic effect of tetramisole may be due to the inhibition of fumarate reductase in sensitive helminths [2, 3]. This hypothesis was based on the following observations. Incubation of *Ascaridia galli* for 15 hours at 37^{0} in an atmosphere of 95 % N_2 and 5 % CO_2 resulted in a decreased succinate production when tetramisole was added to the incubation mixture. 63.4 % inhibition was obtained at a tetramisole concentration of 2.5 μ g/ml. This decreased succinate production is associated with a decrease in the ATP content of the **in vitro** incubated parasites [4].

Evidence has been presented that succinate is one of the major metabolic end-products in several parasites. Succinate is formed from fumarate in the presence of NADH and of the enzyme, fumarate reductase [5]. Since fumarate reductase would appear to be a key enzyme in the energy yielding pathway, we measured the effect of tetramisole on this enzyme. The results presented in Fig. 1 indicate that both optical isomers, i.e. laevo- and dextro-tetramisole, inhibit the NADH-coupled fumarate-succinate system of *Ascaris* muscle. The laevo-isomer proved to be a more potent inhibitor of the enzyme system, a fact compatible with its more potent anthelmintic action. To investigate further the mechanism of the tetramisole-induced inhibition of the fumarate reductase mechanism,

---

* Generic name for the hydrochloride of 2, 3, 5, 6-tetrahydro-6-phenyl-imidazo [2,1b] thiazole.
** levamisole

we isolated particles (R_2) containing succinoxidase activity from muscle strips of **Ascaris suum,** as described by Kmetec and Bueding [6]. Fumarate reductase activity was measured by following the anaerobic oxidation of NADH in the presence of fumarate. The Lineweaver-Burk plots shown in Fig. 2 indicate that tetramisole is a non-competitive inhibitor (K_s = 1.4 x 10^{-4}M; K_i = 4.4 x 10^{-4}M). Incubation of **Ascaris** mitochondria in the presence of ADP, malate and inorganic phosphate results in the formation of ATP [7]. Since ATP production is associated with the fumarate reduction, any inhibition of fumarate reductase would consequently result in a decreased incorporation of inorganic phosphate into ATP. The results shown in Fig. 3 indicate that laevo- and dextro-tetramisole inhibit the malate--induced ^{32}P incorporation into organic phosphate in mitochondria isolated from **Ascaris** muscle by the method of Saz and Lescure [8]. The concentration of laevo- and dextro-tetramisole, needed to obtain 50 % inhibition were 0.9 and 4.3mM respectively. These experiments confirmed that tetramisole is an inhibitor of the energy yielding pathway in **Ascaris.**

Reinecke has shown that tetramisole at a level of 15 mg/kg of body weight was 89 to 100 % effective against all stages of development of **Haemonchus contortus** [9]. Laevo-tetramisole inhibits the fumarate reductase mechanism in third-stage larvae of this nematode [10], and as shown in Table 1, also affects the anaerobic oxidation of NADH in the presence of fumarate in a particulate fraction of fourth-stage larvae. All the results presented so far seem to suggest that the antinematodal activity of tetramisole may be due to inhibition of the fumarate reductase mechanism in helminths. This was confirmed by Prichard in his publication on the mechanism of action of thiabendazole [11]. However, most of these biochemical effects were observed at concentrations higher or after incubation periods longer than those needed to induce contraction of the worms (Van Nueten, this book). This may indicate that the effect of tetramisole on muscle tension is the primary site of chemotherapeutic activity **in vivo.** The tetramisole-stimulated tonus increase in **Ascaris** is followed by a decrease in tension and irreversible paralysis. These two phases may be independent, since the contraction but not the paralysis was inhibited by hexamethonium (Van Nueten, this book).

Based on these observations and on the theories on muscle contraction in higher animals, the following hypothesis for the mechanism of action of tetramisole can be suggested (Fig. 4). The activation of muscle by tetramisole breaks down the myosin-bound ATP to ADP. The next step is normally the rephosphorylation of the resultant bount ADP to bound

ATP. However, our biochemical experiments indicate that in the presence of tetramisole ATP production is inhibited and the rephosphorylation of ADP slowed down or diminished as ATP is exhausted, effects which may result in the formation of actomyosin with consequent muscle stiffness. Although we have at present no direct experimental evidence for this hypothesis, there are a number of observations which seem to lend support to it. Incubation of *Ascaris* in the presence of 10 μg laevo-tetramisole per ml of incubation mixture resulted in complete immobilization of the parasite after only 2 hours, whereas, as shown in the previous communication, it takes only 10 minutes to obtain paralysis when the worms are connected to an isotonic lever with a preload of 1 g. The reaction of the worms to this artificial position may result in an utilization of ATP. Since we have suggested that paralysis is due to an effect on the energy yielding pathway, it is possible that the more rapid response obtained in the latter experiments was the result of a lower energy reserve in the *Ascaris* muscle.

It has also been shown that tetramisole has reversible spasmogenic efffects on guinea pig ileum and rabbit duodenum. Furthermore, although some side-effects were observed in mammals after administration of tetramisole, these always disappear in a few hours. Thus, these **in vitro** and **in vivo** reversible effects are in direct contrast to the irreversible paralysis observed in *Ascaris.* This difference may be attributed to the fact that tetramisole affected the energy production in *Ascaris* muscle but not in mammalian tissues [4].

The experiments of Denham [12] also support our hypothesis. Denham compared the **in vitro** effects of methyridine and tetramisole on the motility and development of *Ostertagia circumcincta.* Both methyridine and tetramisole had a paralysing effect on the exsheated third-stage larvae of this nematode. When the larvae treated with methyridine were washed and incubated without the drug for 18 hours at 37°, all were as active as those in the control tubes. A similar experiment with tetramisole however, showed that after incubation, the washed larvae were moving in a rather sluggish manner as compared with the controls. This difference in recovery may be explained by different mechanisms of action; methyridine is said to act as a neuromuscular blocker [13], whereas tetramisole inhibits the fumarate reductase mechanism and also affects the neuromuscular system of the parasites. The same author has shown that when larvae exposed to methyridine were washed and cultured **in vitro**, they grew as well as the control larvae. Exposure of larvae to tetramisole on the other hand, decreased the growth-rate when compared with controls, and no fifth-stage larvae developed during the test period.

These observations are also considered to lend some support to our hypothesis as it is known that inhibition of energy production resulted in a decreased protein synthesis.

Although our hypothesis lacks direct confirmation, we believe that this proposed mechanism of action explains the different tetramisole-induced effects better than the hypothesis we presented previously [3, 4].

Acknowledgements

The author whishes to express his thanks to Dr. Paul A.J. Janssen for his constant interest; to Mrs. Horemans, Mr. Goossens and Mr. Vermeiren for their technical assistance, and to Mr. & Mrs. Scott for their help in preparing the manuscript.

References.

1. THIENPONT, D., VANPARIJS, O.F.J., RAEYMAEKERS, A.H.M., VANDEN-BERK, J., DEMOEN, P.J.A., ALLEWIJN, F.T.N., MARSBOOM, R.P.H., NIEME-GEERS, C.J.E., SCHELLEKENS, K.H.L. and JANSSEN, P.A.J., Nature **209,** 1084 (1966).
2. BULLOCK, M.W., HAND, J.J. and WALETZKY, E., J. Med. Chem. **11,** 169 (1968).
3. VAN DEN BOSSCHE, H. and JANSSEN, P.A.J., Life Sci. **6,** 1781 (1967).
4. VAN DEN BOSSCHE, H. and JANSSEN, P.A.J., Biochem. Pharmacol. **18,** 35 (1969).
5. SAZ, H.J. and BUEDING, E., Pharmacol. Rev. **18,** 871 (1966).
6. KMETEC, E. and BUEDING, E., J. biol. Chem. **236,** 584 (1961).
7. SAZ, H.J., Comp. Biochem. Physiol. **39B,** 627 (1971).
8. SAZ, H.J. and LESCURE, O.L., Comp. Biochem. Physiol. **30,** 49 (1969).
9. REINECKE, R.K., J.S. Afr. vet. med. Ass. **37,** 27 (1966).
10. VAN DEN BOSSCHE, H., VANPARIJS, O.F.J. and THIENPONT, D., Life Sci. **8,** 1047 (1969).
11. PRICHARD, R.K., Nature **228,** 684 (1970).
12. DENHAM, D.A., Exptl. Parasitol. **28,** 493 (1970).
13. BROOME, A.W.J., Brit. J. Pharmacol. **17,** 327 (1961).

Table 1.
Effect of L-tetramisole upon reduced-NAD: fumarate reductase activity in a particulate fraction of fourth-stage *H. contortus* **larvae**

L-tetramisole (μmoles/ml)	NADH oxidized[a] Absolute (nmoles/min/mg protein)	%
0	8.71 (7.02 - 10.65)	100
0.580	3.42 (2.45 - 4.61)	39.2

[a] number of experiments: 3.

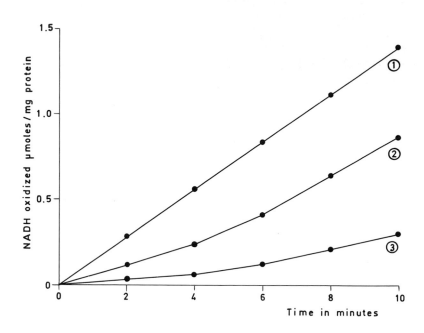

Fig. 1.

Effect of tetramisole on anaerobic oxidation of NADH in the presence of fumarate and a particulate fraction of **Ascaris** muscle. The fraction used was a mitochondrial fraction [8] treated with Tris-buffer (pH 7.4; 0.1M). The incubation mixture consisted of: 2.8 ml Tris-buffer (pH 8.5; 0.04M); 0.1 ml of the particulate fraction (0.84 mg protein); 0.1 ml Tris-buffer (pH 7.4; 0.1M) or 0.1 ml of L- or D-tetramisole (0.44 μmoles/ml of incubation mixture). After 30 min of incubation at 37°, 0.1 ml of fumarate (1.2 μmoles) and 0.02 ml NADH (1.068 μmoles) were added. The reaction was followed at 340 mμ for 10 min at 37° in an atmosphere of nitrogen. 1 = control; 2 = D-tetramisole; 3 = L-tetramisole.

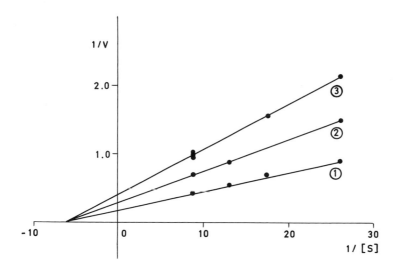

Fig. 2.
Effect of L-tetramisole on NADH oxidation in the presence of fumarate and a R_2-fraction of **Ascaris** muscle. The incubation mixture was the same as that described in Fig. 1. The effect of tetramisole was measured without pre-incubation. 1 = control; 2 = 0.272 μ moles L-tetramisole; 3 = 0.544 μmoles L-tetramisole/ml of incubation mixture.

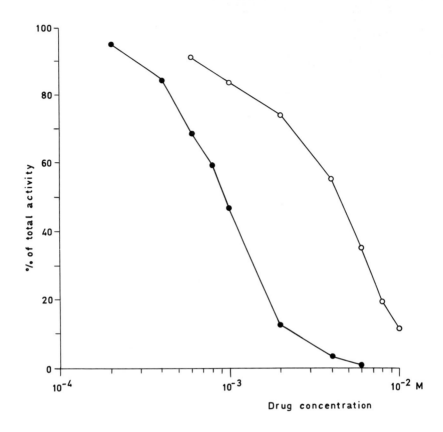

Fig. 3.
Effect of L-(●) and D-(○) tetramisole on malate-induced ^{32}P incorporation into organic phosphate by mitochondria isolated from *Ascaris* muscle. ^{32}P incorporation was measured by the method of Saz [7].

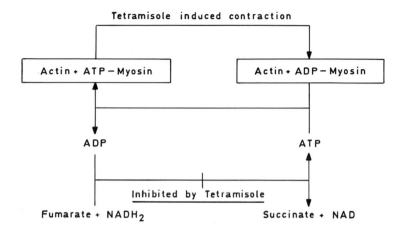

Fig. 4.
Hypothesis for the mechanism of action of tetramisole.

9

RECENT STUDIES ON THE MECHANISM OF ACTION OF BERENIL (DIMINAZENE) AND RELATED COMPOUNDS

B.A. Newton

Medical Research Council Biochemical Parasitology Unit,
The Molteno Institute, University of Cambridge,
Cambridge, CB2 3EE, England

Introduction

The chemotherapy of trypanosomiasis in man and animals is at present dependent upon a relatively small number of synthetic drugs. Drug resistance has been reported to occur against most compounds in current use and no new drugs have been introduced in recent years. These facts emphasise how narrow is the margin of security in the treatment of this disease at the present time and how urgent is the requirement for a new lead upon which to base the development of new chemotherapeutic agents. Such a lead could come from a detailed knowledge of the biochemistry of trypanosomes together with an understanding of the mechanism of action of existing drugs and the nature of drug resistance; work in my own laboratory is directed towards this goal.

Studies of the mechanism of action of existing trypanocidal drugs, using model systems (i.e. trypanosomatid flagellates which can be readily cultured in the laboratory), have revealed the striking fact that, with the exception of arsenicals and suramin, most of the trypanocides in use are potent inhibitors of nucleic acid or protein synthesis. In particular there is good evidence (reviewed by Newton, 1970) that phenanthridine drugs and berenil combine with, and prevent the replication of, deoxyribonucleic acid (DNA). Many synthetic drugs and antibiotics are now known to interact with DNA but the majority of these are too generally toxic to be of any practical value as chemotherapeutic agents; this however is not so for the trypanocides just mentioned. These compounds appear to show a specificity of action at two levels: they are selective for certain protozoan parasites within their host and they may also selectively inhibit the synthesis of extranuclear DNA in the kinetoplast of trypano-somes or in mitochondria of other organisms. At present we do not understand the molecular basis of either of these selective activities, but the present paper describes work done in an attempt to learn something about these effects in the case of berenil (diminazene aceturate; N-1, 3-diamidino-phenyltriazine diaceturate (Fig. 1).

Berenil has been used primarily for the treatment of *Tryapnosoma congolense* and *Trypanosoma vivax* infections of cattle, but limited clinical trials carried out recently in East Africa suggest that the drug may have some value in the treatment of *Trypanosoma gambiense* and *Trypanosoma rhodesiense* infections in man (WHO, 1969). This drug has a number of advantages over other compounds used for the treatment of trypanosomiasis in cattle: it is rapidly excreted, the occurrence of berenil-resistant strains is relatively rare and it is active against strains which have become resistant to phenanthridinium drugs such as homi-dium. It is important to learn the chemical basis of these characteristics.

Preliminary observations

Berenil is rapidly and irreversibly bound by trypanosomes (Bauer, 1958) and ultraviolet microscopy of *Trypanosoma brucei* isolated from rats before and one hour after treatment with a curative dose of the drug revealed a brilliant blue fluorescence in the nucleus and kinetoplast of drug-treated organisms (Newton & Le Page, 1967). Further experiments with *T. brucei* grown in experimental animals or in culture and with *Trypanosoma mega* grown in culture have established that exposure of organisms to low concentrations of berenil results in fluorescence in the kinetoplast only, while in the presence of higher drug concentrations fluorescence occurs first in the kinetoplast and subsequently in the nucleus. The rapid localisation of berenil in DNA-containing organelles of trypanosomes poses a number of questions. Is the drug a selective inhibitor of DNA synthesis? If so, does it produce this inhibition by direct interaction with DNA or by some other means? Does berenil, at low concentrations, exert a selective effect on kinetoplast replication and function? We are now in a position to provide at least partial answers to these questions.

Evidence for the formation of a berenil/DNA complex

(i) **Spectrophotometric studies.** Addition of DNA to a solution of berenil causes a shift in the absorption spectrum of the drug to longer wavelengths, the maximum changing from 370 mm to 380 mm (Fig. 2).

This metachromatic shift can be taken as evidence for the formation of a drug/DNA complex under these particular conditions and use has been made of this effect to measure berenil/DNA binding ratios by the method of Peacocke & Skerrett (1956). Calf thymus DNA was found to bind one molecule of drug for every 4-5 nucleotides; heat denaturation of DNA before the addition of drug was found to double the number of drug-binding sites. Addition of RNA, synthetic homopolymers and a poly-AU copolymer to berenil solutions all produced a spectral shift and it was found that the amount of drug bound by these substances was the same as that bound by heat denatured DNA. No evidence for complex formation between berenil and mono-nucleotides was obtained.

(ii) **Gel filtration studies.** Hummel and Dreyer (1962) described a gel filtration method to detect reversible interactions between macromolecules and substances of low molecular weight. Fairclough & Fruton (1966) used the technique successfully to study peptide/protein interactions and we have adapted it to the study of DNA/drug interactions. The principle is as follows: when DNA is dissolved in a solution of a drug and a complex is formed, the concentration of free drug [c] is reduced by an amount equivalent to the DNA/drug complex. If this solution is placed on a Sephadex G-25 gel column, already equilibrated with drug at concentration [c], and is eluted with further drug at the same concentration, it is found that, as the DNA/drug complex emerges at the excluded volume of the column, the amount of drug in the eluate rises above the equilibrium level, then falls below the base line and finally returns to the equilibrium level. The appearance of a peak followed by a trough in the elution profile provides a criterion of binding of drug to DNA and the amount of drug bound can be calculated from the area under peak or trough. Fig. 3 shows a typical elution profile. The results obtained using this method to study DNA/berenil interactions were in close agreement with results of spectrophotometric studies and indicated a binding of one drug molecule for every four to five nucleotides.

(iii) **Buoyant density studies.** When sedimented in a caesium chloride gradient DNA exhibits a characteristic buoyant density which depends on its conformation, base composition and the presence of various ions (Meselson **et al.,** 1957; Sueoka **et al.,** 1959; Schildkraut **et al.,** 1962; Erikson & Szybalski, 1964; Davidson **et al.,** (1965). In the presence of berenil it was found that the density of DNA was decreased dramatically. This provides additional evidence for the formation of a drug/DNA complex and indicates that the complex is stable at high ionic strength.

Further experiments in which DNA's of differing base composition were used (Table 1) showed that the density shift in the presence of berenil increased with increasing A + T content of DNA, suggesting that the drug may bind preferentially to this particular base pair. In keeping with these results DNA extracted from berenil-treated trypanosomes was found to have a lower buoyant density than DNA from control organisms; furthermore the density of kinetoplast DNA from berenil-treated cells was always decreased to a greater extent than nuclear DNA.

Some characteristics of the berenil/DNA complex

At present the molecular interactions involved in binding berenil to DNA are unknown. Berenil does not increase the viscosity of DNA solutions (Newton, 1967) nor does it cause an uncoiling of the DNA helix (Waring, 1970), thus it seems unlikely that the drug intercalates between adjacent nucleotide pairs in DNA molecules in the manner postulated by Lerman (1961) for acridines and similar planar heterocyclic structures. It has been found that berenil increases the T_m of DNA (i.e. the mid point in the thermal transition of helical to coiled DNA) by $17°C$ (Newton, 1967), but these experiments did not produce any evidence that the drug can cross link complementary DNA strands in a manner similar to the antibiotic mitomycin (Iyer & Szybalski, 1963). The slope of the melting curve of berenil-treated DNA was found to be greater than that of control DNA; this may be due to the preferential binding of drug to A-T base pairs which is indicated by the buyant density studies.

Evidence for selective inhibition of kinetoplast DNA synthesis

Killick-Kendrick (1964), studying the effect of berenil on *Trypanosoma evansi* infections in horses, reported the appearance of dyskinetoplastic organisms. Discussing this observation he favoured the view that berenil treatment selected for naturally occurring dyskinetoplastic organisms by killing only trypanosomes containing kinetoplasts. While this may occur **in vivo** with *T. evansi*, **in vitro** growth experiments with *T. mega* have shown that addition of berenil (3-10 μg/ml) to cultures results in the loss of kinetoplast DNA (judged by acridine orange staining) from 40 percent of the population in one generation time. Clearly under these conditions berenil is not selecting existing dyskinetoplastic organisms from a mixed population; the drug is either inhibiting kinetoplast DNA replication or is modifying this DNA in some way so that it is no longer stained by acridine orange. Comparison of DNA from control and berenil-treated cells by centrifugation in caesium chloride gradients (Fig. 4 and Table 2) showed that less kinetoplast DNA could be extracted from drug-treated than from control organisms.

Further evidence for the selective inhibition of kinetoplast DNA synthesis has been obtained by studying the effect of berenil on the incorporation of 5-bromodeoxyuridine (BUDR) into the DNA of *T. mega.* In control organisms this analogue is incorporated into both nuclear and kinetoplast DNA in place of thymine and the incorporation can be detected as an increase in the buoyant density of the DNA. When organisms were pretreated with berenil for a time sufficient to produce fluorescence in the kinetoplast but not in the nucleus and then transferred to a drug-free medium containing BUDR and incubated for 24 hrs, it was found that the buoyant density of nuclear DNA only was increased. While the selective inhibition of BUDR incorporation into kinetoplast DNA does not necessarily mean that berenil has completely inhibited synthesis of this DNA, this result, together with the observed decrease in extractable kinetoplast DNA and loss of acridine orange stainable material from drug-treated organisms, strongly suggests that this is so.

Relationships between structure and activity

In neutral solution berenil molecules undergo a rearrangement (Fig. 5) which results in the formation of an aminoazo derivative from the original triazine structure (Dr. H. Loewe, personal communication). This breakdown product is without trypanocidal activity and has no effect on the buoyant density of DNA, a finding which suggests that the spacing of the amidino groups of brenil may be critical in the formation of a DNA/drug complex. In an attempt to learn more about the features of berenil which are responsible for the drugs selective activity we have recently examined a number of closely related substances. Structures of two of the most interesting compounds are compared with berenil in Fig. 6, compound 1(C1) was kindly provided by Dr. H. Loewe (Farbewerke Hoechst) and compound 2(C2; M & B.938) by May and Baker Ltd.

In growth experiments with *T. mega* striking differences were found in the minimum growth inhibitory concentrations of these three substances (Fig. 6). Experiments of the type already described for berenil established that C1 and C2 formed complexes with purified DNA and in each case one molecule of drug was bound per 4-5 nucleotides. However, ultraviolet microscopy of drug-treated organisms showed that, in contrast to berenil, treatment with C1 produced no fluorescence in the kinetoplast or nucleus during exposure of organisms to the drug for periods of upto five hours, whereas C2 caused both kinetoplast and nucleus to fluoresce within minutes of its addition to a culture of *T. mega.*

Ultracentrifugation of DNA extracted from organisms which has been grown in the presence of C1 or C2 for 24 hours showed that C1 produced no significant decrease in the buoyant density of either kinetoplast or nuclear DNA whereas C2 decreased the densities of both types of DNA by about the same amount.

Thus, in summary, removal of one nitrogen atom from the triazine bridge of berenil reduces the growth inhibitory activity and the ability of the drug to bind to nuclear and kinetoplast DNA **in vivo**; removal of a second nitrogen atom restores growth inhibitory activity and ability to combine with intracellular DNA, but, in contrast to berenil, C2 does not appear to act selectively on kinetoplast DNA.

An examination of molecular models of these three compounds has shown that removal of one nitrogen atom from berenil reduces the distance between the terminal amidino carbon atoms from 13 Å to 5 Å, whereas removal of a second nitrogen atom results in the molecule opening out again to give a distance of 11 Å between these carbon atoms. A detailed study of the interaction of these compounds with purified nuclear and kinetoplast DNA is now in progress and it is hoped that this work will throw some light on the molecular basis of the selective activity of berenil.

Conclusions

The data presented indicate that berenil resembles certain phenanthridine and acridine drugs in being able to selectively block kinetoplast DNA replication. This action is clearly not the basis of the trypanocidal activity of berenil against bloodstream forms of African trypanosomes; it seems more likely that the drugs action on nuclear DNA is responsible for this. However, I think the rapid action on the kinetoplast is an important attribute of berenil as it will cause a break in the natural life-cycle of the parasite; dyskinetoplastic forms and forms unable to develop functional mitochondria will be unable to develop in the insect vector. An understanding of the chemical basis of the selective activity of drugs such as berenil together with a more detailed knowledge of factors which control the transformation of trypanosomes from trypomastigote to epimastigote forms, may eventually enable us to synthesise a series of drugs which will specifically block this transformation. While these compounds might not cure a trypanosome infection, they could be of value in controlling the spread of the disease by the insect vector and, used in conjunction with other trypanocides, might well reduce the spread of drug resistant variants.

Acknowledgements

The author is indebted to Mrs. I. Hislop and Mrs. R. Holl for skilled technical assistance. He also wishes to thank Messrs May and Baker Ltd. for a gift of M & B.938 and Dr. W.H. Wagner and Dr. H. Loewe of Farbewerke Hoechst for gifts of berenil and compound 2 and for valuable discussions.

References

BAUER, F. (1958). Zbl. f. Bakt. I Orig. **172,** 6045.
DAVIDSON, N., WIDHOLM, J., NANDI, U.S., JENSEN, R., OLIVERA, B.M. and WANG, J.C. (1965). Proc. Natl. Acad. Sci. Wash. **53,** 111.
ERIKSON, R.L. and SZYBALSKI, W. (1964). Virology **22,** 111.
FAIRCLOUGH, G.F. and FRUTON, J.S. (1966). Biochemistry **5,** 673.
HUMMEL, J.P. and DREYER, W.J. (1962). Biochim. Biophys. Acta **63,** 532.
IYER, V.N. and SZYBALSKI, W. (1963). Proc. Natl. Acad. Sci. U.S.A. **50,** 355.
KILLICK-KENDRICK, R. (1964). Ann. trop. Med. Parasit. **58,** 481.
LERMAN, L.S. (1961). J. Mol. Biol. **3,** 18.
MESELSON, M., STAHL, F.W. and VINOGRAD, J. (1957). Proc. Natl. Acad. Sci. U.S.A. **43,** 587.
NEWTON, B.A. (1967). Biochem. J. **105,** 50P.
NEWTON, B.A. (1970). Adv. Pharmacol. Chemotherap. **8,** 149.
NEWTON, B.A. and LE PAGE, R.W.F. (1967). Biochem. J. **105,** 50P.
PEACOCKE, A.R. and SKERRETT, J.N.H. (1956). Trans. Faraday Soc. **52,** 261.
SCHILDKRAUT, C.L., MARMUR, J. and DOTY, P. (1962). J. Mol. Biol. **4,** 430.
SUEOKA, N., MARMUR, J. and DOTY, P. (1959). Nature, Lond. **183,** 1429.
WARING, M.J. (1970). J. Mol. Biol. **54,** 247.
W.H.O. (1962). Techn. Rep. Ser. No. 434.

Table 1.
Effect of Berenil on the buoyant density of DNA.

DNA (10 μg/ml), in Tris-buffer (0.02 M, pH 8.5) containing 0.15 M NaCl and 0.1 M EDTA + or − Berenil (30 μg/ml), was dialysed (24 hr) against Tris/EDTA/NaCl. 0.23 ml samples were then well mixed with 0.83 ml CsCl (1.710 g/ml final density) and a suitable marker DNA added. Solutions were centrifuged at 44,000 r.p.m. for 20 hrs at 20° in an analytical ultracenttifuge.

DNA Source	Moles % G + C	Buoyant Density (g/ml)		Decrease in Denisty (mg/ml)
		Control	+ Berenil	
Micrococcus lysodeikticus	72	1.732	1.723	11
Escherichia coli	49	1.708	1.689	19
Streptococcus faecalis	35	1.694	1.665	29
Clostridium perfringeus	29	1.688	1.656	32

Table 2.
Effect of Berenil (100 μg/ml) on growth and DNA content of *Trypanosoma mega*

	Control			+ Berenil (100 μg/ml)		
Time (Hrs)	Cell Count x 10⁻⁶	Kineto-plast DNA % of total	Nuclear DNA % of total	Cell Count x 10⁻⁶	Kineto-plast DNA % of total	Nuclear DNA % of total
0	4.1	24.0	76	4.2	23.6	76.4
24	9.4	24.8	75.2	7.1	16.2	83.8
48	19.5	24.4	75.6	9.2	11.8	88.2

Organisms were grown in a peptone/yeast extract/haemin medium. DNA was extracted as described in legend to Fig. 4. Relative amounts of nuclear and kinetoplast DNA were calculated from microdensitometer tracings of bands obtained in caesium chloride gradients.

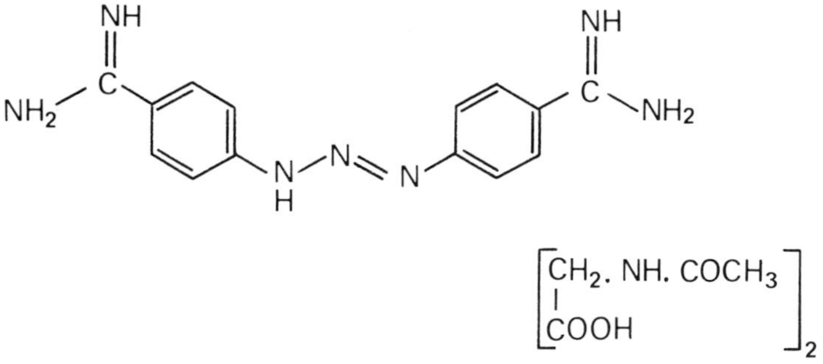

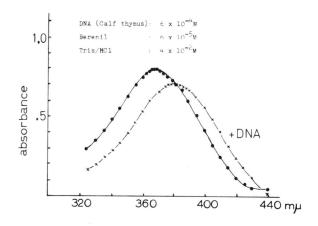

Fig. 1.

Berenil (Diminazene; N-1, 3-Diamidino-phenyltriazine Diaceturate).

Fig. 2.

Effect of DNA on the absorption spectrum of berenil.

● — ● Berenil (6×10^{-5} M) in Tris/HCl buffer (4×10^{-2} M; pH 8.5)

x — x plus calf thymus DNA (6×10^{-4} M).

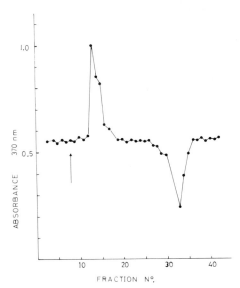

Fig. 3.
Elution profile obtained from a Sephadex G-25 column equilibrated with berenil (2×10^{-5} M) in Tris/HCl buffer (4×10^{-2} M; pH 8.5). The DNA/berenil mixture (1 molecule of drug/4 nucleotide residues) was added at a time indicated by the arrow. Fractions collected every 10 min. Flow rate 43 ml/hr.

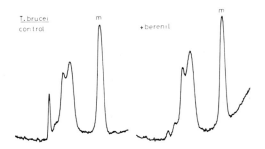

Fig. 4.
Effect of berenil on the DNA content of Trypanosoma brucei. Organisms were grown in a trypotose-casein-liver digest-blood medium in the presence or absence of berenil (6×10^{-5} M) for 24 hrs. Cells were harvested by centrifugation, washed in a buffered salts solution, suspended in Tris buffer (0.02 M, pH 8.5) containing 0.15 M NaCl and 0.1 M EDTA, and lysed by addition of sodium lauryl sulphate (final concn. 1.0 % w/v). When lysis was complete the preparation was treated with pronase (1 mg/ml) for 2 hr at 37° C. Preparations were dialysed overnight against Tris/EDTA/NaCl before centrifugation at 44,000 r.p.m. for 20 hrs at 20° C. m = marker DNA ρ = 1.731.

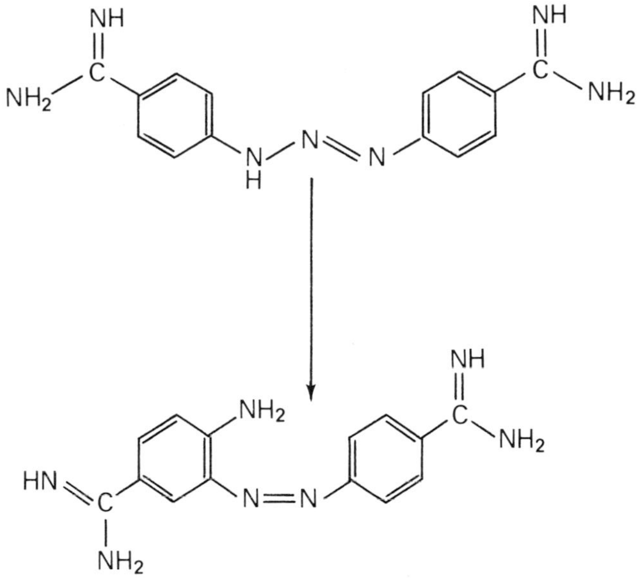

Fig. 5.
Breakdown of berenil in neutral solution to form an 0-aminoazo derivative. (Based on a personal communication from Dr. H. Loewe).

M.I.C

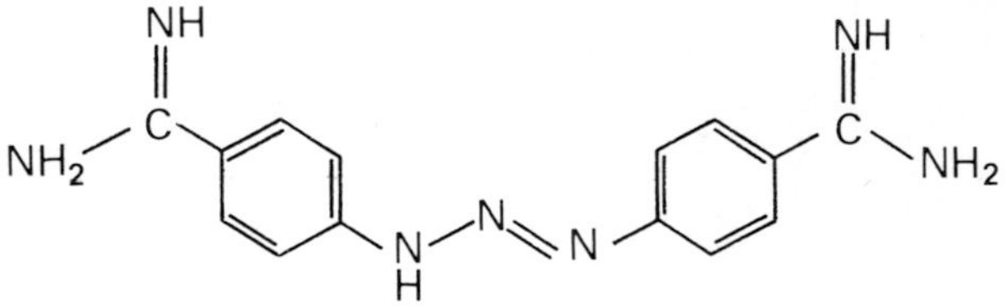

BERENIL

10^{-5} M

COMPOUND 1

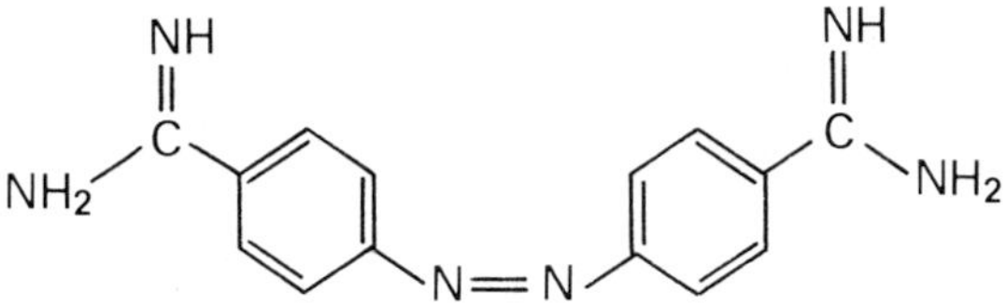

2×10^{-4} M

COMPOUND 2

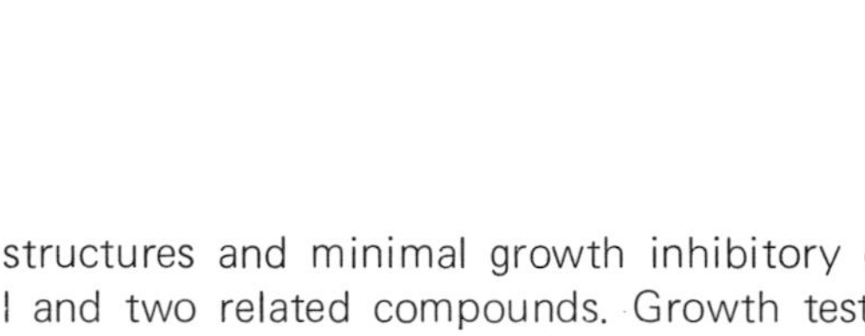

4×10^{-6}

Fig. 6.
Comparison of structures and minimal growth inhibitory concentrations (MIC) of berenil and two related compounds. Growth tests were performed with *Trypanosoma mega* in a peptone/yeast extract/haemin medium. MIC = lowest concentration in which no growth occurred after 72 hrs. In all experiments an initial inoculum of 2×10^6 organisms/ml was used. Berenil and Compound 1 were obtained from Farbewerke Hoechst and Compound 2 (M & B.938) from Messrs May and Baker Ltd.

10

BIOCHEMICAL EFFECTS OF THE ANTHELMINTIC DRUG MEBENDAZOLE

H. Van den Bossche*
Department of Comparative Biochemistry
Janssen Pharmaceutica - Research Laboratories
Beerse - Belgium

Introduction

Mebendazole is the generic name for methyl 5(6)-benzoyl-2-benzimidazole carbamate (Fig. 1). It is an off-white to slightly yellow powder, freely soluble in formic acid, soluble in benzaldehyde, sparingly soluble in dimethyl sulphoxide (DMSO) and insoluble in water, alcohol, ether and chloroform.

Mebendazole has a broad spectrum anthelmintic acitivity (Thienpont, unpublished data) affecting not only gastro-intestinal nematodes but also a nematode living in the trachea, *Syngamus trachea* (Table 1). It is interesting to note that this drug is highly effective against *Strongyloides, Trichuris, Trichinella, Ancylostoma, Necator, Enterobius* and *Ascaris*, all parasites affecting man. Mebendazole also exhibits cestocidal activity against all species investigated until now (Table 2).

From parasitological studies it may be concluded that mebendazole has a broad sprectrum of activity against helminths, both those in aerobic environments, such as *Syngamus,* and those in anaerobic environments such as *Enterobius.* Its mechanism of action cannot therefore lie in the inhibition of the oxygen uptake.

Mebendazole is a rather slow-acting drug, indicating a possible interference with or inhibition of the uptake of exogenous glucose into the nematode. The consequent increased utilization of endogenous carbohydrate reserves would eventually lead to an inadequate energy supply and result in death of the nematode. In the present study, **in vitro** and **in**

* This investigation was supported by Grant No. D 1/4 - 1644 from the "Instituut tot Aanmoediging van het Wetenschappelijk Onderzoek in Nijverheid en Landbouw (IWONL)".

vivo experiments were designed to test the hypothesis that mebendazole inhibits glucose transport into the parasite. A comparison was also made between the effects of mebendazole with those of other benzimidazoles.

Materials and Methods

Adult female *Ascaris suum* and *Syphacia muris* were obtained as previously described [1, 2]. *Ascaridia galli* and *Syngamus trachea* were collected from artificially infected chickens and turkeys respectively, and *Hymenolepis nana* from naturally-infected mice.

A. Glucose uptake studies.

1. In vitro experiments: *A.suum* were incubated in 100 ml glucose salt medium (0.12M NaCl; 0.005M KCl; 0.001M $CaCl_2$; 0.001M $MgCl_2$; 0.005M potassium phosphate buffer (pH 7.4); 0.045M $NaHCO_3$; 0.016M glucose; 0.1 mg streptomycin and 40 x 10^3 units penicillin G). Mebendazole was dissolved in DMSO and added to this medium in which two female worms were incubated at $37°C$ for 24 hours in an atmosphere of 95 % N_2: 5 % CO_2. After incubation, the worms were transferred to a drug-free basic medium. Control worms were incubated in a glucose salt medium containing similar quantities of DMSO (final concentration: 0.1 %). At the end of each incubation period, the glucose content of the medium was determined by the anthrone method [3]. *S. muris* were placed in conventional Warburg manometric vessels containing 5 ml salt medium [2] to which glucose had been added to a final concentration of 0.005M. Mebendazole dissolved in dimethylformamide (DMF) was also added to the medium. Controls were similarly set up with equivalent quantities of DMF (final concentration: 0.5 %). All worms were incubated at $37°C$ for 5 hours in an atmosphere of 95 % N_2: 5 % CO_2 after which glucose content of the medium was determined as mentioned above.

To determine the effect of mebendazole on the distribution of radioactivity in the organs (body wall, intestine, reproductive system) and pseudocoelomic fluid of *A.suum,* the worms (1 female/100 ml medium) were incubated in the glucose salt medium to which 50 nC glucose-C14 had been added (specific activity: 3mC/mM; Radiochemical Centre, Amersham, U.K.). Incubation was for 24 hours at $37°C$ in an atmosphere of 95 % N_2: 5 % CO_2. The various organs were dissected out and homogenized (Ultra-Turrax) in water. One ml of medium, pseudocoelomic fluid, and of homogenate was added to 15 ml of scintillator solution (Insta-Gel, Packard), and assayed for radioactivity using a Packard 3310

Tri-Carb Liquid Scintillation Spectrometer. Correction for quenching was applied by internal standardization.

2. In vivo experiments. Mebendazole was fed to *S.trachea*-infected turkeys for 24 hours at levels of 0.05 and 0.1 % of the diet. Glucose-C14 was given in the drinking water at 100 μC/l. After 24 hours the turkeys were slaughtered, the worms collected, washed thoroughly in physiological saline, dissected and washed again to remove the blood present in the intestine. The intestine, body wall, reproductive system of all the worms collected from one turkey were combined, weighed in tared scintillation counting vials and digested according to Mahin **et al.** [4]. Fifteen ml of Insta-Gel was added and the radio-activity determined. Control values were obtained from worms collected from turkeys receiving no drug.

B. Maltose uptake studies.

1. In vitro experiments. The effect of mebendazole on the maltose uptake by different organs of *A.suum* was studied under the same experimental conditions as described for the glucose uptake studies, with the exception that maltose (final concentration: 0.0088M) was used in place of glucose, 50 nC maltose-C14 (specific activity: 7 mC/mM, Radiochemical Centre) was added to 100 ml of the maltose salt medium.

2. In vivo experiments. Mebendazole was fed to *A-galli*-infected chickens at levels of 0.004 and 0.01 % of the diet. Maltose-C14 was given in the drinking water at 50 μC/l. After 24 hours the chickens were slaughtered and the worms collected, washed and dissected. The intestines were washed thoroughly, combined with the other organs and treated further as described for the experiments with *S. trachea.*

C. Effect of mebendazole on the glycogen content.

1. In vitro experiments. *A.suum* and *S.muris* were incubated as described for the glucose uptake experiments with the exception that DMSO was added instead of DMF. At the end of the incubation period (48 hours for *Ascaris,* 5 hours for *S.muris)* the *Ascaris* worms were dissected. The body walls (muscle + cuticle) were weighed and digested in 20 ml KOH 30 % at 100° C for 30 min. *S.muris* were collected and similarly digested in 1 ml of KOH 30 %. Glycogen was precipitated with ethanol and determined by the anthrone method as used in a previous study [2].

2. In vivo experiments. Rats and chickens infected with *S.muris* and *A.galli* respectively were treated orally with a single dose of mebendazole. Rats and chickens were killed 24 and 26 hours respectively after treatment and the worms collected. 100 pinworms or 2 *Ascaridia* were

141

digested in 1 ml of KOH 30 % as previously described and the glycogen content was similarly determined. Mice infected with *H.nana* were treated orally with two doses of mebendazole (50 mg/kg, given 12 hours apart). The animals were killed 24 hours after the last dose, and the worms collected, washed with 0.85 % saline, weighed and treated as described for *S.muris*. Turkeys, infected with *S.trachea* were treated with mebendazole for 24 hours at a level of 0.1 % of the diet. At the end of the treatment period the turkeys were slaughtered, the worm pairs collected from the trachea, transferred to 0.85 % saline, washed several times and weighed. One worm pair was digested in 1 ml of KOH 30 % as described for *S.muris.*

D. Enzyme assays.

Chickens infected with *A.galli* were slaughtered 26 hours after treatment with a single oral dose of mebendazole, (40 mg/kg). The worms were collected, dissected and the intestinal tracts removed. These were homogenized in 10 volumes of 0.25 M mannitol (pH 7.4) using a Potter-Elvehjem homogenizer with teflon pestle. The homogenates were then used for maltase and hexokinase determinations. Maltase activity was determined by the formation of glucose from maltose, final concentration of 0.0055 M in a 0.05 M maleate buffer at pH 6, according to Borgström and Dahlquist [5] . Glucose formed after an incubation period of 60 min. at 37° C was determined by the Glucostat method (Worthington Biochemical Corporation). Hexokinase activity was determined according to Walker and Parry [6] .

Rats infected with *S.muris* were slaughtered 24 hours after treatment with a single oral dose of mebendazole (2.5 mg/kg). The worms were collected as described above and 100 worms were homogenized in 1 ml of a medium (0.15 M KCl and 0.0016 M $KHCO_3$). The homogenate was centrifuged for 5 min. at 1000 g and the supernatant collected. Hexokinase activity was determined as described above. To determine glycogen phosphorylase a and a + b activity in *A.suum,* the worms were homogenized in a medium containing 20 % glycerol in 0.05 M glycylglycine buffer (pH 7.4), 1 g worm being used per 3 ml of medium. Enzymatic activity was determined according to Bueding and Fisher [7] .

E. ATP determination.

Ascaris worms were incubated as described for the glucose uptake experiments with the exception that they were previously starved for 24 hours. After an incubation period of 24 hours the worms were homogenized in ice-cold 8 % perchloric acid, 1 worm per 20 ml of homogenizing medium. The mixtures were centrifuged at 10,000 g for 10 min. and measured volumes of supernatant were adjusted with K_2CO_3 (3.75 M) to a pH of 6.5 and allowed to stand for 10 min. in an ice bath. The insoluble potassium perchlorate was removed by centrifugation at 5000 g for 10 min. ATP was determined in aliquots of the supernatants as proposed by Bueding **et al.** [8]. The glucose concentration of the medium was determined by the Glucostat method.

F. Protein concentration.

Protein was determined by the Folin method of Lowry **et al.** [9].

Results and Discussion.

The effect of mebendazole upon the glucose uptake by *A.suum* and *S.muris* is presented in Tables 3 and 4. These results indicate that concentrations of mebendazole lower than those affecting the motility of either *Ascaris* or *Syphacia,* significantly inhibit the rate of anaerobic glucose uptake **in vitro.** Exposure of *Ascaris* worms for 24 hours to mebendazole and their transferrence to a drug-free medium illustrated the persistent inhibition of glucose uptake, strongly indicative of an irreversible inhibitory effect. Although control worms were similarly incubated with drug solvent, it was thought possible that DMSO or DMF changes the intestinal cells and consequently facilitates the mebendazole-induced effects. With this point in mind, **in vivo** experiments were conducted on turkeys infected with *S.trachea.* As shown in Table 5, an inhibition of glucose uptake was again observed, a significant decrease in radioactivity being seen at a mebendazole concentration of as little as 0.05 % of the diet. The results for the **in vitro** uptake of C14-labelled sugars and the distribution of radioactivity in the different organs of *A.suum* are presented in Table 6 (glucose) and table 7 (maltose). In both cases, the results indicate a significant inhibition in the rate of C14 - uptake in the different organs. There is, however a difference in the amount of radioactivity removed from the medium. With maltose as substrate the radioactivity of the medium decreased by only 22 % as compared with 60 % when glucose was used. Mebendazole did not affect the maltase activity in *A.galli* and *A.suum* intestinal cells. However, since mebendazole unquestionably inhibits the uptake of C14 from maltose, it may be concluded that the uptake of glucose metabolised from maltose at the brush border membrane is also inhibited. The fact that the radioactivity in the experimental medium differed by only 22 % from the control would suggest that this metabolised glucose is re-excreted into the medium.

The effect of mebendazole on maltose utilization was further confirmed **in vivo.** As shown in Table 8, a significant decrease in radioactivity was observed when mebendazole was fed for 24 hours to chickens infected with *A.galli.*

The foregoing experiments indicated that mebendazole interferes with glucose uptake. Inhibition of glucose uptake can result either from an inhibition of glucose catabolism or from an inhibition of glucose entry and/or transport. If no inhibition of glucose-6-phosphate catabolism can be shown, inhibition of glucose uptake must result in an increased utilisation of reserve glycogen. As shown in Table 9, a marked decrease in the endogenous glycogen of *Ascaris* and *Syphacia* results from the **in vitro** mebendazole-induced inhibition of glucose uptake. This effect was confirmed **in vivo.** The results presented in Table 10 indicate a significant decrease (P ⟨ 0.0005) in glycogen content of adult *S.muris* 24 hours after feeding a single oral dose of 0.63 mg/kg of body weight to the rat host. This effect on glycogen stores was also found to be dose -related.

Reduction of glycogen stores was also noted in *A.galli* 26 hours after treating the chicken host with a single oral dose of mebendazole (40 mg/kg), and in *S.trachea* when the drug was administered over 24 hours to a turkey host. The marked decrease in glycogen content of *H.nana* from a mouse host, killed 24 hours after the second of two doses of mebendazole (50 mg/kg) given 12 hours apart, is consistent with the drug's cestocidal activity (Table 11). Whereas expulsion of *S.muris, A.galli* and *S.trachea* takes almost three days, **in vivo** glycogen depletion was noted 24 to 26 hours after treatment with mebendazole indicating that the biochemical properties of the drug are not secondary to its anthelmintic effect.

The results presented in Table 12 indicate that exposure of *A.suum* to mebendazole concentrations which inhibit glucose uptake, also produced a significant decrease (P = 0.0073) in the body wall concentration of ATP. Greater glucose uptake may have been due to the starvation of the worms prior to incubation. It was thought possible that inhibition of hexokinase may affect glucose uptake and consequently increase glycogen utilization. We therefore studied the effect of mebendazole on hexokinase activity. Chickens infected with *A.galli* were slaughtered 26 hours after a single oral dose of mebendazole (40 mg/kg body weight). The results presented in Table 13 indicate that glucose phosphorylation is unaffected in the intestine a similar result being recorded with homogenates of *S.muris* collected from rats 24 hours after oral treatment with mebendazole (2.5 mg/kg body weight). The derangement of glucose uptake and glycogen reserves induced by mebendazole cannot therefore be attributed to an inhibition of glucose phosphorylation.

Phosphorylase a and a + b activity in *Ascaris,* 24 hours after incubation with mebendazole (1 μg/ml of incubation mixture), was unaffected. Thus glycogen depletion cannot be related to an effect of mebendazole on phosphorylase activity, as has been discribed for niridazole [7].

The benzimidazoles are known as potent uncouplers of oxidative phosphorylation in mammalian mitochondria [10], and we have shown (H. Van den Bossche, this book) that they also inhibit the malate-induced phosphorylation in *Ascaris* mitochondria. We therefore designed experiments to show the effect of mebendazole on phosphorylation in mitochondria of *Ascaris* muscle. Fifty percent inhibition of malate-induced phosphorylation was only obtained using mebendazole concentrations up to 9 x 10^{-3}M, corresponding to approximately 2.6 mg/ml of incubation mixture. However, 50 % inhibition of the **in vitro** glucose uptake by *Ascaris* is achieved with only a few micrograms of mebendazole, indicating that its anthelmintic effect is not due to an interference in the energy yielding pathway.

All the experiments performed to date suggest that it is the direct inhibition of glucose uptake which is responsible for glycogen depletion in parasites, ultimately leading to decreased generation of ATP so essential for survival and replication. Recently, Beames [11] has shown that the intestinal cells of *Ascaris* possess the ability to move hexose molecules from the luminal to the pseudocoelomic fluid against a concentration gradient. Fairbairn and Passey [12] have however, shown that the glucose concentration of the pseudocoelomic fluid is very low. Thus as Beames [11] pointed out, there may always be a favorable gradient for the diffusion of sugar from the intestinal cells to the body fluid. Therefore it seems plausable that mebendazole interferes with the diffusion of glucose from the luminal to the pseudocoelomic fluid, rather than with active transport. Experiments designed to test this hypothesis are at present in progress.

It is interesting to note that several benzimidazoles possess anthelmintic activity. A number of these compounds, i.e. parbendazole, cambendazole, thiabendazole and some mebendazole analogues, R 17147 and R 18986, have been examined with regard to their effects on glycogen utilization. (The formulae of these benzimidazoles are given in Fig. 1). As shown in Fig. 2 parbendazole, cambendazole and R 17147 similarly deplete the glycogen content of *Syphacia* 24 hours after oral treatment of the rat hosts. The figure also illustrates the greater potency of mebendazole in stimulating glycogen utilization. The effects of cambendazole and parbendazole are very similar. These results are compatible with the finding that mebendazole has an anthelmintic activity against *S.muris* several times greater than parbendazole (Thienpont and Vanparijs, unpublished data). The anthelmintic activity of the latter drug would appear to be comparable with that of cambendazole [13]. The mebendazole analogue R 17147 is about four to five times less active than mebendazole and is also less effective in inducing glycogen utilization. This was further confirmed when the glycogen content of *A.galli* was measured 26 hours after oral treatment of the chicken host with a single dose of either mebendazole or R 17147. These results are presented in Table 14 and indicate that the glycogen content was only slightly affected by R 17147. The four benzimidazoles investigated until now (mebendazole, parbendazole, cambendazole, R 17147) are substituted in position 5 and have a carbamate function. Thiabendazole, however, is not substituted in this position and has no carbamate function. The glycogen content of *S. muris,* 24 hours after oral treatment of the rat host with a single dose of thiabendazole (5 mg/kg body weight), was only slightly affected by the drug (Table 15). Moreover, a decrease of only 31.5 % was noted after the rats were treated orally with two high doses of 50 mg/kg, given 12 hours apart, and slaughtered 24 hours after the second dose. This may indicate, that either substitution in position 5 or a carbamate function is required for the benzimidazole to affect glycogen utilization in helminths. The fact that R 18986 [14] did not affect the glycogen content of the rat pinworm (Table 15) suggests that it is the carbamate function which is essential. Prichard [15] has shown that thiabendazole inhibits the fumarate reductase mechanism in adult *Haemonchus contortus.* This difference in the mode of action may explain why thiabendazole affected glycogen content only slightly. Experiments done until now suggest that structural modifications may result not only in increased anthelmintic activity but also in another mode of action.

References

1. VAN DEN BOSSCHE, H. and Janssen, P.A.J., Life Sci. **6,** 1781 (1967).

2. VAN DEN BOSSCHE, H., SCHAPER, J. and BORGERS, M., Comp. Biochem. Physiol. **38,** 43 (1971).

3. VAN MUUSTER, R.J.J., Ned. Tijdschr. Geneesk. **96,** 1345 (1952).

4. MAHIN, D.T. and LOFBERG, R.T., Anal. Biochem. **16,** 500 (1966).

5. BORGSTROM, B. and DAHLQUIST, A., Acta Chem. Scand. **12,** 1997 (1958).

6. WALKER, D.C. and PARRY, M.J., in "Methods in Enzymology" (W.A. Wood, Ed;) vol. 9 p. 381. Academic Press, New York (1966).

7. BUEDING, E. and FISHER, J., Mol. Pharmacol. **6,** 532 (1970).

8. BUEDING, E., KMETEC, E., SWARTZWELDER, C., ABADIE, S. and SAZ, H.J., Biochem. Pharmacol. **5,** 311 (1961).

9. LOWRY, O.H., ROSEBROUGH, N.J., FARR, A.L. RANDALL, R.J., J. Biol. Chem. **193,** 265 (1951).

10. JONES, O.T.H. and WATSON, W.A., Biochem. J. **102,** 564 (1967).

11. BEAMES, C.G. Jr., J. Parasitol. **57,** 97 (1971).

12. FAIRBAIRN, D. and PASSEY, R.F., Exptl. Parsitol. **6,** 566 (1957).

13. HOFF, D.R., FISHER, M.H., BOCHIS, R.J., LUSI, A., WAKSMUSKI, F., EGERTON, J.R., YAKSTIS, J.J., CUCKLER, A.C. and CAMPBELL, W.C., Experientia **26,** 550 (1970).

14. BRUGMANS, J.P., THIENPONT, D.C., VAN WIJNGAARDEN, I., VANPARIJS, O.F., SCHUERMANS, V.L. and LAUWERS, H.L., JAMA **217,** 313 (1971).

15. PRICHARD, R.K., Nature **228,** 684 (1970).

Acknowledgements

The author wishes to express his thanks to Dr. Paul A.J. Janssen for his constant interest; Mrs. Horemans, Mr. Goossens and Mr. Vermeiren for their technical assistance, and to Mr & Mrs Scott for their help in preparing the manuscript.

Table 1.
Anthelmintic activity of mebendazole against nematodes [a]

RHABDIASIDEA		*Strongyloides*
TRICHURIDEA		*Trichuris*
		Capillaria
		Trichinella
STRONGYLIDEA	*STRONGYLIDAE*	*Strongylus*
	ANCYLOSTOMIDAE	*Ancylostoma*
		Uncinaria
		Necator
	CYATHOSTOMIDAE	*Trichonema*
		Oesophagostomum
		Chabertia
	TRICHOSTRONGYLIDAE	*Trichostrongylus*
		Cooperia
		Hyostrongylus
		Ostertagia
		Haemonchus
		Nematodirus
	SYNGAMIDAE	*Syngamus*
OXYURIDEA	*OXYURIDAE*	*Oxyuris*
		Enterobius
		Syphacia
	HETERAKIDAE	*Heterakis*
ASCARIDIDEA		*Ascaris*
		Parascaris
		Toxascaris
		Toxocara
		Ascaridia

[a] Thienpont **et al.**, unpublished data.

148

Table 2.
Anthelmintic activity of mebendazole against cestodes [1]

CYCLOPHYLLIDEA	*ANOPLOCEPHALIDAE*	*Moniezia*
	DILEPDIDIDAE	*Dipylidium*
	HYMENOLEPIDIDAE	*Hymenolepis*
	DAVAINEIDAE	*Raillietina*
	TAENIIDAE	*Taenia hydatigena*
		Hydatigera
		taeniaeformis

[a] Thienpont et al., unpublished data.

Table 3.
Effect of mebendazole on the exogenous glucose uptake by A.suum [a]

Drug concentration µg/ml	Glucose uptake [b] mg/24 hours/g worm		
	1st day	2nd day	3rd day
0	16.31 ± 5.47 (20)	14.41 ± 4.07 (18)	13.09 ± 1.02 (17)
0.0125	14.58 ± 3.46 (3)	9.81 ± 4.85 (3)	13.24 ± 1.12 (3)
0.0250	10.03 ± 3.19 (4)	6.90 ± 4.53 (4)	6.73 ± 5.45 (4)
0.0500	8.50 ± 2.84 (4)	4.38 ± 0.65 (4)	3.83 ± 1.10 (4)
0.1000	6.17 ± 2.00 (4)	6.42 ± 2.10 (4)	5.27 ± 2.24 (4)
0.2500	3.32 ± 1.15 (6)	3.95 ± 1.72 (6)	4.19 ± 0.67 (4)

[a] **Ascaris** were incubated for 24 hours in a medium containing mebendazole. The worms were then transferred to the drug-free medium which was used throughout for the control worms.

[b] Mean value ± S.D. followed by the number of determinations in brackets.

Table 4.
Effect of mebendazole on the exogenous glucose uptake by S. muris

Drug concentration μg/ml	Glucose uptake [a] μg/5 hours/100 worms	%
0	836.6 ± 173.24 (8)	100
5	476.2 ± 142.51 (9)	55.14
8	48.7 ± 7.57 (3)	5.64

(a) Mean value ± S.D. followed by the number of observations in brackets.

Table 5.
Effect of mebendazole on the uptake of glucose-C14 by S.trachea after treatment of the turkey host with the drug.

Drug concentration % of the diet	Radioactivity counts/min/g worm
0	4445.7 ± 1446 (9)
0.05	1462.5 ± 258 (4)
0.1	1243.4 ± 254 (7)

Table 6
Effect of mebendazole on the uptake of radioactivity by the different organs of A.suum after incubation of the worms in a medium to which C14-labelled glucose was added.

	Specific activity [a] counts/min/g tissue		Decrease %
	controls	mebendazole [b]	
Muscle	1156 ± 895	172 ± 96	85.12
Intestine	969 ± 295	85 ± 24	91.23
Reproductive system	905 ± 565	109 ± 54	87.96
Pseudocoelomic fluid	1319 ± 602	508 ± 258	61.49
Uptake from medium	2993 ± 1084	1190 ± 412	60.24

(a) Mean value of five experiments ± S.D.

(b) 0.25 μg of mebendazole/ml of incubation mixture

Table 7.
Effect of mebendazole on the uptake of radioactivity by the different organs of A.suum after incubation of the worms in a medium to which C14-labelled maltose was added.

	Specific activity [a] counts/min/g tissue		Decrease %
	controls	mebendazole [b]	
Muscle	554 ± 297	156 ± 115	71.84
Intestine	960 ± 636	180 ± 139	81.25
Reproductive system	752 ± 572	171 ± 93	77.26
Pseudocoelomic fluid	2315 ± 613	818 ± 338	64.67
Uptake from medium	4613 ± 991	3601 ± 750	21.94

(a) Mean value of 4 experiments ± S.D.

(b) 0.25 μg of mebendazole/ml of incubation mixture

Table 8.
Uptake of C14-labelled maltose by A.galli after treatment of the chicken host with mebendazole.

Drug concentration % of the diet	Radioactivity [a] counts/min/g worm
0	1905 ± 274 (6)
0.004	1183 ± 326 (5)
0.01	412 ± 93 (7)

(a) Mean value $\pm$ S.D. followed by the number of determinations in brackets.

Table 9.
Effect of mebendazole on the endogenous glycogen content.

Species	Drug concentration μg/ml	Glycogen content [a] μg glycogen/100 worms	%
S. muris	0	295.9 ± 47.59 (6)	100
	5	196.1 ± 29.93 (4)	66.27
		mg glycogen/g body wall	
A. suum [b]	0	144.1 ± 27.13 (4)	100
	0.1	92.4 ± 19.98 (5)	64.12

(a) Mean value $\pm$ S.D. followed by the number of observations in brackets
(b) Body wall = muscle + cuticle

Table 10.
Glycogen content of S.muris 24 hours after oral treatment of the rat host with mebendazole.

Mebendazole mg/kg body weight	Glycogen content[a] μg/100 worms
0	383.4 ± 56.5 (55)
0.63	252.7 ± 12.6 (6)
1.25	211.9 ± 10.8 (8)
2.50	179.7 ± 32.5 (8)
5.00	161.6 ± 14.9 (4)
10.00	129.0 ± 8.4 (5)

(a) Mean value ± S.D. followed by the number of determinations in brackets.

Table 11.
Glycogen content of A.galli, S.trachea and H.nana after oral treatment of their hosts with mebendazole.

Species	Mebendazole	Glycogen concen-tration[a] μg/mg worm	%
A. galli	0	32.55 ± 3.45 (43)	100
	40 mg/kg	19.09 ± 1.00 (9)	58.65
S. trachea	0	9.00 ± 1.27 (20)	100
	0.1 % of the diet	2.42 ± 0.64 (19)	26.86
H. nana	0	69.64 ± 22.58 (18)	100
	2 x 50 mg/kg	28.57 ± 11.65 (19)	41.02

(a) Mean values ± S.D. followed by the number of determinations in brackets.

Table 12.
Effect of mebendazole on the glucose uptake and ATP-concentration after incubation of A.suum with the drug.

Mebendazole µg/ml of medium	Glucose uptake[a] mg/g worm/24 hours	ATP content[a] µmoles/g worm
0	35.20 ± 3.86 (6)	1.09 ± 0.12 (6)
1	16.38 ± 1.21 (6)	0.89 ± 0.05 (5)

(a) Mean value ± S.D. followed by the number of determinations in brackets.

Table 13.
The hexokinase activity in intestinal cells of *A.galli* and *S.muris* after oral treatment of their hosts with mebendazole.

Species	Mebendazole (mg/kg body weight)	Hexokinase activity (nmoles/min/mg protein)
A. galli (intestine)	0	39.4 ± 20.0 (9)
	40	39.4 ± 7.6 (6)
S. muris (whole worm)	0	160.0 ± 48.0 (10)
	2.5	180.0 ± 45.0 (11)

Table 14.

Glycogen content of _A.galli_ 26 hours after oral treatment [a] of the chicken host with either mebendazole or R 17147.

Drug	Glycogen concentration (mg/g worm) [b]	P
—	32.55 ± 3.45 (43)	
Mebendazole	19.09 ± 1.00 (9)	‹ 0.0001
R 17147	28.59 ± 4.07 (20)	=0.0001

(a) 40 mg drug/kg of body weight

(b) Mean value ± S.C. followed by the number of determinations in brackets.

Table 15.

Glycogen content of _S.muris_ 24 hours after oral treatment [a] of the rat host with thiabendazole or R 18986.

Drug	Glycogen content μg/100 worms[b]
—	383.4 ± 56.51 (55)
Thiabendazole	318.9 ± 45.89 (4)
R 18986	402.4 ± 15.51 (5)

(a) 5 mg drug/kg of body weight

(b) Mean value ± S.D. followed by the number of determinations in brackets.

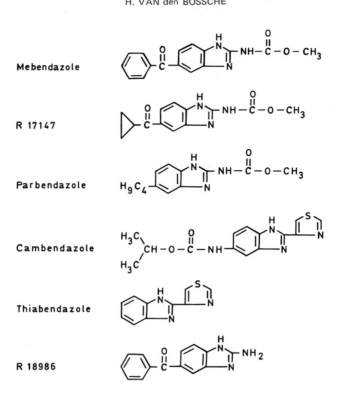

Fig. 1.
Structural formulae of the benzimidazoles used in this study.

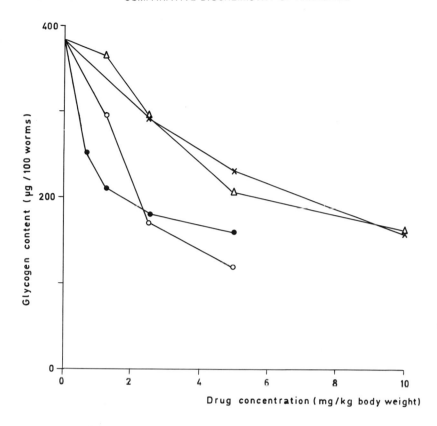

Fig. 2.
Glycogen content of *S.muris* 24 hours after oral treatment of the rat host with Mebendazole (●), R 17147 (○), Parbendazole (x) or Cambendazole (▲).

11

KINETOPLAST DNA. STRUCTURE AND FUNCTION

M. Steinert and Suzanne Van Assel

Département de Biologie Moléculaire, Faculté des Sciences
Université Libre de Bruxelles
67, rue des chevaux, B 1640 - Rhode-St-Genèse, Belgium.

A great deal of confusion has occured in connection with the nature and function of the kinetoplast, and this confusion is reflected by the large number of different words which have been used to specify this peculiar organelle. I personnally (M.S.) largely contributed to this confusion by successively using at least three different names and I am now asking myself whether it should not be suitable to shift to a fourth one. This meeting being devoted to comparative biochemistry, I would like to consider, very briefly, the following questions: how can kinetoplast DNA be compared with typical mitochondrial DNA in other cells and, is it reasonable to assume that kinetoplast DNA may perform the same functions as mitochondrial DNA?

Let us first mention some common characters. Like most mitochondrial DNAs, kinetoplast DNA is, at least partially, in the form of closed circles [1, 5]. Like mitochondrial DNA, k-DNA is not associated with basic proteins [6] and its high sensitivity to intercalating dyes in the living cell is now well documented (see ref. 7 for a review of this subject). However, in some respects, k-DNA differs greatly from mitochondrial DNA. For instance, kinetoplast DNA is present in relatively much higher quantities than mitochondrial DNA in most cells, excepting perhaps yeast. All the k-DNA complement of the trypanosome is concentrated in a single body, the kinetoplast, which is always and very characteristically situated behind the basal body of the flagellum. As regards synthesis, mitochondrial DNA usually replicates continuously, throughout the whole cell cycle, whereas k-DNA is synthesized during the S phase only, together with nuclear DNA [8, 9 and 10]. But, perhaps, what makes k-DNA look so different from typical mitochondrial DNA is its apparently highly complex molecular structure. As Dr. Newton recalled yesterday, kineto- plast DNA behaves in a very abnormal way in CsCl gradient: it bands extremely rapidly and forms a very sharp and narrow band [11]. The reason for this has been recently discovered in at least three different laboratories [12, 14] : highly purified k-DNA, in the presence of a suitable fluorescent dye, appears, in the fluorescence microscope, in the form of clumps of constant size and shape (fig. 1). Quantitative estimations of the DNA contents of each clump show, in fact, that each one of these particles is a whole kinetoplast. So, in the intact kinetoplast, most probably all the minicircles are linked together to form these unique very large structures which can be extracted without being disrupted. In what precise way are all these circles (and perhaps also linear segments of DNA) linked together is not known, although quite a few of them appear to form catenanes (fig. 2) [2, 3] .

This closely packed DNA is also characterized by a polyphasic melting curve. This rather abnormal melting behavior seems to be due to topological constraint, as a normal transition is observed on sonicated k-DNA samples (fig. 3).

So, no clear answer to our question, of wether kinetoplast DNA is really homologous to mitochondrial DNA, can be obtained from this compari- son of structural characters. Obviously, a definite answer should come from comparative studies of the biological function of these DNAs.

There is now ample support, from both genetic and biochemical eviden- ce, for the idea that mitochondrial DNA contains the necessary informa- tion for the synthesis of mitochondrial r-RNA, for mitochondrial t-RNA and, most probably also for the coding of a very small number of polypeptide chains. No such evidence has ever been obtained with regard to kinetoplast DNA, although we know, from old and recent work on akinetoplastic trypanosomes, that the kinetoplast, like mitochondrial DNA in yeast, appears to be involved in the biogenesis of the respiring mitochondrion (see ref. 7 for a review).

Now, in order to perform the same genetic function as mitochondrial DNA, kinetoplast DNA should have the same informational capacity, in other words it should contain a unique sequence of similar length. At first glance, this seems not to be the case, as we have seen that kinetoplast DNA appears to be made, essentially, of very small circles which are about ten times smaller than the DNA circles extracted from vertebrate mitochondria. However, the linear segments which have been found in k-DNA could perhaps provide more genetic information. The informational capacity of a DNA can be determined by measuring its kinetics of reassociation after melting [15, 16]. Figure 4 presents a typical result of some preliminary work in this direction. Pure k-DNA has been prepared from *Crithidia luciliae* as described previously [12] and sonicated in order to obtain fragments which sediment as 8 S particles in alkaline solution. After being melted at 100°C, the DNA is allowed to reassociate at 56°C in a 0.08M phosphate buffer. The diphasic kinetics of the reaction indicate that at least two distinct sequences of different length are present. Although no precise computation of sequence size could be made from these preliminary data, it seems reasonably clear that the rapidly annealing fraction corresponds to the minicircles of 0.7 μ (fig. 2) and that the kinetic complexity of the slower component is much higher and might compare with the size of the mitochondrial genome in yeasts [17].

So it is quite reasonable to predict that, in the near future, a typical mitochondrial DNA function will be discovered and ascribed to the kinetoplast. No doubt the next problem to be solved will be the biological significance of this rather exceptional example of redundancy in mitochondrial DNA.

This work has been supported by the World Health Organization (Geneva) and by the "Fonds de la Recherche Fondamentale Collective" (Brussels). We thank Dr. Pamela Malpoix for kind assistance in revising the English text.

References

1. RIOU, G. and PAOLETTI, C., J. Mol. Biol., **28,** 377 (1967).
2. RIOU, G. and DELAIN, E., Proc. Natl. Acad. Sc. U.S., **62,** 210 (1969).
3. SIMPSON, L. and DA SILVA, A., J. Mol. Biol., **56,** 443 (1971).
4. LAURENT, M. and STEINERT, M., Proc. Natl. Acad. Sc. U.S., **66,** 419 (1970).
5. RENGER, H.C. and WOLSTENHOLME, D.R., J. Cell Biol., **47,** 689 (1970).
6. STEINERT, M., Exp. Cell Research, **39,** 69 (1965).
7. SIMPSON, L., "The kinetoplast of the hemoflagellates". In Intern. Rev. Cytology, **32,** (1971), in press.
8. STEINERT, M. and STEINERT, G., J. Protozool., **9,** 203 (1962).
9. VAN ASSEL, S. and STEINERT, M., Exp. Cell Research, **65,** 353 (1971).
10. SIMPSON, L. and BRALY, P., J. Protozool., **17,** 511 (1970).
11. DU BUY, H.G., MATTERN, C.F.T. and RILEY, F.L., Science, **147,** 754 (1965).
12. LAURENT, M., VAN ASSEL, S. and STEINERT, M., Biochem. Biophys. Res. Comm., **43,** 278 (1971).
13. WILLIAMSON, D.H., STUART, K., GUTTERIDGE, W.E. and BURDETT, I., 11th. Seminar on Trypanosomiasis Research, London, 12 Nov. 1970.

14. SIMPSON, L., personnal communication.
15. WETMUR, J.G. and DAVIDSON, N., J. Mol. Biol., **31,** 349 (1967).
16. BRITTEN, R.J. and KOHNE, D.E., Science, **161,** 529 (1968).
17. HOLLENBERG, C.P., BORST, P. and VAN BRUGGEN, E.F.J., Biochim. Biophys. Acta, **209,** 1 (1970).
18. KLEINSCHMIDT, A.K., LANG, D., JACHERTS, D. and ZAHN, R.K., Biochim. Biophys. Acta, **61,** 857 (1962).

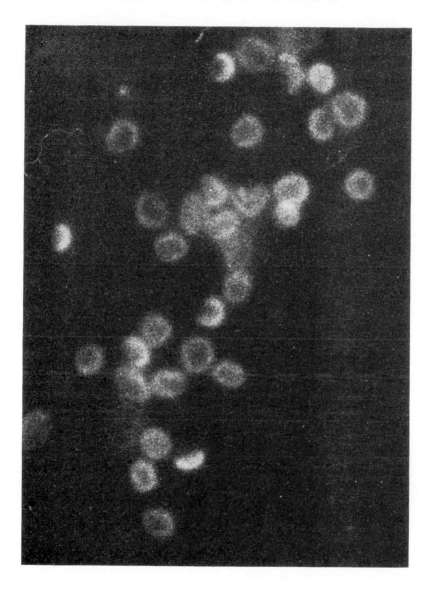

Fig. 1.
Highly purified k-DNA, extracted from *Crithidia luciliae,* as observed under the fluorescence microscope in the presence of 40 μg/ml ethidium bromide.

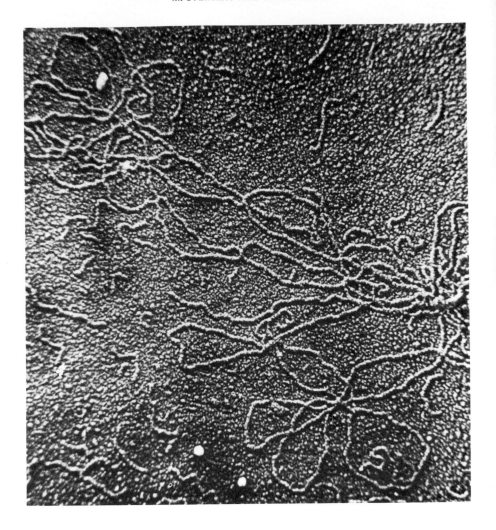

Fig. 2.
Kinetoplast DNA from *C. luciliae.* The sample has been briefly sonicated and prepared for electron microscopy as described by Kleinschmidt et al. [18]. The large clumps are partially dissociated and the fragments seem to be made of catenated circular monomers, about 0.7 μ in contour length.

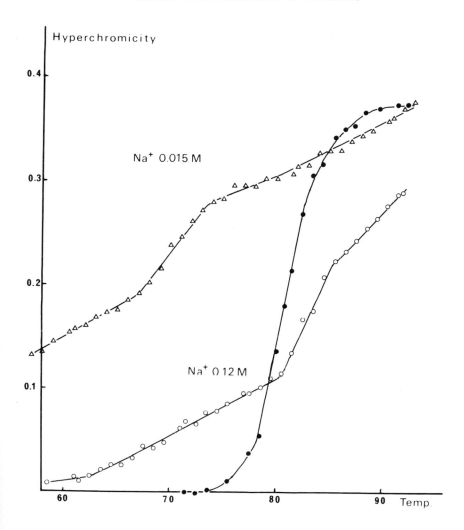

Fig. 3.
Melting profiles of purified *C. luciliae* k-DNA.
(○): intact k-DNA, heated in 0.08M phosphate buffer.
(▲): intact k-DNA, heated in 0.01M phosphate buffer.
(●): sonicated k-DNA, heated in 0.08M phosphate buffer.

165

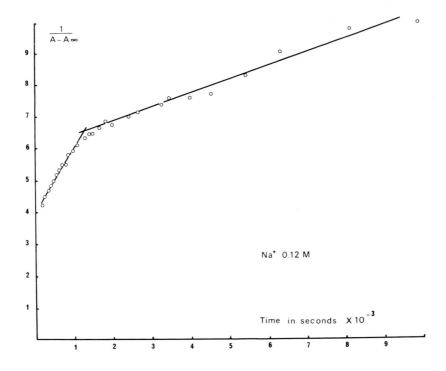

Fig. 4.
Second-order rate plot for the reassociation of *C. luciliae* k-DNA.

12

ORGANIZATION OF THE DNA IN THE KINETOPLAST OF TRYPANOSOMATIDAE

Delain, E., Brack, Ch., Lacome, A. and Riou, G.

Unités de Microscopie Electronique et de Pharmacologie Moléculaire
Institut Gustave-Roussy - 94 - Villejuif - France

Introduction

Trypansomes are parasitic flagellates characterized by a unique mito-chondrion. A specialized region of this organelle, called the kinetoplast, contains a large amount of DNA. The kinetoplast is localized at the basis of the flagellum.

The mitochondrial DNA (mtDNA) of all animal cells studied up to now has been shown to consist of double-stranded covalently closed circular molecules with a contour lenth of abour 5 μ (reviewed by Borst et al[1] and Rabinowitz et al[2]); in contrast, the mtDNA of some other eukaryote cells (higher plants and yeasts) is linear. Circular DNA molecules have been isolated from the kinetoplast of many species of Trypanosomatidae (*Leishmania - Crithidia-Trypanosoma* [3-9]). These mo-lecules are much smaller than the mtDNA of higher animals, their contour length varying from 0.2 to 0,8 μ, depending on the trypanosome species.

One characteristic feature of the kinetoplast is its particularly high content of DNA (kDNA). This kDNA is so concentrated that it can be detected by light microscopy, either by specific DNA staining or by fluorescence. In ultrathin sections it exhibits a characteristic and regular structure. The kDNA is therefore a very suitable material to study the physico-chemical properties of circular DNA, its replication, and the trypanocidal effect of DNA interacting drugs at the cellular and the molecular level.

Technical approach

Preparation of kinetoplast DNA (kDNA)

Different techniques have been described for the preparation of purified kDNA. After chemical extraction, the total cell DNA can be further fractionated into different components; but it is not possible to ascertain the kinetoplastic origin of the different satellite DNAs. The best way to obtain pure kDNA would be to extract it from isolated and purified kinetoplasts.

Isolation of kinetoplasts

The trypanosomes were disrupted either by sonication *(T. cruzi)* or with a Potter homogenizer *(T. equiperdum).* A homogeneous suspension of trypanosomes, 100 mg per ml in mannitol 0.21 M, sucrose 0.07 M, in buffer solution (Tris 0.001 M, EDTA 0.0001 M, pH 7.5) was sonicated with a Branson sonifier, in an ice bath. After 10-15 sec. 80-90 % of the trypanosomes were broken. DNase was added to hydrolyze the nuclear DNA in standard conditions. However, *T. equiperdum* kDNA did not resist to the DNase treatment.

The disrupted cells are layered on the top of a discontinuous sucrose gradient (1.75 M, 1.30 M, 1 M, 0.75 M sucrose in the same buffer), and centrifuged in a Beckman ultracentrifuge rotor SW 25, for 1 hour at 22 000 r.p.m at 4°C. Most of the kinetoplasts banded between the 1.30 M and the 1.75 M sucrose layers. They were washed and recentrifuged in 0.15 M NaCl. 0.015 M Na$_3$ citrate (SSC). Fig. 1 shows a micrograph of purified kinetoplasts. Their intact double membrane has preserved the characteristic paracrystalline structure of kDNA from the DNase action. The disadvantage of this technique is the relatively small yield of purified kinetoplasts. The kDNA was observed in the electron microscope, either after osmotic disruption or after chemical extraction from purified kinetoplasts.

Extraction and purification of kDNA

Total cell DNA was extracted with chloroform-isoamylic alcohol according to the method of Marmur [10]. The choice of the extraction method has proved to be very important for the extraction of kDNA. In previous work, where we had used the phenol method of Kirby [11] for the preparation of *T. equiperdum* DNA, we selectively lost the kDNA with the buoyant density of $\rho = 1.691$ g/ml (Riou et al [9]).

The kDNA was fractionated from the total DNA by several methods. We used complex formation of DNA with heavy metal such as mercury. This method previously described by Davidson et al[12] is based on the large changes in the buoyant densities of DNAs subjected to caesium sulfate density gradient centrifugation after binding mercuric ions[3]. The amount of mercuric ions bound to DNA is directly proportional to its A-T content. This fractionation procedure is only efficient if the base composition of the kDNA and the nuclear DNA are different.

Another ultracentrifugation method for the fractionation of kDNA consists in the centrifugation of the total DNA in CsCl with a high resolutive rotor (fixe-angle rotor 50 or 40.3, Beckman ultracentrifuge). In this case, the greater the difference in the base compositions is, the better is the resolution of the different DNA bands.

The kDNA can be further purified in an ethidium bromide (EB)-CsCl gradient as described by Radloff et al[13]. EB is a trypanocidal drug which binds to the DNA. The amount of dye fixed to the DNA depends on its molecular form. Covalently closed circles bind less dye than linear or nicked circular DNA molecules, and can be fractionated after centrifugation in an EB-CsCl gradient.

Electron microscopy

Osmotic lysis of isolated kinetoplasts

The procedure was similar to that of Kleinschmidt et al[14] adapted to mitochondria by Nass[15]. A suspension of kinetoplasts in SSC, containing cytochrome c, was spread on an hypophase of distilled water. The DNA-containing cytochrome c film was picked up on carbon coated grids and shadowed with platinum while rotating (30 mg 6° 10 cm).

Spreading of the chemically purified kDNA

The microversion of the spontaneous adsorption method of Lang and Mitani[16] was used with the following modifications. Droplets of about 15 μl of a solution containing a final concentration of DNA: 0,5 μg/ml; cytochrome c: 15 μg/ml; ammonium acetate 0.05 M, were placed on a Teflon plate. A small beaker containing 37 % formaldehyde was placed beside it, and both were covered with a Petri dish. After 30-60 minutes diffusion time, the DNA-containing protein film was picked up, dried in 95 % ethanol and shadowed.

Organization of the DNA in the kinetoplast of Trypanosomes

After osmotic lysis of purified kinetoplasts of *T. cruzi* (Figs 2-3) and *T. equiperdum* (Figs 4-5), we observed large patches of kDNA, more or less well expanded. Usually, the central region of such patches is very dense and badly resolved, but at the periphery, the individual kDNA molecules can be distinguished as being complexly imbricated. A magnification of the kDNA molecules observed at the periphery of one patch is shown on Figs 2-3. Many circular DNA molecules seem to be attached to each other, forming figures of catenanes(↔) or figures of 8 (↮). These kDNA associations are often surrounded by many extended free circles (Fig. 2). The tension of the molecules is due to the spreading on distilled water. The kDNA circles of *T. equiperdum* are smaller (0.3 μ) than those of *T. cruzi* (0.5 μ) but their organization is similar (Fig. 4-5).

When the osmotic lysis is performed on an hypophase of distilled water containing 1 μg/ml of EB, many molecules are supertwisted (Fig. 6). The supertwisted state of circular DNA molecules from osmotically disrupted kinetoplasts strongly suggests that these molecules are covalently closed **in situ.**

The DNA spreading technique described in this report permits to obtain kDNA patches with a diameter of 10-20 μ, in which molecules are easily distinguishable at some places. Fig. 7 shows a large kDNA-association of approximately 10x14 μ, which is assumed to be most of the DNA content of one single kinetoplast. On well resolved portions of such large kDNA associations, we can see rare linear molecules (Fig. 8 →), circles which are apparently catenated (↔), and circles attached to each other only by one point (↮). In addition, long branched chains of "linear" molecules regularly crossing over from a complex network to which circles are attached (Fig. 8).

The chains and the circles which are attached by only one point are morphologically similar to the figure of 8 and to the oligomeric figures of 8 already described in **Leishmania tarentolae** kDNA by Simpson et al[6] and can be compared to the fused molecules described by Clayton et al.[17] in mtDNA molecules of leukemic cells. It is possible that all these associations are actually composed of such fused molecules: one strand of the double helix consisting of circular units, the other one being a long linear polymeric strand, which links all these units to each other. In fact, the topological catenation of circular molecules has never been proved.

Trypanocidal drugs and kDNA organization

We have described the arrangement of the kDNA molecules, showing that in *T. cruzi* this extranuclear DNA is probably very highly repetitive and in an unusually high concentration. The experiments with the drugs mentioned below give further evidence for the existence of a meso-morphous state of this kDNA. The absence of histone-like proteins in the kinetoplast as in the mitochondria, and the selective permeability of the mitochondrial membrane, could explain the particular sensibility of kDNA to some drugs.

The DNA-interacting drugs can be roughly grouped into two classes: the intercalating and the non intercalating ones[18]. We have tried to describe the specific effects of these two types of substances on the organization **in vivo** of the kDNA of *T. cruzi.*

The characteristic double-row arrangement in untreated *T. cruzi* kDNA (Fig. 9) rapidly disappears when the trypanosomes are grown in the presence of ethidium bromide or daunomycine. These two substances are known to intercalate between the base pairs of the DNA, and to produce a torsion in the double helix, leading to a modified tertiary structure of covalently closed circular molecules. Ethidium bromide (Fig. 10) and daunomycine (Fig. 11) both induce the rearrangement of the kDNA into very characteristic arches (→). This structure resembles the organization of the DNA in the chromosomes of Dinoflagellates.

When *T. cruzi* was treated with the non intercalating trypanocide hydroxystilbamidine and berenil, their kDNA was not organized into arches, but rearranged into very different structures. Ultrathin sections (Fig. 12) and kDNA cytochrome preparations (Fig. 13) have demonstra-ted that the kDNA of these treated trypanosomes consists of long imbricated "sausage-like" fibrillar bodies[19, 20].

We have shown that EB induces the formation of circular oligomeric forms of kDNA[4]. In berenil treated trypanosomes the kDNA contains a relatively high proportion of double-branched circular molecules. The behaviour of these circular replicating molecules in an EB-CsCl gradient have shown that their two template strands are covalently closed.

The study of the DNA organization in the kinetoplast leads to different concepts on the topology of such circular molecules. In fact it is quite impossible to relate these molecular associations to the classical models of associations of circular DNA (catenated or fused). The comparison of the effects of many different drugs may help to a better understanding of the real organization of the kDNA, and also of the way these drugs interact with circular DNA.

References

1. BORST, P. and KROON, A.M., Intern. Rev. Cytol. **26,** 107 (1969).
2. RABINOWITZ, M. and SWIFT, H., Physiol. Rev. **50,** 376 (1970).
3. RIOU, G. and PAOLETTI, C., J. Mol. Biol. **28,** 377 (1967).
4. RIOU, G. and DELAIN, E., Proc. Nat. Acad. Sci. **62,** 210 (1969).
5. LAURENT, M. and STEINERT, M., Proc. Nat. Acad. Sci. **66,** 419 (1970).
6. SIMPSON, L. and DA SILVA, A., J. Mol Biol. **56,** 443 (1971).
7. RENGER, H.C. and WOLSTENHOLME, D.R., J. Cell Biol. **47,** 689 (1970).
8. RENGER, H.C. and WOLSTENHOLME, D.R., J. Cell Biol. **50,** 533 (1971).
9. RIOU, G., LACOME, A., DELAIN, E., BRACK, Ch. and PAUTRIZEL, R., C.R. Acad. Sci. (1971).
10. MARMUR, J., J. Mol. Biol. **3,** 208 (1961).
11. KIRBY, K.S., Biochem. J. **66,** 495 (1957).
12. DAVIDSON, N., WIDHOLM, J., NANDI, U.S., JENSEN, R., OLIVERA, B.M. and WANG, J.C., Proc. Nat. Acad. Sci. **53,** 111 (1965).
13. RADLOFF, R., BAUER, W. and VINOGRAD, J., Proc. Nat. Acad. Sci. **57,** 1514 (1967).
14. KLEINSCHMIDT, A.K. and ZAHN, R.K., Z. Naturforsch. **14b,** 730 (1959).
15. NASS, M.M.K., Proc. Nat. Acad. Sci., **56,** 1215 (1966).
16. LANG, D. and M. MITANI, Biopolymers **9,** 373 (1970).
17. CLAYTON, D.A., DAVIS, R.W. and VINOGRAD, J., J. Mol. Biol. **47,** 137 (1970).
18. WARING, M., J. Mol. Biol. **54,** 247 (1970).
19. DELAIN, E., BRACK, Ch., RIOU, G. and FESTY, B., J. Ultrastruct. Res. **37,** 200 (1971).
20. BRACK, Ch., DELAIN, E., RIOU, G. and FESTY, B. (in press).

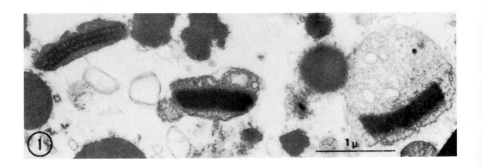

Fig. 1.

Isolated kinetoplasts of *T. cruzi* obtained by sonication of trypanosomes and purification in sucrose gradient (x 26,000).

Fig. 2.
Isolated kinetoplasts of blood stream forms of *T. cruzi* were osmotically lyzed. At the periphery of a large DNA patch, the individual circular molecules can be discerned. Note the numerous free molecules on the upper part of the picture (x 15,000).

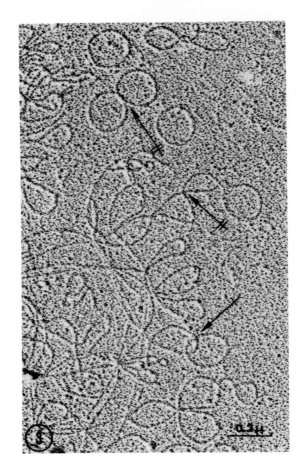

Fig. 3.
Detail of Fig. 2 shows the imbrication of circular molecules. They are
either catenated (↦) or fused (↦) (x 65,000).

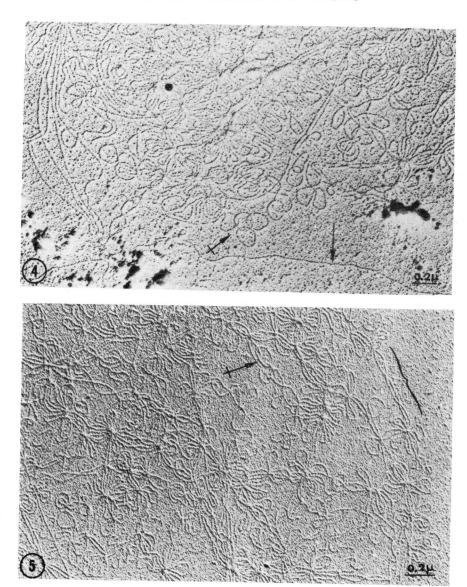

Fig. 4 and 5.

Preparations of kDNA of *T. equiperdum* obtained after osmotic lysis of isolated kinetoplasts. The small circles are imbricated like catenanes (↔), or linked to form chains (bracket). Some long linear molecules can be seen (→) (x 50,000).

175

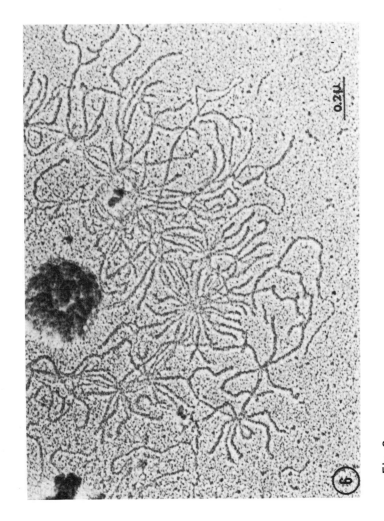

Fig. 6.
Kinetoplasts of *T. cruzi* were lyzed on distilled water containing 1 μg/ml of ethidium bromide. The circular molecules are tightly supertwisted (x 65,000).

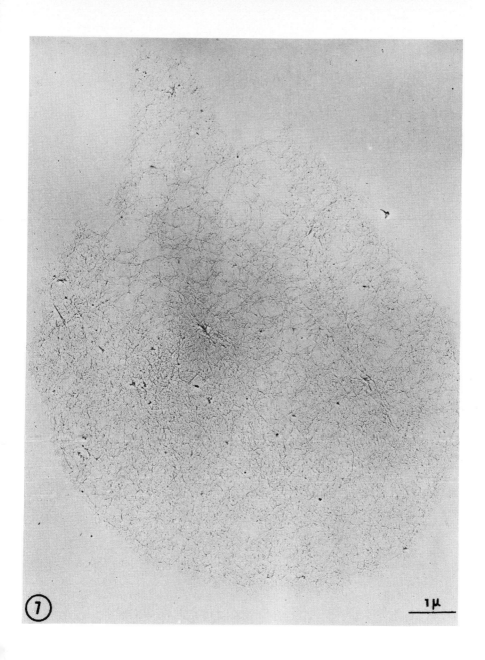

Fig. 7.
Large patch of kDNA of *T. cruzi,* spread after chemical extraction by the
"microdrop method". Such a preparation is considered to correspond to
the kDNA content of one single kinetoplast (x 12,000).

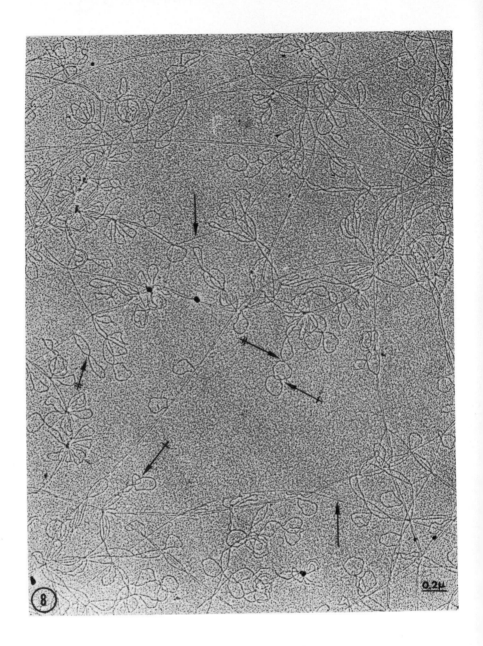

Fig. 8.
Detail of a kDNA association prepared and spread as in Fig. 7. The numerous chains are very extended. Single or connected molecules are linked to these chains. The circles are either catenated (↔) or fused (↮). Some long linear molecules are marked by → (x 40,000).

178

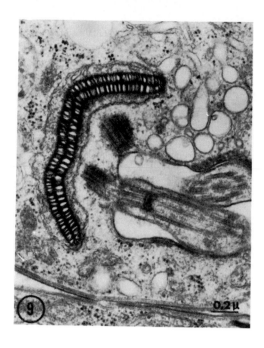

Fig. 9.
Typical morphology of the kinetoplast in thin sections of *T. cruzi*. The kDNA is organized in a regular double-layered pattern (x 35,000).

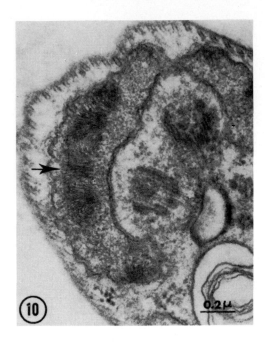

Fig. 10.
T. cruzi treated with 0.5 μg/ml of ethidium bromide for 15 h. The kDNA is rearranged into arches (→) (x 50,000).

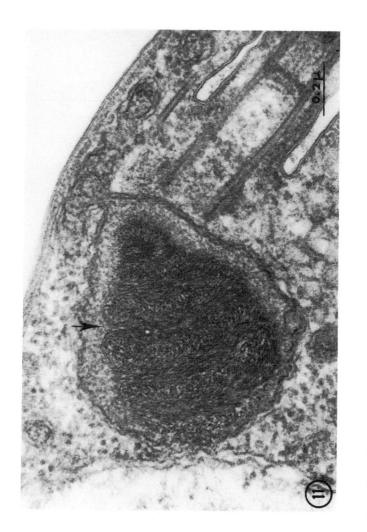

Fig. 11.
T. cruzi treated with 15 μg/ml of daunomycine for 3 days. The kDNA is similarly organized into arches (x 75,000).

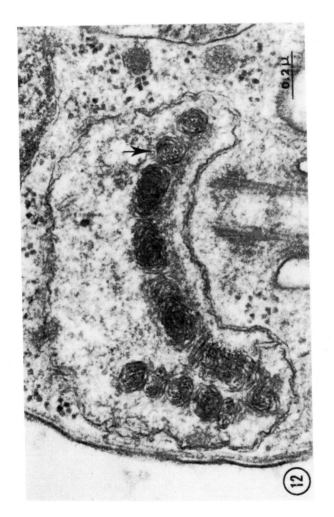

Fig. 12.
T. cruzi treated with 5 μ g/ml of hydroxystilbamidine for 2 days. The long "sausage-like" fibrillar structures are cross sectioned, and have a circular profile (→) (x 68,000).

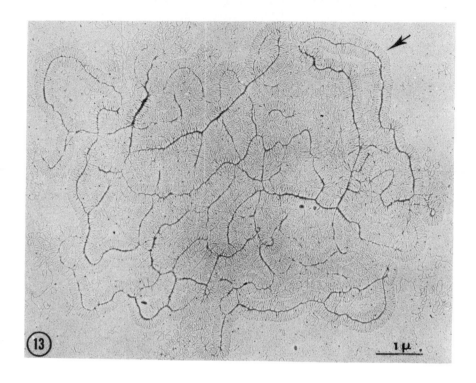

Fig. 13.

kDNA chemically extracted from *T. cruzi* treated for 4 days with 2 μg/ml of berenil. Long lampbrush-like structures (ref. 20) consist of hundreds of circular molecules which are attached to a central axis. Such structures (→) can be related to the fibrillar bodies described in Fig. 12 (→) (x 15,000).

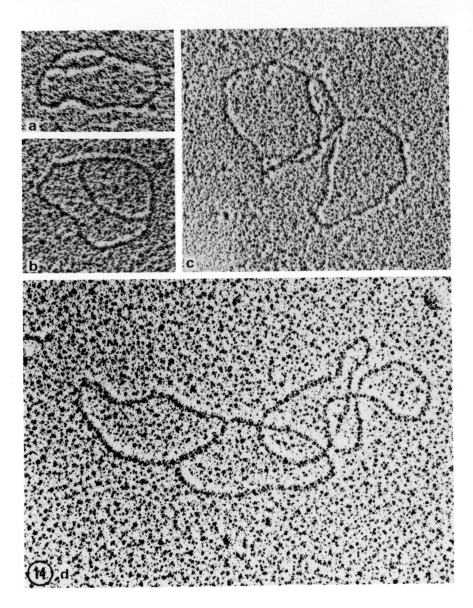

Fig. 14.
Replicating molecules observed in the kDNA of berenil treated *T. cruzi.*
a) and b) are free replicating circles at different stages of replication.
c) is a replicating molecule which is attached to another circle by one of the replicated segments.
d) shows a more complex association where a replicating molecule seems to be catenated with two other circles (x 200,000).

13

DNA OF KINETOPLASTIDAE: A COMPARATIVE STUDY

B.A. Newton and J.K. Burnett
Medical Research Council Biochemical Parasitology
Unit, The Molteno Institute, University of Cambridge,
Cambridge, CB2 3EE, England.

Introduction

It is now well established that DNA isolated from kinetoplastid flagellates can be separated into major and minor components by isopycnic ultracentrifugation in caesium chloride. Also, there is good evidence, for some species, that one minor component originates from the kinetoplast. The unique characteristics of kinetoplast DNA, which will be discussed by Dr. Steinert (chapter 11) and Dr. Riou (chapter 12) in this symposium, have attracted many workers in recent years and in the enthusiasm for detailed examination of these beautifully arranged minicircles it is not surprising that studies to date have been largely restricted to DNA from species of flagellates which can be easily grown in the laboratory.

In the course of studies on the mechanism of action of trypanocidal drugs in our laboratory we observed that the buoyant densities of kinetoplast-DNA and nuclear DNA isolated from salivarian and stercorarian trypanosomes and from insect flagellates differed significantly; we also found that DNA from some species of trypanosomes contained more than one satellite, (Newton & Burnett, 1971; Newton, 1971). These findings encouraged us to extend our survey and during the last four years we have been gradually accumulating information about the DNA of flagellates representing the major genera and sub-genera of the order **Kinetoplastida,** Honigberg, 1963. The aim was not simply to gain information which might aid the taxonomist, although it was clearly of interest to learn, for example, whether any differences could be detected in the DNA from the morphologically identical species, *Trypanosoma brucei,* *Trypanosoma rhodesiense* and *Trypanosoma gambiense,* and whether a comparison of buoyant densities of DNA from mono- and digenetic flagellates would shed any light on phylogenetic relationships. We were also interested in the answer to more specific questions, such as: can differences be detected in the DNA of culture and bloodstream forms of *brucei*-group trypanosomes? and does the DNA of pleomorphic and monomorphic strains of the same species differ? The investigation is still far from complete and this contribution is a report of work in progress, however we believe some interesting data is now beginning to emerge.

Extraction procedures and the possibility of selective loss of DNA satellites

Equilibrium density gradient centrifugation in caesium chloride permits small quantities of native DNA (ca 1 μg) to be examined for the presence of minor components and it is therefore a most useful analytic procedure. However, the results obtained by this procedure are only of value in comparative studies if one can be sure that the extraction methods used liberate the total DNA from organisms being studied: this is extremely difficult to establish and it is a point which has frequently been ignored in studies of DNA heterogeneity. In the course of developing techniques for the extraction of DNA from trypanosomes and related flagellates (Newton, 1967) it became clear that the kinetoplast DNA could be partially or totally lost by the use of standard deproteinization and phenol extraction procedures (Sevag, 1938; Kirby, 1957; Marmur, 1961). A similar selective loss of the d(A-T) satellite from *Gecarcinus lateralis* (land crab) DNA was reported by Skinner & Triplett (1967) and it is clear that such losses led Riou & Pautrizel (1969) to identify incorrectly a minor component of *T. equiperdum* and *T. gambiense* DNA as kinetoplast DNA. The procedure used by the latter workers resulted in a complete loss of kinetoplast DNA (ρ = 1.693) from preparations of *T. equiperdum* and an almost complete loss of satellite B (Fig. 1 and Table 1) which is characteristic of *brucei*-group trypanosomes (Fig. 2). Our own work has shown that even centrifugation of whole cell lysates (prepared by detergent treatment in the presence of EDTA and followed by ribonuclease digestion) in caesium chloride cannot be relied upon to reveal all the DNA components in a particular organism. Fig. 3 shows that k-DNA is only released from *Crithidia oncopelti, Crithidia fasciculata* and *Trypanosoma lewisi* in a form that will band in a caesium chloride gradient if lysates are treated with pronase before centrifugation. However, pronase treatment is not necessary for the release of k-DNA from

186

all species of trypanosomes: the banding patterns of DNA in lysates of *T. brucei* and *T. cruzi* for example, are unaltered by pronase treatment. These findings, together with recent reports of DNA fractions in mammalian cells which are not extracted by normal procedures (Penn & Suwalski, 1969; Burden, 1971), stress the need for caution in interpreting results of equilibrium centrifugation studies on DNA extracts.

Satellite DNA's of kinetoplastid flagellates

Table 1 records buoyant densities of DNA components extracted from twenty-three species of flagellates representing six genera and six subgenera. The recorded values represent the mean of at least four determinations (standard deviation $\pm$ 0.001). Considering this data the following points are of particular interest.

(i) **Kinetoplast DNA.** All species studied, with the exception of *T. evansi* (SAK strain), which is a dyskinetoplastic strain, contain a DNA satellite which bands rapidly in caesium chloride gradients (within 30 mins of commencing centrifugation at 44,000 r.p.m.). This is a characteristic of DNA extracted from isolated kinetoplast (DuBuy, Mattern & Riley, 1966; Newton, 1967) and for this reason the rapidly banding satellite has been called kinetoplast DNA in Table 1; it should be noted however that kinetoplasts have so far been isolated from only four or five of the species listed. The buoyant densities of the fast banding satellite range from 1.689-1.703 g/ml; these values correspond to DNA base compositions ranging from 29-43 moles percent guanine + cytosine, but, in view of the unique banding properties and tertiary structure of kinetoplast DNA, base ratios calculated from buoyant densities may be of doubtful significance. Estimates of the amount of kinetoplast DNA calculated by integration of areas under the peaks on microdensitometer tracings of DNA bands in caesium chloride gradients, and expressed as a percentage of total cell DNA, range from 9-10 % in *T. brucei*-group trypanosomes to 28 % in *T. vespertilionis*. The stercorarian trypanosomes contain approximately twice as much kinetoplast DNA as the salivarian species.

(ii) **DNA satellites of brucei-group trypanosomes.** The DNA's of *T. brucei, T. rhodesiense, T. gambiense, T. evansi* and *T. equiperdum* are characterised by the presence of satellite B (Table 1), density 1.697-1.702 g/ml, which bands between the main DNA component and kinetoplast DNA. This satellite accounts for about 30 % of the total DNA of these organisms. It is not yet known whether satellite B is of nuclear origin; alkaline denaturation experiments and centrifugation in caesium chloride-ethidium bromide gradients (Radloff et al., 1967) suggest that the DNA of this satellite is not composed of closed circular molecules. Since a satellite of density 1.697-1.702 g/ml has not been detected in DNA from other salivarian or stercorarian trypanosomes it is interesting to speculate whether this DNA carries information required to control some of the unique changes in structure and metabolism which *brucei*-group trypanosomes undergo in the course of their development cycle in mammalian and insect hosts. Experiments to investigate this possibility are planned.

(iii) **Satellite A.** Of the species examined only *Crithidia oncopelti* and *Trypansoma congolense* contained a DNA satellite which was lighter than the kinetoplast. In *C. oncopelti* satellite A (Table 1 ρ = 1.694) is believed to be associated with a discrete cytoplasmic organelle, the bipolar body (Newton & Horne, 1957; Marmur, Cahoon, Shimura & Vogel, 1963). It has been suggested (Gill & Vogel, 1963) that this organelle is an endosymbiotic bacterium but the evidence upon which this suggestion is based is, in the opinion of one author (Newton, 1968), open to considerable criticism. It is not yet known whether the light satellite (ρ = 1.693) in *T. congolense* is of nuclear or cytoplasmic origin, but its existance provides a clear cut difference between the DNA's from the several strains of *T. congolense* and *T. vivax* examined.

(iv) **Satellite C.** Three species *(T. gambiense, T. theileri* and *T. dionisii)* were found to contain a satellite which was heavier than the main DNA component (satellite C in Table 1 ρ = 1.708-1.719). Again we do not know the location of this material in the cell but its existance in *T. gambiense* is of particular interest since it provides, for the first time, a means of distinguishing this species of trypanosome from the morphologically identical species *T. brucei* and *T. rhodesiense.* Like satellite B, there is no evidence that satellites A and C are composed of closed circular DNA molecules. The high buoyant density of satellite C suggests that it is composed of a G + C-rich DNA; similar satellites have been observed in many other cell types and this DNA has been reported to hybridize strongly with ribosomal RNA (Coudray, Quetier and Guille, 1970).

The DNA of culture forms and bloodstream forms of trypanosomes

Dyskinetoplastic strains of brucei-group trypanosomes (either naturally occurring or drug-induced) are unable to infect tsetse or grow in culture and they do not undergo the changes in ultrastructure, antigenicity and metabolic activity which occur when bloodstream forms of normal strains transform from long slender trypomastigotes into epimastigotes. This fact has led to the suggestion that kinetoplast DNA carries information required for this transformation and that it is vital for normal cyclical development of these flagellates. However, when pleomorphic strains of *T. brucei* are maintained over prolonged periods by syringe passage in laboratory animals they become monomorphic (i.e. they lose their ability to transform from trypomastigotes to epismastigotes) but they retain an apparently normal Feulgen positive kinetoplast. In view of this it was clearly of interest to compare DNA's extracted from bloodstream forms of pleomorphic and monomorphic strains of *T. brucei* with DNA from culture forms. All three preparations were found to give similar banding patterns in caesium chloride gradients and no significant differences were detected in the relative amounts of individual satellites. In contrast to these results Renger & Wolstenholme (1970), have reported that DNA extracted from bloodstream forms of *T. lewisi* contains a main band (ρ = 1.707), a kinetoplast band (ρ = 1.699) and a heavy satellite (ρ = 1.721) whereas DNA from culture forms lacks the heavy satellite and contains a main component with a density of 1.711 g/ml. The quite large change in density of the main DNA component is surprising, but the authors do not comment on it. Experiments in our own laboratory have failed to reveal similar differences between DNA's from bloodstream and culture forms of *T. lewisi* (Molteno strain) or *T. cruzi*.

Phylogenetic relationships & DNA buoyant densities

Many authors have been tempted to speculate on the evolution of the family Trypanosomatidae (Baker, 1963; Hoare, 1967; Woo, 1970) and the phylogeny of African trypanosomes has been discussed in detail (Ashcroft, 1959; Ormerod, 1961; Hoare, 1970). It is not our intention to embarque on a lengthy discussion of these topics in the present paper, but it does seem wothwhile asking the question: does any obvious pattern of relationships emerge from our data on the buoyant densities of DNA from kinetoplastid flagellates? Studies of DNA from a wide range of cell types support the view that DNA base analysis is an important taxonomic aid and show that, in general, organisms which are closely related genetically or by the criteria of numerial taxonomy have DNA base compositions which are similar. In Fig. 4 flagellates have been grouped on the basis of buoyant densities of either their kinetoplast DNA or their nuclear DNA. Considering the former grouping it will be seen that the range of ρ values for a particular group of organisms is small (the range corresponds to a variation in G + C content of about 4 %) and that there is a progressive decrease in density from values corresponding to a high G + C content in the kinetoplast DNA of crithidia, through intermediate values in the stercoraria to densities corresponding to a high A + T content in the kinetoplast DNA of *brucei*-group trypanosomes. On this basis it can be seen that *T. congolense* and *T. vivax* fall within the stercorarian group rather than the *brucei*-group; this is, to some extent, in keeping with what we know of the intermediary metabolism of these organisms. In contrast, grouping organisms according to nuclear DNA buoyant densities results in a spread of stercorarian trypanosomes over almost the whole range of ρ values observed for nuclear DNA. Flagellates belonging to the sub-genus *Trypanosoma* are still closely grouped but *T. congolense, T. lewisi* and *T. cruzi* occur in the same group.

On the basis of these findings it is tempting to speculate whether the observed progressive increase in A + T content of kinetoplast DNA reflects a gradual evolutionary development of this group of organisms from monogenetic flagellates of insects through the stercorarian parasites to the *brucei-* group of salivaria which, Hoare (1948) believes, evolved relatively recently. As yet we know nothing about the information content of kinetoplast DNA. Laurent & Steinert (1970) have pointed out that the contour length of the circular kinetoplast DNA molecules may be species specific: the contour length of *Leishmania tarentolae* kinetoplast DNA is only 0.22 μm (Simpson & Da Silva, 1971) compared with values of 0.40 μm, 0.45 μm and 0.75 μm for the kinetoplast DNA's from *T. lewisi, T. cruzi* and *T. mega* respectively (Renger & Wolstenholme, 1970; Riou & Delain, 1969; Laurent & Steinert, 1970). Is there a relationship between contour length and information content and, if so, will the more sophisticated *brucei*-group trypanosomes be found to contain larger circular kinetoplast DNA molecules? There is now evidence for the presence of linear DNA in the kinetoplast of some trypanosomes (Laurent & Steinert, 1970; Riou et al., this symposium); it may be that the relative amounts of circular and linear DNA's will be found to vary from species to species.

Acknowledgements

The authors wish to thank Mrs. I. Hislop, Mrs. R. Holl and Mr. G.E. Mewis for highly skilled technical assistance at various stages of this work. They are also indebted to the many people who have provided strains of trypanosomes. This investigation received financial support from the Medical Research Council, the Overseas Development Administration and the World Health Organisation.

References

ASHCROFT, M.T. (1959). Trop. Dis. Bull. **56**, 1073.
BAKER, J.R. (1963). Expl. Parasit. **13**, 219.
BURDON, R.H. (1971). Biochem. Soc. Proc. (in press).
COUDRAY, Y., QUETIER, F. & GUILLE, E. (1970). Biochim. Biophys. Acta **217**, 259.
DUBUY, H.G., MATTERN, C. & RILEY, F. (1966). Biochim. Biophys. Acta **123**, 298.
GILL, J.W. & VOGEL, H.J. (1963). J. Protozool., **10**, 148.
HOARE, C.A. (1967). Adv. Parasitol. **5**, 47.
HOARE, C.A. (1970). In: The African Trypanosomiasis p. 3. Ed. H.W. MULLIGAN, George, Allen & Unwin; London.
KIRBY, K.S. (1959). Biochim. Biophys. Acta **36**, 117.
LAURENT, M. & STEINERT, M. (1970). Proc. Nat. Acad. Sci. Wash. **66**, 419.
MARMUR, J. (1961). J. Mol. Biol. **3**, 208.
MARMUR, J., CAHOON, M.E., SHIMURA, Y. & VOGEL, H.J. (1963). Nature Lond. **197**, 1128.
NEWTON, B.A. (1967). J. gen. Microbiol. **48**, iv.
NEWTON, B.A. (1968). An Rev. Microbiol. **22**, 109.
NEWTON, B.A. (1971). Trans. R. Soc. trop. Med. Hyg. **65**, (in press).
NEWTON, B.A. & BURNETT, J.K. (1971). Trans. R. Soc. trop. Med. Hyg. **65**,243.
NEWTON, B.A. & HORNE, R.W. (1957). Expl. Cell Res. **13**, 563.
ORMEROD, W.J. (1967). J. Parasitol. **53**, 824.
PENN, N.W. & SUWALSKI, (1969). Biochem. J. **115**, 563.
RADLOFF, R., BAUER, W. & VINOGRAD, J. (1967). Proc. Nat. Acad. Sci. Wash. **57**, 1514.
RENGER, H.C. & WOLSTENHOLME, D.R. (1970). J. Cell Biol. **47**, 689.
RIOU, G. & DELAIN, E. (1969). Proc. Nat. Acad. Sci. Wash. **62**, 210.
RIOU, G. & PAUTRIZEL, R. (1969). J. Protozool. **16**, 509.
SEVAG, M.G., LACKMAN, D.B. & SMOLENS, J. (1938). J. Biol. Chem. **124**, 425.
SIMPSON, L. & DA SILVA, A. (1971). J. Mol. Biol. **56**, 443.
SKINNER, D.M. & TRIPLETT, L.L. (1967). Biochem. Biophys. Res. Commun. **28**, 892.
WOO, P.T.K. (1970). Nature, Lond. **228**, 1059.

Table 1.

Buoyant Densities of DNA's from Kinetoplastid Flagellates (Order: KINETOPLASTIDA; Suborder: TRYPANOSOMATINA).

Flagellates, washed free of growth media or blood constituents were suspended in Trisbuffer (0.02 M pH 8.5) containing 0.15 M NaCl and 0.1 M EDTA. Cells were lysed by addition of sodium lauryl sulphate (final conc. 1.0 % w/v) and, when lysis was complete, treated with pronase (1 mg/ml) for 2 hr at 37° C. Preparations were dialysed overnight against Tris/EDTA/NaCl before ultracentrifugation in caesium chloride.

Genus, Subgenus & Species		Buoyant density[a]				
		Satel-lite A	Kine-toplast DNA	Satel-lite B	Main DNA Com-ponent	Satel-lite C
Genus: LEPTOMONAS	*L. collosoma*	—	1.699	—	1.716	—
	L. mesogramma	—	1.701	—	1.719	—
Genus: HERPETOMONAS	*H. muscarum*	—	1.698	—	1.717	—
Genus: BLASTOCRI-THIDIA	*B. culicis*	—	1.696	—	1.717	—
Genus: CRITHIDIA	*C. fasciculata*	—	1.703	—	1.716	—
	C. oncopelti	1.694	1.703	—	1.712	—
Genus: LEISHMANIA	*L. tarentolae*	—	1.701	—	1.718	—
	L. enrietti	—	1.702	—	1.721	—
Genus: TRYPANOSOMA						
Subgenus: Megatrypanum	*T. melophagium*	—	1.695	—	1.702	—
	T. theileri	—	1.696	—	1.701	1.708
Subgenus: Herpetosoma	*T. lewisi*	—	1.698	—	1.706	—
	T. musculi	—	1.698	—	1.706	—

[a] Recorded values for ρ represent the mean of at least four determinations (standard deviation ± 0.001).

(Table 1)

Genus, Subgenus & Species		Buoyant density[a]				
		Satel-lite A	Kine-toplast DNA	Satel-lite B	Main DNA Com-ponent	Satel-lite C
Subgenus:						
Schizotrypanum	*T. cruzi*	—	1.698	—	1.709	—
	T. dionisii	—	1.700	—	1.712	1.717
	T. vespertilionis	—	1.698	—	1.718	—
Subgenus:						
Duttonella	*T. vivax*	—	1.696	—	1.713	—
Subgenus:						
Nannomonas	*T. congolense*	1.693	1.697	—	1.708	—
Subgenus:						
Trypanozoon	*T. brucei*	—	1.691	1.702	1.707	—
	T. rhodesiense	—	1.690	1.701	1.707	—
	T. gambiense	—	1.690	1.702	1.707	1.719
	T. evansi (NS)	—	1.689	1.702	1.707	—
	T. evansi (SAK)	—	—	1.702	1.709	—
	T. equiperdum	—	1.693	1.701	1.706	—

[a] Recorded values for ρ represent the mean of at least four determinations (standard devration ± 0.001).

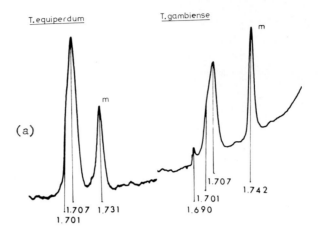

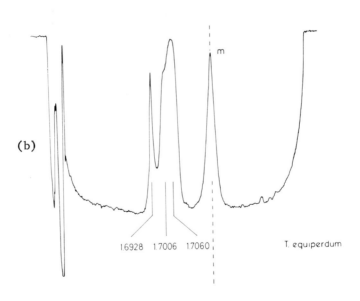

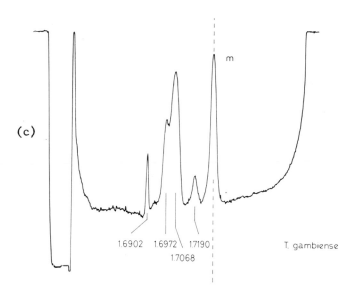

Fig. 1.

Microdensitometer tracings of DNA bands obtained by analytical ultra-centrifugation in caesiumchloride (44,000 r.p.m. for 20 hrs at 20° (a) Data from Riou and Pautrizel (1969) showing banding patterns obtained when DNA from *T. equiperdum* and *T. gambiense* is extracted by a phenol procedure. (b) and (c) Banding patterns of DNA obtained from the same species of trypanosomes by the procedure detailed in the legend to Table 1. m = marker DNA (ρ = 1.731).

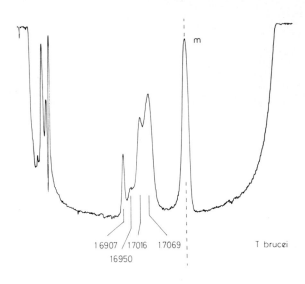

16907 /17016 17069   T brucei

16950

Fig. 2.
DNA from *T. brucei.* Method of preparation and conditions of ultracentrifugation as stated in Table 1 and Fig. 1. m = marker DNA ($\rho = 1.731$).

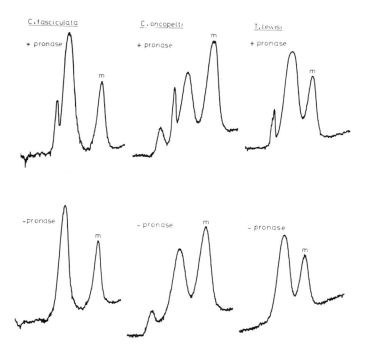

Fig. 3.
Effect of pronase treatment of DNA release from flagellates. Cell lysates prepared as in Table 1. There is a selective loss of the fast-banding kinetoplast DNA when cell lysates are not treated with pronase before ultracentrifugation in caesium chloride. m = marker DNA (ρ = 1.731).

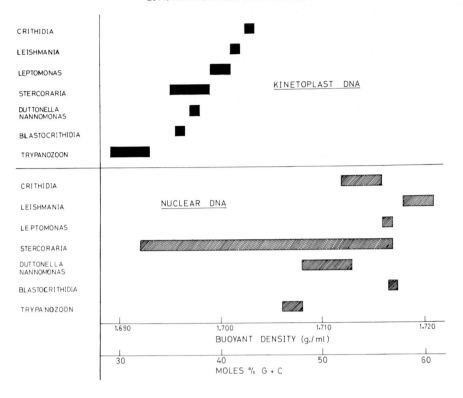

Fig. 4.
Comparison of the buoyant densities of kinetoplast and nuclear DNA from flagellates six genera.

14

SOME STUDIES ON THE DNA OF *PLASMODIUM KNOWLESI*

W.E. Gutteridge* and P.I. Trigg
Division of Parasitology,
National Institute for Medical Research, Mill Hill,
London NW7 IAA.

Introduction

These investigations of the DNA of the blood stages of *Plasmodium knowlesi* were prompted by the many indications in the literature which suggested that the ultimate lethal action of two of the most successful antimalarial drugs, chloroquine and pyrimethamine, was an interference with either the replication or transcription of the genetic information contained within the malarial parasite. It was hoped that these drugs would aid the elucidation of the pathways involved in DNA metabolism in the malarial parasite and that such an elucidation would in turn indicate not only new targets for chemotherapeutic attack but also how the already known targets could be better attacked.

Analysis of DNA in the ultracentrifuge.

Isopycnic ultracentrifugation in CsCl of extracts of the blood schizont stages of *P. knowlesi* obtained from infected rhesus monkeys revealed two DNA components (Fig. 1). Both were judged to be DNA by their sensitivity to deoxyribonuclease. The major component had a density of 1.697 g/cm^3, equivalent to 37 % G + C, while the minor one, which accounted for about 5 % of the total DNA, had a density of 1.679 g/cm^3 (19 % G + C). On heat denaturation, the major component increased in density by about 0.015 g/cm^3, indicating that it was native DNA. Too little DNA was present in the minor component to determine if a similar shift occurred.

* Present address: Biological Laboratory, University of Kent, Canterbury, Kent.

In contrast to our primate malarial species, extracts of two rodent species, *P. berghei* and *P. vinckei*, often used as model species in laboratory investigations of malaria, showed only single DNA components in CsCl gradients. These both had densities of 1.683 g/cm^3 (24 % G + C) which also increased by about 0.015 g/cm^3 on heat denaturation. However, extracts of *P. falciparum* the only human malarial species we were able to study, gave results identical to those for *P. knowlesi,* validating the use of the latter species in our laboratory investigations.

We have no firm evidence about the intracellular location of the two DNA components in *P. knowlesi.* However, by analogy with other organisms and from our observation that only the nucleus in this organism can be stained by the Feulgen technique (Gutteridge & Trigg, in preparation), we are sure that the major component represents nuclear DNA. The minor component might arise from the mitochondrial-like organelles which have been described in mammalian malarial parasites (Rudzinska & Trager, 1968). The trypanosomal equivalent of mitochondrial DNA is located within the kinetoplast. In preparations made by gentle detergent lysis, this kinetoplast DNA is obtained as a rapidly sedimenting large complex (Williamson, Stuart, Gutteridge & Burdett, 1971). Indications of this property are provided by its ability to form a sharp band in analytical CsCl gradients within 30 minutes. The minor DNA component of *P. knowlesi* does not form a sharp band and takes about 10 hours to appear in analytical CsCl gradients (Fig. 2). It is unlikely therefore that it exists as large complexes. Further studies are contemplated, however, since its base composition is extremely low (19 % G + C) and since it might be the intra-mitochondrial binding site of the 8-aminoquinolines which are known to concentrate within the malarial mitochondria (Aikawa & Beaudoin, 1970) and can bind to DNA **in vitro** (Whichard, Morris, Smith & Holbrook, 1968).

Isolation and properties of nuclear DNA

DNA was prepared from frozen schizont stages of *P. knowlesi* using a method based on that described by Marmur (see Gutteridge, Trigg & Williamson, 1971). Analysis in the ultracentrifuge showed that the minor DNA component was lost during this procedure, so that the method yielded pure nuclear DNA.

The thermal denaturation kinetics of this DNA were typical of most double-stranded deoxyribonucleic acids (Fig. 3). The degree of hyperchromicity was 30 % and the mid-point of the melting curve (T_m) was 85° C. This T_m is equivalent to a base composition of 38 % G + C, a value close to that obtained for the main DNA component by measurement of density in CsCl gradients. This agreement indicates that it is very unlikely that there is an unusual base in the DNA. The absorbance of the solution of the DNA hardly diminished when the peak denaturation was followed by cooling, indicating that the DNA does not rapidly renature. It is most likely, therefore, to be composed of linear molecules.

Our data indicated that the nuclear DNA of *P. knowlesi* is composed of typical double-stranded linear molecules with no unusual bases and a base composition of about 37 % G + C. Thus, there is little or no difference in physical properties between primate malarial DNA and DNA from mammalian cells, since mammalian DNA has a base composition in this range. This concusion led us to a detailed consideration of the possibility that the main DNA component in our preparations of *P. knowlesi* was host monkey DNA and that only the minor component represented parasite DNA. The evidence which ruled out this explanation of our results has been described elsewhere (Gutteridge, Trigg & Williamson, 1971).

Binding of chloroquine to nuclear DNA

There have been many investigations of the ability of chloroquine (Fig. 4) and related 4-aminoquinolines to bind to DNA isolated from a variety of organisms and cells (e.g. Cohen & Yielding, 1965; Hahn, O'Brien, Ciak, Allison & Olenick, 1966). Such investigations have led to the suggestion that these drugs exert their antimalarial action by binding to the DNA of the malarial parasite and thus interfering with its functioning. No investigator however, has tested this by using DNA isolated from a malarial parasite. It has thus been difficult to judge the significance of these studies, particularly as even the basic properties of DNA from a malarial parasite were unknown until recently. However, sufficient nuclear DNA remained from the experiments described above for a study of its ability to bind chloroquine **in vitro.**

It was found that in the presence of increasing concentrations of native malarial DNA, the absorption spectrum of chloroquine exhibited a strong hypochromism in the rang 325-350 nm small red shifts of the absorption maxima and an isosbestic point at 300 nm (Fig. 5). The same quantities of heat denatured malarial DNA and native mammalian DNA produced similar effects. The fluorescence emission of the drug at its peak of 380 nm was quenched up to about 5-fold by native or heat-denatured malarial DNA and again, native mammalian DNA produced a similar effect (Fig. 6). Chloroquine itself protected native malarial (Table 1) and mammalian DNA from thermal denaturation to a similar degree. These data are consistent with the binding of chloroquine to malarial DNA and indicate that the binding affinity is of the same order as that for mammalian DNA.

These results were by no means unexpected once the similarity of the physical properties of DNA from primate malarial parasites and mammalian cells was established. They mean that although our own data provide no indication as to the exact nature of the interaction of chloroquine and malarial DNA, further studies with the more readily available mammalian and bacterial DNAs should yield results which would be significant for primate malaria.

Until recently, it was generally accepted that chloroquine exerted its antimalarial action by binding to the DNA of the malarial parasite and thus interfering with the functioning of the macromolecule, though the evidence for this conclusion has come mainly from studies with bacteria (Hahn **et al.,** 1966; Schellenberg & Coatney, 1961; Polet & Barr, 1968). Our demonstration of the binding of chloroquine to malarial DNA is in agreement with this hypothesis but it must be emphasized that it points only to a potential mechanism which can occur once the drug has penetrated into the parasite. This is particularly important since recent studies with the electron microscope suggest that the food vacuole, where haemoglobin breakdown to amino acids is thought to occur, is the first organelle in the parasite to be affected when the host is treated with the drug. This is followed by changes in the nucleus and nucleolus (Macomber, Sprinz & Tousimis, 1967; Warhurst & Hockley, 1967). Alterations in the food vacuole result in clumping and expulsion of the phagosomes and it has been suggested that the resulting amino acid starvation is the primary means whereby chloroquine is effective against the trophozoite stages of the malarial parasite (Howells, Peters, Homewood & Warhurst, 1970).

It was clear that only a study of the effects of chloroquine on the metabolism of the malarial parasite would distinguish between an action through a binding to DNA or an action through an interference with amino acid production. The former should lead to a selective inhibition of nucleic acid synthesis and the latter to a selective inhibition of protein synthesis. In fact, all four processes that we were able to study were inhibited to a similar extent (Table 2). Our data thus appear to rule out an action solely through amino acid starvation, a result which was not unexpected since even if haemoglobin degradation was affected, amino acids would still be obtained from the blood and by **de novo** biosynthesis (see Polet, Brown & Angel, 1969). Our data also appear to rule out an effect caused solely by a binding to DNA since besides nucleic acid synthesis, protein synthesis and respiration were also affected. The cessation of protein synthesis could be related either to the partial breakdown of ribosomal RNA that occurs within the parasite within 4 hours of the injection of chloroquine into an infected animal (Warhurst & Williamson, 1970) or to pigment clumping causing amino acid starvation or to a combination of the two effects. Respiration is presumably affected by yet another mechanism. Other metabolic processes, such as lipid biosynthesis, which we cannot as yet study satisfactorily **in vitro,** could also be inhibited since chloroquine can bind to phospholipids (Williamson & Warhurst, unpublished observation). However, we can conclude that the growth inhibitory action of chloroquine includes at least four separate effects, though only one of them, respiration, would be malariacidal. This deduction endorses our earlier conclusions from the DNA binding experiments that the selectivity of the action of chloroquine cannot reside at the level of its intracellular binding sites. It is difficult to see how every one of these could be distinct from its equivalent site in mammalian cells. In this connection, it is interesting to note that it now seems well established that erythrocytic stages of malarial parasites, unlike mammalian cells, can actively concentrate chloroquine 100-1000-fold (Polet & Barr, 1968; Fitch, 1969). It is most likely that the selectivity resides at the level of drug uptake.

Origins of purines and pyrimidines for DNA synthesis

Experiments with *P. knowlesi* in culture using radioactive pyrimidines (thymine, thymidine, deoxycytidine, cytidine, uracil, uridine) indicated that none were utilised for nucleic acid synthesis, though the pyrimidine precursor, orotic acid, was utilised to a limited extent (Gutteridge & Trigg, 1970). Absence of pyrimidines from culture media did not affect the growht of parasites (Trigg & Gutteridge, 1971) so that it seems likely that *P. knowlesi* can synthesise the pyrimidine ring.

By contrast, experiments with radioactive purines indicated that all those tested (adenine, adenosine, deoxyadenosine, guanine, guanosine and hypoxanthine) were incorporated into nucleic acids. We could not detect incorporation of ^{14}C-formate and parasites in culture media without added purine did not grow as well as parasites did in control cultures. These data suggested that *P. knowlesi* is unable to synthesise the purine ring and is dependent on an exogenous source of purine for DNA and RNA synthesis. This conclusion further suggested that the parasite would be vulnerable to inhibition by purine analogues.

A study of the effects of cordycepin (Fig. 4) on cultures of *P. knowlesi* **in vitro** and on *P. berghei* **in vivo** indicated that this was so (Trigg, Gutteridge & Williamson, 1971). Concentrations of drug as low as 10^{-6}M affected growth **in vitro** after a lag period of about 4 hours. Single doses of 50 mg/kg and above were active against *P. berghei,* but complete cure was not obtained even at the highest dose tested (200 mg/kg). Toxic effects were observed in infected animals at doses of 200 mg/kg and above. It might be profitable to synthesise analogues of cordycepin since one might be made which is more active against the malarial parasite but less toxic to the host. Such an analogue would prove a particularly useful drug since its mode of action would be different from that of any currently used antimalarial drug and thus problems of cross-resistance should not arise.

Periodicity of nuclear DNA synthesis

The intraerythrocytic cycle of *P. knowlesi* in the rhesus monkey is highly synchronous and takes 24 hours. It thus appeared to be an ideal system in which to study the periodicity of DNA synthesis in a malarial parasite. However, initial experiments involving the incubation of parasites with radiotracers in culture yielded equivocal results (Gutteridge & Trigg, 1970). When ^{3}H-adenosine was used as tracer, DNA synthesis appeared to be confined to the ring and trophozoite stages, whereas when ^{3}H-orotic acid was used, DNA synthesis appeared to occur mainly during the schizont stage (Fig. 7). It was concluded therefore that the final resolution of the problem must await the development of a technique to measure parasite DNA in infected blood without any form of incubation **in vitro.**

Since then, we have developed a technique using an MSE type A low speed zonal rotor which allows the preparation of large batches of "free" parasites essentially uncontaminated by host white cells (Gutteridge, Trigg & Cover, 1971). These have been assayed for DNA by conventional techniques. Preparations of both late-trophozoites and schizonts contained similar amounts of DNA but preparations of ring stages contained much less DNA (Table 3). This suggests that DNA synthesis occurs mainly during the ring and trophozoite stages of the intra erythrocytic cycle of *P. knowlesi* and that its synthesis is paralleled by the incorporation of ^{3}H-adenosine.

This conclusion led to a detailed study of the incorporation of ^{3}H-adenosine into cultures consisting initially of late-trophozoite stage parasites (Fig. 8). Incorporation occurred until schizogony began, it ceased during this period and began again as soon as parasites had invaded new red cells. Thus it can be concluded that the S phase occurs during the ring and trophozoite stages of the intra-erythrocytic cycle with the G1 and G2 phases, if they occur at all, being of very short duration and occurring at the young ring and late trophozoite stages respectively. Schizogony consists of a "multiple mitosis" which occurs without further DNA synthesis and yields a 16-nuclear schizont.

Action of pyrimethamine on DNA synthesis

Pyrimethamine (Fig. 4) is a potent inhibitor of the dihydrofolate reductase from *P. knowlesi* (Gutteridge & Trigg, 1971). It seemed likely therefore that it would cause an inhibition of DNA synthesis since this enzyme in bacteria and mammalian cells is concerned with the conversion of uridylate to thymidylate. Furthermore, it seemed likely that the inhibition would be specific to DNA synthesis since other pathways involving folate metabolism for purine (see above) and amino acid (Polet & Conrad, 1968) biosynthesis are not present in *P. knowlesi.* Some evidence for this has already been presented (Schellenberg & Coatney, 1961).

We found that as **in vivo** (McGregor & Smith, 1952), pyrimethamine acted **in vitro** only against the schizont stages of *P. knowlesi.* Growth was normal in cultures containing initially ring stage parasites until schizogony began, but parasites then assumed an abnormal appearance and segmentation and reinvasion did not occur. The minimum concentration of drug required to produce this effect was 10^{-9}M which is the same concentration as that required to inhibit markedly the dihydrofolate reductase of the organism (see Gutteridge & Trigg, 1971). However, we know now that DNA synthesis in *P. knowlesi* occurs during the ring and trophozoite stages so that if the drug acts through an inhibition of DNA synthesis, it would be these stages and not the schizont stage that would be most susceptible to its action. It was possible, though, that effects on DNA synthesis would not result in morphologically detectable alterations until the time for nuclear division was reached.

We found that:

1) drug concentrations as high as 10^{-5}M did not affect the incorporation of ^{3}H-adenosine into DNA of cultures containing initially ring-stage parasites in a 16 hour incubation during which time parasites grew to the late trophozoite stage;

2) the drug did not affect the incorporation of ^{3}H-adenine into DNA of cultures consisting initially of late-trophozoite stage parasites until after

morphological damage could be seen and reinvasion was complete in control cultures (Fig. 9);

3) significant segmentation and reinvasion could be prevented in cultures by adding drug **after** schizogony had begun (Table 4);

4) trophozoites isolated from a monkey treated with pyrimethamine (5 mg/kg as the isethionate) 16 hours previously when the parasites were ring-stages contained normal amounts of DNA and other macromolecules (Table 5). These data are not consistent with the hypothesis that the ultimate effect of pyrimethamine on *P. knowlesi* is an inhibition of DNA synthesis.

There can be no doubt that the effect of pyrimethamine is mediated through an inhibition of the malarial dihydrofolate reductase. Ferone, Burchall and Hitchings (1969) observed a positive correlation between the binding of pyrimethamine and three dihydrotriazines by the dihydro-folate reductase of *P. berghei* and the activity of these compounds **in vivo** against infections of the organism. This led them to suggest that the potent inhibition of this enzyme by pyrimethamine is the basis of its chemotherapeutic action. We found that there is also an absolute correlation between the concentrations of pyrimethamine required to inhibit dihydrofolate reductase of *P. knowlesi* 50 % and the minimum concentration required to affect development **in vitro.** Later, Ferone (1970) demonstrated the occurrence of an altered dihydrofolate reducta-se in a strain of *P. berghei* resistant to pyrimethamine. Thus, if inhibition of DNA synthesis does not occur, our results can be most easily explained if there is some other metabolic process in the malarial parasite, requiring a fully functioning dihydrofolate reductase, which is only required during the process of schizogony. However, acceptance of this explanation makes it necessary to further postulate either that thymidylate for DNA synthesis is synthesized by a pathway not involving dihydrofolate reductase or that pyrimethamine penetrates to the malarial parasite only during schizogony. We consider the second possibility more likely since the permeability of the host red cell membrane could increase as the parasite grows, fills and ultimately distends the red cell. Such an alteration has been suggested as one reason why late tropho-zoites and schizonts, but not rings and young trophozoites, can be agglutinated with immune serum (Brown, Brown and Hills, 1968) and would explain the anomaly of the time course of ^{3}H-orotic acid incorporation (Fig. 7). It would require pyrimethamine in a radioactive form to test this experimentally and this is not available to us at present. Either possibility, however, would fit in well with our studies on the periodicity of DNA synthesis and with the electron microscope studies of

Aikawa and Beaudoin (1968) on the effects of pyrimethamine on *P. gallinaceum*. This last group found that the drug interferred at metaphase of the schizont nuclear division, rather than at prophase when chromosome replication takes place and thus any inhibition of DNA synthesis would be expected to become apparent.

Conclusions

Our investigations have yielded some information about the basic physical properties of the DNA of *P. knowlesi.* They have provided indications only, however, about the metabolic pathways involved in its metabolism, partly because chloroquine and pyrimethamine proved not to be specific inhibitors of DNA replication or transcription in *malarial parasites.* Nevertheless, our investigations have indicated some future studies which might shed light on areas of malarial metabolism particularly vulnerable to inhibition by drugs. These include:

1) Isolation and analysis of "mitochondrial" DNA and investigations of its ability to bind 8-aminoquinolines;

2) synthesis and testing of other analogues of adenosine;

3) examination in detail of the mechanism of the selective uptake of chloroquine;

4) determination of the methyl donor in the synthesis of thymidylate;

5) study of the cytology and biochemistry of the nuclear divisions which occur during schizogony;

6) investigation of the integrity of the red cell membrane during the intraerythrocytic cycle.

In addition, our data on the action of pyrimethamine illustrate the dangers of using solely the incorporation of radiotracers into a malaria parasite in culture as a screen for potential antimalarial drugs (see Canfield, Altstatt & Elliot, 1970) or in immunological studies (see Cohen, Butcher & Crandall, 1969). Any system which would miss the antimalarial activity of a compound as effective as pyrimethamine should be used with discretion.

Acknowledgements

The analyses of DNA in the ultracentrifuge were carried out in conjunction with Dr. D.H. Williamson, those of the binding of chloroquine to nuclear DNA with Dr. P.M. Bayley and those on the effects of cordycepin with Dr. J. Williamson. Financial support was received from the World Health Organization. We thank Dr. F. Hawking for much advice and encouragement during the early stages of these investigations.

References

AIKAWA, M. & BEAUDOIN, R.L. (1968). J. Cell Biol., **39,** 749.

AIKAWA, M. & BEAUDOIN, R.L. (1970). Exptl. Parasitol., **27,** 454.

BROWN, I.N., BROWN, K.N. & HILLS, L.A. (1968). Immunology, **14,** 127.

BURTON, K. (1956). Biochem. J., **62,** 315.

CANFIELD, C.J., ALTSTATT, L.B. & ELLIOT, V.B. (1970). Am. J. Trop. Med. Hyg., **19,** 905.

COHEN, S., BUTCHER, G.A. & CRANDALL, R.B. (1969). Nature, **223,** 368.

COHEN, S.N. & YIELDING, K.L. (1965). J. Biol. Chem., **240,** 3123.

FERONE, R. (1970). J. Biol. Chem., **245,** 850.

FERONE, R., BURCHALL, J.J. & HITCHINGS, G.H. (1969). Mol. Pharmacol., **5,** 49.

FITCH, C.D. (1969). Proc. Nat. Acad. Sci., **64,** 1181.

GUTTERIDGE, W.E. & TRIGG, P.I. (1970). J. Protozool., **17,** 89.

GUTTERIDGE, W.E. & TRIGG, P.I. (1971). Parasitology, **62,** 431.

GUTTERIDGE, W.E., TRIGG, P.I. & BAYLEY, P.M. (1971). Parasitology, in the press.

GUTTERIDGE, W.E., TRIGG, P.I. & COVER, B. (1971). Trans. R. Soc. Trop. Med. Hyg., **65,** in the press.

GUTTERIDGE, W.E., TRIGG, P.I. & WILLIAMSON, D.H. (1971). Parasitology, **62,** 209.

HAHN, F.E., O'BRIEN, R.L., CIAK, J., ALLISON, J.L. & OLENICK, J.G. (1966). Milit. Med., **131,** (Suppl) 1071.

HOWELLS, R.E., PETERS, W., HOMEWOOD, C.A. & WARHURST, D.C. (1970). Nature, **228,** 625.

MACOMBER, P.B., SPRINZ, H. & TOUSIMIS, A.J. (1967). Nature **214,** 937.

MARMUR, J. & DOTY, P. (1962). J. Mol. Biol., **5,** 109.

MCGREGOR, I.A. & SMITH, D.A. (1952). Brit. Med. J., **1,** 730.

POLET, H. & BARR, C.F. (1968). J. Pharmacol. Exp. Therapeut. **164,** 380.

POLET, H. & CONRAD, M.E. (1968). Proc. Soc. Exp. Biol. Med., **127,** 251.

POLET, H., BROWN, N.D. & ANGEL, C.R. (1969). Proc. Roy. Soc. Exp. Biol. Med., **131,** 1215.

RUDZINSKA, M.A. & TRAGER, W. (1968). J. Protozool., **15,** 73.

SCHELLENBERG, K.A. & COATNEY, G.R. (1961). Biochem. Pharmacol., **6,** 143.

TRIGG, P.I. & GUTTERIDGE, W.E. (1971). Parasitology, **62,** 113.

TRIGG, P.I., GUTTERIDGE, W.E. & WILLIAMSON, J. (1971). Trans. R. Soc. trop. Med. Hyg., **65,** 514.

WARHURST, D.C. & HOCKLEY, D.J. (1967). Nature, **214,** 935.

WARHURST, D.C. & WILLIAMSON, J. (1970). Chem. Biol. Interactions, **2,** 89.

WHICHARD, L.P., MORRIS, C.R., SMITH, J.M. & HOLBROOK, Jr., D.J. (1968). Mol. Pharmacol., **4,** 630.

WILLIAMSON, D.H., MOUSTACCHI, E. & FENNELL, D. (1971). Biochim. Biophys. Acta, **238,** 369.

WILLIAMSON, D.H., STUART, K., GUTTERIDGE, W.E. & BURDETT, I. (1971). Trans. R. Soc. trop. Med. Hyg., **65,** 246.

Table 1.

Effect of chloroquine on the thermal denaturation of DNA from *P. knowlesi.* Cuvettes contained 3.0 ml of 5×10^{-3}M Tris buffer, pH 7.5, and 40 μg of DNA. Chloroquine present as the diphosphate. From data of Gutteridge, Trigg & Bayley (1971).

Drug Conc^n (xM)	T_m ($^\circ$C)
—	55
10^{-5}	64
3×10^{-5}	80

Table 2.

Action of chloroquine on macromolecular biosynthesis in *P. knowlesi* over 16 hours in culture. Mill Hill medium (Trigg & Gutteridge, 1971) was used supplemented with 100 μc of adenosine-T(G) and 10 μc of L-isoleucine-C^{14}(U) in each flask. Parasites were incubated from the ring to the trophozoite stage and then complete flasks were harvested. 0.5 ml was used to prepare smears. 1.0 ml was precipitated with 0.6M perchloric acid and the centrifugal supernatant fluids estimated for lactate using lactate dehydrogenase. 3.0 ml was harvested by centrifugation, washed, precipitated with trichloracetic acid and fractionated to yield DNA, RNA and protein fractions (Trigg, Gutteridge & Williamson, 1971). Data of Gutteridge, Trigg & Bayley (1971).

Drug Conc^n (xM)	% inhibition of incorporation into or production of:			
	DNA	RNA	Protein	Lactate
10^{-4}	91	92	90	79
10^{-5}	86	87	88	86
10^{-6}	89	88	86	79
10^{-7}	20	12	2	7
10^{-8}	0	0	0	10
10^{-9}	2	5	6	7

Table 3.
DNA contents of the intraerythrocytic stages of *P. knowlesi.* "Free"-parasite preparations were made from infected monkey blood by a technique involving the use of sucrose gradients and immune lysis (see Gutteridge, Trigg & Williamson, 1971). These were counted in a haemocytometer, precipitated with cold perchloric acid and then extracted with hot acid to yield a hot perchloric acid soluble fraction. This was estimated for DNA with the diphenylamine reagent (Burton, 1956).

Stage	No. of determinations	Mean value of DNA (μg/10^8 organisms)	Standard deviation	Significance of difference
Ring	1	0.76	—	—
Trophozoite	5	11.2	$\pm$ 5.3	0.1
Schizont	4	17.5	$\pm$ 2.9	

Table 4.
Effect of pyrimethamine (10^{-9}M) on the growth of *P. knowlesi* in culture. The incubation was started once the parasite had reached the G2 and division phases. From the data of Gutteridge & Trigg (1971).

	Control					+ Pyrimethamine				
Time	Parasites /10^4 RBC	R	T	S	Ab	Parasites /10^4 RBC	R	T	S	Ab
0	106	—	22	78	—	106	—	22	78	—
4	434	74	2	22	—	139	22	—	12	66
25	596	—	82	14	2	131	10	28	27	35

Key = R, ring; T, trophozoite; S, schizont; A, abnormal form.
The figures represent the % of the parasites present at the various stages. Gametocyte counts are not included so that the figures do not add up to 100%

Table 5.

Nucleic acid and protein content of trophozoites of *P. knowlesi* from a monkey treated with pyrimethamine. Drug as the isethionate was given intravenously at a dose of 5 mg/kg when the parasites were at the ring stage and the red cell parasitaemia of the peripheral blood was 40 %. The monkey was bled 17 hours later when the parasites had reached the late trophozoite stage (overall parasitaemia of blood 22 %). A "free" parasite preparation was made and estimated for DNA, RNA and protein.

	$\mu g/10^8$ parasites	
	Control	**Drug-treated**
DNA	11.2 ± 5.3	11.0
RNA	60 ± 23	47
Protein	1035 ± 315	1380

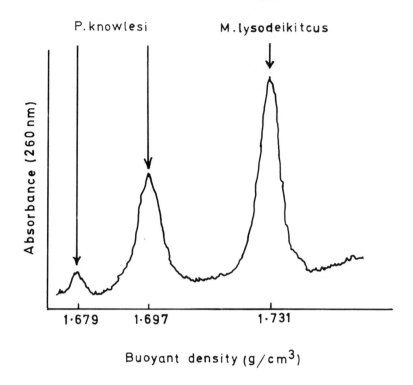

Fig. 1.

Densitometer trace of U.V. photograph of CsCl gradient of DNA from *P. knowlesi* and DNA from *Micrococcus lysodeikticus* used as marker. "Free" parasite preparations were made from infected monkey blood by a technique involving the use of sucrose gradients and immune lysis (see Gutteridge, Trigg & Williamson, 1971). These preparations were free from contaminating host blood cells and platelets. They were frozen, disrupted by passage through an Eaton press and examined by analytical ultracentrifugation in CsCl gradients (Williamson, Moustacchi & Fennell, 1971). The picture was taken at 20 hours when the gradient had come into equilibrium Data of Gutteridge, Trigg & Williamson (1971).

Fig. 2.
Negative prints of U.V. photographs of CsCl gradient of DNA from *P. knowlesi* + DNA from *M. lysodeikticus.* "Free" parasite preparations, made as in Fig. 1, were lysed directly with sodium lauryl sulphate and examined in the ultracentrifuge (see Williamson, Stuart, Gutteridge & Burdett, in preparation). Pictures were taken after 7, 8, 9 and 10 hours (left to right).

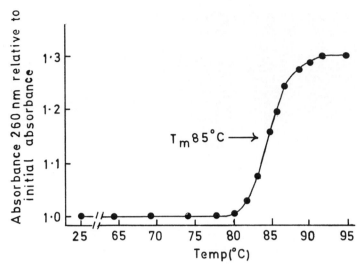

Fig. 3.
Melting curve of nuclear DNA from *P. knowlesi.* The culture contained ∿
40 µg DNA dissolved in 3 ml of SSC (Marmur & Doty, 1962). Data of
Gutteridge, Trigg & Williamson (1971).

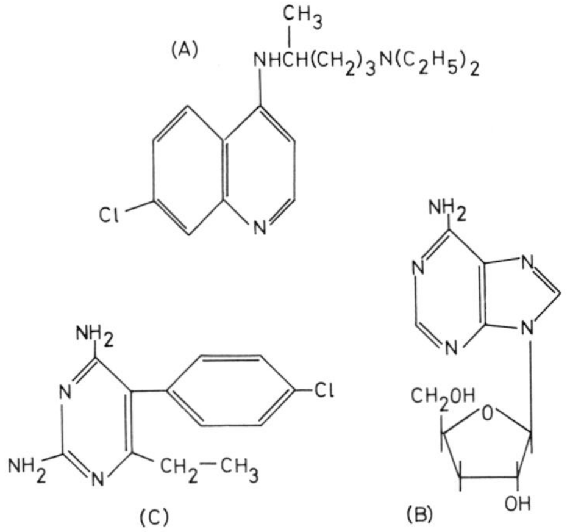

Fig. 4.
Chemical structures of chloroquine (A), cordycepin (B) and pyrimethamine (C).

214

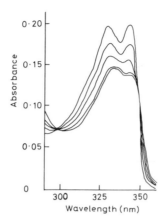

Fig. 5.
Effect of increasing concentration of native DNA from *P. knowlesi* on the absorption spectrum of chloroquine (10^{-5}M) in 3 ml of 5×10^{-3}M Tris buffer, pH 7.5. From top of bottom, the amounts of DNA added were 0, 4, 8, 12 and 16 μg. Data of Gutteridge, Trigg & Bayley (1971).

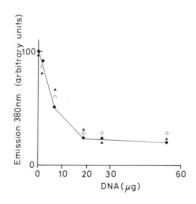

Fig. 6.
Effect of increasing concentrations of DNA on the emission at 380 nm of a solution of chloroquine (6×10^{-6}M) in 5×10^{-3}M Tris buffer pH 7.5 excited by light at 330 nm. Key: ●—●, native nuclear DNA from *P. knowlesi;* ○ — ○, heat denatured nuclear DNA from *P. knowlesi;* ▲—▲, native rat liver nuclear DNA. Data of Gutteridge, Trigg & Bayley (1971).

215

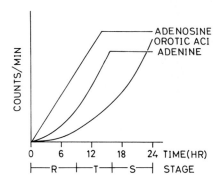

Fig. 7.
Diagram of time courses of incorporation of radiotracers into DNA of *P. knowlesi* in culture. Diagram is compounded from Fig. 1, 4 and 6 of Gutteridge & Trigg (1970). Key: R, ring; T, trophozoite; S, schizont.

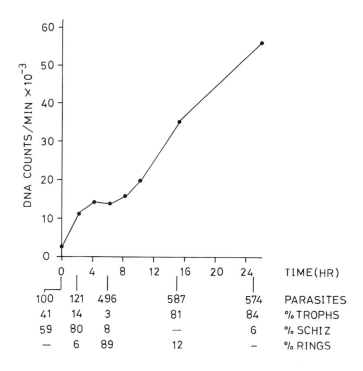

	100	121	496		587		574	PARASITES
	41	14	3		81		84	% TROPHS
	59	80	8		—		6	% SCHIZ
	—	6	89		12		—	% RINGS

Fig. 8.

Time course of the incorporation of ^{3}H-adenosine into the DNA of *P. knowlesi* in culture. Mill Hill medium (Trigg & Gutteridge, 1971) was used supplemented with 100 μc of adenosine-T(G) in each flask. The incubations were initiated with late-trophozoite stage parasites and a complete flask was harvested at each time point. 0.5 ml was used to prepare smears and 4.0 ml was harvested by centrifugation, washed, precipitated with trichloracetic acid and fractionated to yield DNA fractions (Gutteridge & Trigg, 1970). Parasite numbers are for 10^4 red cells. Gametocyte and abnormal parasite counts are not included so that the percentages do not all add up to 100 %. Data of Gutteridge & Trigg, in preparation.

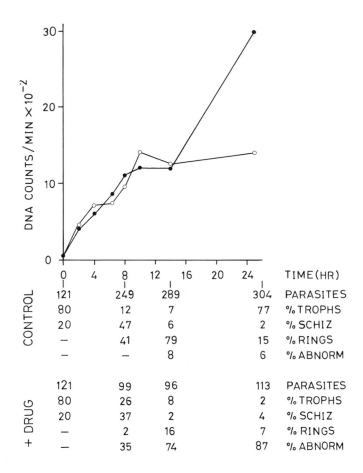

CONTROL					
TIME(HR)	0	8	12	24	
121		249	289	304	PARASITES
80		12	7	77	% TROPHS
20		47	6	2	% SCHIZ
–		41	79	15	% RINGS
–		–	8	6	% ABNORM

+ DRUG				
121	99	96	113	PARASITES
80	26	8	2	% TROPHS
20	37	2	4	% SCHIZ
–	2	16	7	% RINGS
–	35	74	87	% ABNORM

Fig. 9.

Time course of the effect of pyrimethamine (10^{-9}M) on the incorporation of ^{3}H-adenine into the DNA of *P. knowlesi* in culture. Medium 199 was used supplemented with 100 μc of adenine-2,8-T in each flask. The incubations were initiated with late trophozoite stage parasites and a complete flask was harvested at each time point. The procedure then followed that described in Fig. 8. Key: ●—●, control; ○—○, + pyrimethamine. Parasite numbers are for 10^4 red cells.

15

DIHYDROFOLATE REDUCTASES IN PARASITIC PROTOZOA AND HELMINTHS*

Julian J. Jaffe, Department of Pharmacology
The University of Vermont College of Medicine
Burlington, Vermont 05401, U.S.A.

It is generally believed that during the course of evolution toward a life of obligate parasitism, invading organisms gradually underwent metabolic adaptations, with accompanying morphological and physiological changes, in response to the physicochemical composition of their particular environmental niches within the host. Such successful adaptation by parasites to restricted environments has almost invariably been achieved by sacrificing the versatility of various aspects of their metabolism and physiology. Moreover there is evidence to indicate that even when parasite and host possess indentical biochemical pathways subserving a particular area of metabolism, the analogous enzymes therein may differ quantitatively in terms of fine structure, geometry, and surface charge. [1] Identification of such biochemical and physiological peculiarities offers the opportunity for exploitation by appropriate chemical agents, provided the differences between analogous features of parasite and host are broad enough to favor selective toxicity.

An excellent example supporting this view is provided by Bueding and his colleagues [2, 3], who ascribed the antischitosomal activity of trivalent antimonial compounds to an 80-fold greater sensitivity of schistosomal phosphofructokinase to inhibition by these agents than the analogous mammalian enzyme. In this case, the consequences of this weak selectively toxic action at the enzyme level are further amplified at the physiological level because of the much greater dependence of the

* The author's personal contributions were made possible by support from the National Cancer Institute, NIH (Grant CA-08114-06) and the United States-Japan Cooperative Medical Science Program administered by the National Institute of Allergy and Infectious Diseases, NIH, HEW (Grant AI-08261-03).

schistosomes than the mammalian host upon phosphorylative glycolysis for energy generation.

Obligate parasites characteristically have high rates of growth and cell division during one or more phases of their life cycles. Because it has been established that folic acid (folate) derivatives play an important role in the synthesis of nucleic acids and proteins by a wide variety of organisms, and because certain chemical compounds which disturb folate metabolism were found empirically to inhibit the growth **in vivo** of *Plasmodium*[4], *Toxoplasma*[5], and *Eimeria*[6] and to affect larval development **in utero** of *Litomosoides carinii* **in vivo**[7], increased interest has developed to compare the characteristics of folate metabolism in various parasites and their vertebrate hosts, with particular attention being devoted to the enzyme dihydrofolate (DHF) reductase, which catalyzes the converseion of folate to its metabolically active forms.

The metabolically active forms of folate are those in which positions 5, 6, 7, and 8 of the pyrazine portion of the pteridine moiety are reduced. Six tetrahydrofolate (THF) coenzymes have been discovered which act as acceptors of various one-carbon units and which participate in reactions involving interconversions among several amino acids, initiation of peptide chain synthesis, and synthesis **de novo** of purine nucleotides and the pyrimidine deoxyribomononucleotide, thymidylic acid (d TMP). In the course of all these reactions except the synthesis **de novo** of d TMP, THF is regenerated and is again available for reconversion to its various coenzymic forms. The major **de novo** biosynthetic pathway leading to d TMP involves the methylation of deoxyuridylate (dUMP), this reaction being catalyzed by thymidylate synthetase with N^5, N^{10}-methylene THF serving as donor of the single carbon substituent as well as the required hydrogen atoms[8]. As a consequence of the reaction, DHF is regenerated and must be reduced to THF in the presence of DHF reductase before it can again participate in one or more metabolic pathways.

Thus, DHF reductase is of considerable importance because it can mediate the formation of THF by reduction of exogenous folate (if particular cells are able to assimilate the preformed molecule); and because it mediates the reduction of endogenously synthesized DHF (either as a by-product of dTMP synthesis **de novo** or by way of, presumably, guanosine triphosphate, p-aminobenzoate, and glutamate[8]. The vulnerability of DHF reductase as a target in antiparasite chemotherapy will depend upon a number of complex interacting factors, such as the comparative properties of this enzyme (including differential sensitivity to inhibitors) in parasite and host, the comparative interrelation of metabolites and inhibitior in terms of transport across membranes, the relative importance of the enzyme in parasite and host for the maintenance of intracellular THF levels, pool size of distal products of THF-mediated reactions, and so on.

Presence of DHF Reductase in Parasites

DHF reductase activity has now been detected in crude extracts of plasmodia, [9, 11] crithidia, [12] bloodstream and culture forms of salivarian and stercorarian trypanosomes [10, 13, 14] , and adult *Schistosoma mansoni* [15], *Nippostrongylus brasiliensis* [16], and four species of filariae [15]. As indicated in Table I, there is a considerable range in the levels of DHF reductase reported to be present in these parasites. Such variation may be due in part to the efficiency of different extraction procedures but also may reflect differing physiological states of the organisms when harvested. In the case of *Trypanosoma (Schizotrypanum)cruzi,* it is noteworthy that the level of this enzyme in culture forms is much higher than in the mammalian bloodstream forms. With respect to plasmodia, the claim by Walter **et al.** [10] that the specific activity of DHF reductase in 10^{10} *P.chabaudi* triphozoites is only one-sixth that in a comparable number of schizonts should be accepted with reservation, in the light of Bahr's finding [17] that the mean dry mass of *P. berghei* schizonts (and presumably their protein content) is approximately 6 times that of the merozoites plus trophozoites.

The data in Table 1 also indicate that four species of adult filariae *(Dirofilaria immitis, Litomosoides carinii, Dipetalonema witei* and *Onchocerca volvulus)* are much richer sources of DHF reductase than the closely related strongylid nematode parasite, *N. brasiliensis,* as well as the trematode *S. mansoni*. Crude extracts of female *D. immitis* and *L. carinii* contain more than twice as much DHF reductase per mg protein as those of the males. It is unlikely that the higher enzyme content of these female filariae is due to contributions from embryos within them, since extracts of 10^7 *D. immitis* microfilariae are devoid of DHF reductase activity [15].

General Properties

All the DHF reductases of parasite origin closely resemble the analogous enzyme from mammalian sources in their strong preference for DHF over folate (F) as substrate and NADPH over NADH as cofactor. Indeed, within the pH range of 6.0 - 7.5, F was unable to replace DHF as a substrate for the DHF reductases from plasmodia, [9, 11] trypanosomatid flagellates, [10, 12, 14] and *N. brasiliensis*, [16] and was instead in most cases a weak inhibitor of the enzyme. However, when equimolar amounts of F were substituted for DHF in reaction systems at physiological pH containing DHF reductases from *S. mansoni,* four filarial species, or rat liver, the reaction in all cases proceeded at approximately 10 % of the maximal rates [15] . Since it has been found that DHF reductases from a variety of cells that can assimilate exogenous F can also catalyze the reduction of F to THF at a low rate, the above findings might suggest that those parasites possessing DHF reductases incapable of reducing F rely exclusively upon synthesis **de novo** of F-containing cofactors. While there is strong indirect evidence that plasmodia, [18, 19] *Toxoplasma* [5, 20] and *Eimeria* [6, 21] are so specialized, no conclusions can yet be drawn concerning the biosynthetic capabilities of other protozoan and helminthic parasites for F-containing cofactors.

Except in the case of DHF reductases from crithidia [12] and *N. brasiliensis* [16] , NADH can partially replace NADPH as cofactor for the analogous enzyme from the other parasites as well as from mammalian cells, the reaction generally proceeding at 10-30 % of the maximal rate when NADPH is present.

The relationship between the rate of the reaction and the concentration of either substrate or cofactor, expressed as the apparent affinity constants (Km) for DHF and NADPH, is shown in Table 2. With respect to Km values for DHF, those of the protozoan and helminthic DHF reductases fall between the high extreme reported for the analogous bacterial enzymes $(2 \times 10^{-5}M)$ [22] and the low extremes reported for those of chicken liver $(2-5 \times 10^{-7}M)$ [23, 24] and rat liver $(2 \times 10^{-7}M)$. [25] With respect to Km values for NADPH, those of the plasmodial DHF reductases, being relatively low $(1-5 \times 10^{-6}M)$ could be distinguished, with but one exception *(T. (T.)brucei,* $5 \times 10^{-6}M$), [14] from those of the other protozoan and helminthic DHF reductases $(1-2.8 \times 10^{-5}M)$. However, the NADPH Km values for the plasmodial DHF reductases are comparable to those reported for the analogous enzymes from avian [24] and mammalian [26, 27] sources $(1.7-5 \times 10^{-6}M)$.

The DHF reductases from parasites belonging to different genera, and in the case of *Plasmodium,* from 3 species within the genus, differ quantitatively in their sensitivity to changes in pH. For example, the pH optimum of DHF reductase from *P. berghei* [9] is in the narrow range 7.0 - 7.2, while that of the *P. knowlesi* enzyme [11] is broader (pH 6.0 - 7.5) and that of the *P. chabaudi* enzyme [10] quite sharply defined (pH 5.8). The DHF reductases of trypanosomotid flagellates [12, 13] generally maintain essentially maximal activity over a broad pH range (6.0-7.0; 6.0-7.7) as does the analogous enzyme from *N. brasiliensis* [16] (pH 7.5-8.5). The pH-activity profiles of DHF reductases from *S. mansoni* and filariae were not ascertained, but these enzymes also appear not to be very sensitive to changes in pH. However, all presently characterized DHF reductases from parasites exhibit single pH optima, in contrast to the analogous enzymes from two vertebrate sources (birds and mammals), which show two pH optima. [24, 27, 28]

A sharp dichotomy was found between the estimated molecular weights of protozoan DHF reductases on the one hand and metazoan DHF reductases on the other. The molecular weights of the enzymes from *P. berghei,* [9] *P. knowlesi* [11] and *P. lophurae* [29] are in the range 150,000-200,000, and those from *C. oncopelti, T. (S.)cruzi,* and *T. (T.)brucei rhodesiense* are in the range 100,000-200,000. [30] By contrast, the molecular weights of DHF reductases from *S. mansoni,* [15] *N. brasiliensis,* [16] *D. immitis,* [15] *L. carinii,* [15] *D. witei,* [15] and *O. volvulus* [15] as well as those from avian [24, 30] and mammalian sources [30, 32] are all around 20,000. Considering the possibility that the *P. berghei* DHF reductase might be an aggregate of subunits, Ferone **et al.** [9] attempted to alter the position of exit of the enzyme from Sephadex G-100 columns by carrying out chromatography in the presence of 0.2 M KCl, 4 M urea, 0.1 M 2-mercaptoethanol, or 50 μM DHF plus 0.04 M 2-mercaptoethanol. Neither the agents which are known to break hydrogen bonds or disulfide linkages nor the substrate itself was able to decrease significantly the molecular weight of this particular enzyme. This strongly suggests that the *P. berghei* DHF reductase, and presumably the other protozoan DHF reductases so far investigated, exist naturally as protein molecules of 100,000-200,000 molecular weight. If this be true, it is one of two outstanding distinguishing general biochemical properties among the presently characterized DHF reductases of protozoan and metazoan origin, the other being the 20-hour half life of the *P. berghei* enzyme in the presence of urea, [9] in contrast to half-lives in the order of minutes for the analogous enzymes from avian [33] and mammalian [31] sources.

Sensitivity to Folate Analogs and Certain Antiparasitic Drugs

Although all DHF reductases are characterized by extreme sensitivity to inhibition by 4-amino analogs of folate [8] (Table 3), Burchall and Hitchings [22] were the first to point out that DHF reductases of bacterial and mammalian origin could be distinguished by their differential sensitivity to certain 2,4-diaminopyrimidines and related low molecular weight heterocycles. Characteristic drug inhibitor profiles have since been described for the analogous enzymes from plasmodia, [9, 11] crithidia, [12] trypanosomes, [14] *S. mansoni* [15] *N. brasiliensis,* [16] and filariae, [15] and most of these data are summarized in Table 3, together with similar data from studies of mammalian DHF reductases, [9, 15, 22] for purposes of comparison. The original papers should be consulted for further information.

An outstanding feature of the drug inhibitor profile of plasmodial DHF reductases is their much greater sensitivity to the 2,4-diaminopyrimidines pyrimethamine and trimethoprim than the analogous mammalian enzyme. The comparative drug inhibitor profiles of DHF reductases from one salivarian *(T. (T.)brucei rhodesiense)* and one stercorarian *(T. (S.)cruzi)* trypanosome which appear in Table 3 represent a more complete study of other members of these two sections of the genus *Trypanosoma* [14]. It was found that the drug inhibitor profiles of 5 species of salivarian trypanosomes were closely similar and could be clearly distinguished from those of 2 species of stercorarian trypanosomes. Such similarities and differences between a particular protein in organisms belonging to the same genus may reflect phylogenetic kinship and divergencies. By contrast, the drug inhibitor profiles of DHF reductases from culture forms and bloodstream forms of the same species are indistinguishable [14].

Although DHF reductases from salivarian trypanosomes, like those from plasmodia, are much more sensitive to inhibition by trimethoprim than is the analogous mammalian enzyme, the differential sensitivity of the trypanosomal enzymes to this inhibitor is not as great as that of the plasmodial enzymes. Moreover, the trypanosomal DHF reductases are much less sensitive than are the plasmodial enzymes to inhibition by pyrimethamine, another 2,4-diaminopyrimidine, BW 60-212, and especially the dihydrotriazine, BW 57-43.

The drug inhibitor profiles of crithidial DHF reductases more closely resemble those of stercorarian rather than those of salivarian trypanosomes, an observation which also may reflect a closer phylogenetic relationship between these two groups of trypanosomatid flagellates.

As could be expected, the helminthic DHF reductases are also very sensitive to inhibition by two close analogs of folate, methotrexate and a related 2,4-diaminoquinazoline (CCNSC 529,861). [35] It is of interest that although methotrexate and CCNSC 529,861 are roughly equipotent on a molar basis against the schistosomal (as well as the mammalian) enzyme, the latter compound is significantly more potent against all four filarial DHF reductases, especially that of *O. volvulus.* The schistosomal and filarial DHF reductases are relatively insensitive to inhibition by the 2,4-diaminopyrimidines pyrimethamine, trimethoprim and BW 60-212. Interestingly, quite marked differences were found in the sensitivity of the DHF reductases from the four filarial species and that of the closely related nematode *N. brasiliensis* to pyrimethamine and BW 60-212, although their sensitivity to trimethoprim is mutually closely similar.

The N-nitroso derivative of a 6-dichlorobenzyl-substituted 2,4-diamino-quinazoline (NO-Q) is significantly more potent than the parent compound (H-Q) against the DHF reductases of *S. mansoni, D. immitis, L. carinii* and *D. witei* (as well as *T. (T.)brucei rhodesiense*). By contrast the DHF reductase of *O. volvulus* does not exhibit differential sensitivity to the two homologs, a property it shares with the analogous mammalian enzyme.

Suramin, a complex substitued area of the aminonaphtalene-sulfonic acid type, was found to be a weak inhibitor of DHF reductases from *S. mansoni, D. immitis, L. carinii,* and *D. witei* (as well as *T. (T.brucei rhodesiense)* and an even weaker inhibitor of mammalian DHF reductase, in terms of the concentration required for 50 % inhibition of optimal activity. However, the DHF reductase from *O. volvulus* is much more sensitive to inhibition by suramin (ID_{50} value of $2 \times 10^{-6}M$), being thus 35 times more sensitive to this inhibitor than the analogous mammalian enzyme in terms of the ID_{50}.

In view of McCullough and Bertino's [36] admonition that purified DHF reductases should be used in measuring inhibition produced by analogs, it should be noted that the drug sensitivities of such enzymes from various sources [15, 30] were measured following their partial purification by passage through Sephadex G-200, and in every case the drug sensitivities did not differ significantly from those of the enzymes from the corresponding crude extracts.

The conclusions which might be drawn from such drug inhibitor studies are that the differential sensitivity of plasmodial DHF reductases to pyrimethamine and trimethoprim and that of salivarian trypanosomal DHF reductases to trimethoprim are sufficiently large as to suggest the basis for effective chemotherapeutic action of these agents against the corresponding parasitoses. Yet in practice, positive correlations were observed between the concentrations of pyrimethamine and trimethoprim required to inhibit the DHF reductases of the parasite by 50 % and the minimum concentration required to inhibit development **in vitro** and **in vivo** only in the case of plasmodia [9, 11] . Indeed, trimethoprim and other antifolates so active against the DHF reductases of salivarian trypanosomes have practically no activity against the intact organisms **in vitro** or **in vivo** [14]. In the case of the helminths, Gutteridge **et al.**[16] recently stated that "the prospects of using dihydrofolate reductase inhibitors in the chemotherapy of helminths infections are not encouraging". Yet it is noteworthy that methotrexate was found to be toxic against developing larvae of a nematode parasite of insects, *Neoaplectana glaseri*, **in vitro** [37] and to inhibit embryonal development **in utero** of *L. carinii* in cotton rats [7].

Such uncertainties regarding the correlation between the degree of differential sensitivity of an isolated parasite DHF reductase to particular antifolate compounds and the ability of such agents to affect the growth and development of the parasite **in vivo** raises questions concerning the physiological roles of THF coenzymes and the relative importance of DHF reductase in maintaining THF levels in different groups of parasites.

Involvement of THF Coenzymes in Parasite Metabolism

As has already been mentioned, the general biological functions of THF coenzymes, elucidated from studies of bacteria, yeasts, and avian and mammalian cells, involve the transfer of one-carbon units in interconversions among several amino acids, initiation of peptide chain systhesis, and in synthesis **de novo** of purine nucleotides and dTMP.

It is well established that in free-living organisms glycine and serine are readily interconverted in the presence of serine transhydroxymethylase, with N^5, N^{10}-methylene THF serving as the hydroxymethyl transferring agent [8]. While protozoan and metazoan parasites are able to assimilate both these amino acids preformed from host sources [38, 40] no one yet reported the presence in any parasite of an operative THF-requiring glycine-serine interconverting system.

Investigations of bacterial, avian, and mammalian metabolism have revealed that the synthesis **de novo** of methionine proceeds by way of the methylation of homocysteine, the methyl donor being 6-methyl THF or its triglutamate congener [8, 41]. The 5-methyl group is formed by enzymatic reduction of N^5, N^{10}-methylene THF (or its triglutamate congener). Two types of methyl transfer reactions have been described, not necessarily occurring in the same organism. One requires cobalamin and S-adenosylmethionine, while the other functions independently of these cofactors but has an absolute requirement for the triglutamate form of 5-methyl THF.

While presently available evidence suggests that most parasites depend primarily, if not exclusively upon preformed methionine supplied by their hosts [38, 40, 42, 44] , it has been reported [45] that *P. berghei* has the ability to synthesize methionine **de novo**, such production accounting for at least 20 % of the total methionine content of this parasite. Also, various species of *Crithidia* apparently can synthesize methionine from homocysteine [46]. The nature of the methylating system in both cases remains obscure. In fact, not enough is yet known about the general amino acid metabolism of parasites or the fine structure of their protein synthetic machinery to make feasible at this time any meaningful assessment of the relative importance of the pertinent THF-mediated reactions.

On the other hand, sufficient information about certain facets of parasite nucleic acid metabolism is available to warrant the conclusion that for most, if not all, parasites of vertebrates, the N^5, N^{10}-methylene THF-mediated synthesis **de novo** of dTMP is probably the most crucial for their normal growth and the maintenance of their high reproductive capacity. This conclusion is based on the finding that plasmodia, [47, 48] trypanosomes, [46, 49] *S. mansoni,* [50, 51] and perhaps other parasitic helminths [52, 53] have extremely limited abilities to synthesize purine nucleotides **de novo,** whereas it has been established that plasmodia, [47, 54] trypanosomes, [49] *S. mansoni* and *D. immitis* [15] can synthesize all their pyrimidine nucleotides **de novo**. It is noteworthy that plasmodia have a absolute dependence upon synthesis **de novo** of dTMP [10, 54] whereas trypanosomes [10, 49] and helminths [15] are capable of phosphorylating exogenously supplied thymidine. Nevertheless, even in these latter cases, it is doubtful that the extremely limited amounts of thymidine in mammalian host extracellular fluids [55] would be sufficient **per se**, when converted to dTMP and thence to thymidine-5'-triphosphate (dTTP) by way of the salvage pathway, to meet the needs of these parasites for this vital DNA precursor and allosteric activator of purine ribonucleoside diphosphate reductase [56].

Thus, although other possible roles of THF coenzymes in parasites have not yet been elucidated, there is strong evidence to suggest that N^5, N^{10}-methylene THF is required by many, if not all, parasites for production of dTMP for DNA synthesis, and that the relative importance of DHF reductase in the regeneration of this THF coenzyme would be inversely proportional to the size of intracellular THF pools, which might fluctuate in response to physiological requirements. Further investigations of the folate metabolism of parasites are needed before we can solve the riddle of why salivarian trypanosomes are not affected by trimethoprim [13] and why the effects of pyrimethamine and trimethoprim upon *P. knowlesi* are only expressed at the schizont stage of this parasite [11].

Finally, the discovery that suramin can inhibit DHF reductase deserves additional comment. Suramin was found to be curative against *L. carinii* infections in the multimammate rat, *Mastomys natalensis* (Dr. G. Lammler, personal communication), and it is established as one of the drugs of choice against *Onchocera volvulus* infections in man[57] and also early *T.(T.)brucei gambiense* and *T.(T.)brucei rhodesiense* infections in man[58]. The precise modes of its antiparasitic actions remain uncertain. Although this drug was found to inhibit a number of enzymes **in vitro** including various oxidases and kinases, suramin is not a general enzyme poison and most of the non-proteolytic enzymes tested were unaffected by the drug[59]. Hexokinase was the most sensitive of the enzymes tested, being inhibited by 50 % at a concentration of suramin in the range of 2×10^{-5}M.

It can be seen in Table 3 that the DHF reductases of schistosomal and filarial (as well as trypanosomal) origin are among the most sensitive enzymes, of all those tested, to inhibition by suramin. Especially noteworthy is the sensitivity of the *O. volvulus* reductase, which was inhibited by 50 % in the presence of 2×10^{-6}M suramin. Even after taking into account the extensive plasma protein binding of suramin[60], this concentration of unbound drug is likely to be maintained following the usual dose regimen (1 gram intravenously once weekly for 5 weeks) administered to patients suffering from onchocerciasis. Thus it is conceivable that the clinical efficacy of suramin against this important filarial infection may be due in part to its ability to inhibit onchocercal DHF reductase.

References

1. HITCHINGS, G.H. in Drugs, Parasites and Hosts (Eds. L.G. GOODWIN and R.H. NIMMO-SMITH), p. 196. Little Brown and Co., Boston (1962).
2. SAZ, H.J. and BUEDING, E. Pharmacol. Rev. **18,** 871 (1966).
3. BUEDING, E. J. Gen. Physiol. **33** 475 (1950).
4. HILL, J. in Experimental Chemotherapy, Vol. 1 (Eds. R.J. SCHNITZER and F. HAWKING), p. 513. Academic Press, New York (1963).
5. EYLES, D.E. and COLEMAN, N. Antibiot. Chemother. **5,** 529 (1955).
6. LUX, R.E. Antibiot. Chemother. **4,** 971 (1954).
7. HAWKING, F. WORMS, M.J. J. Parasitol. **53,** 1118 (1967).
8. BLAKLEY, R.L. in The Biochemistry of Folic Acid and Related Pteridines, p. 231. American Elsevier Publ. Co., Inc., New York (1969).
9. FERONE, R., BURCHALL, J.J. and HITCHINGS, G.H. Mol. Pharmacol. **5,** 49 (1969).
10. WALTER, R.D., MUHLPFORDT, H. and KONIGK, E. Z. Tropenmed. Parasitol. **21,** 347 (1970).

11. GUTTERIDGE, W.E. and TRIGG, P.I. Parasitology **62,** 431 (1971).
12. GUTTERIDGE, W.E., McCORMACK, J.J. JR., and JAFFE, J.J. Biochim. Biophys. Acta **178,** 453 (1969).
13. JAFFE, J.J. and McCORMACK, J.J. JR., Mol. Pharmacol. **3,** 359 (1967).
14. JAFFE, J.J., McCORMACK, J.J. JR., and GUTTERIDGE, W.E. Exp. Parasitol. **25,** 311 (1969).
15. JAFFE, J.J., McCORMACK, J.J. and MEYMERIAN, E. Biochem. Pharmacol. in the press (1971).
16. GUTTERIDGE, W.E., OGILVIE, B.M. and DUNNETT, S.J. Int. J. Biochem. **1,** 230 (1970).
17. BAHR, G.F. Mil. Med. (Suppl.) **131,** 1064 (1966).
18. HITCHINGS, G.H. Clin. Pharmacol. Therap. **1,** 570 (1969).
19. FERONE, R. and HITCHINGS, G.H. J. Protozool. **13,** 504 (1966).
20. SANDER, J. and IDTVEDT, T.M. Acta Path. Microbiol. Scand. (B) **78,** 664 (1970).
21. McMANUS, E.C., OBERDICK, M.T. and CUCKLER, A.C. J. Protozool. **14,** 379 (1967).
22. BURCHALL, J.J. and HITCHINGS, G.H. Mol. Pharmacol. **1,** 126 (1965).
23. OSBORN, M.J., FREEMAN, M. and HUENNEKENS, F.M. Proc. Soc. Exp. Biol. Med. **97,** 429 (1958).
24. KAUFMAN, B.T. and GARDINER, R.C. J. Biol. Chem. **241,** 1319 (1966).
25. WANG, D. and WERKHEISER, W.C. Fed. Proc. **23,** 324 (1964).
26. MORALES, D.R. and GREENBERG, D.M. Biochim. Biophys. Acta **85,** 360 (1964).
27. FREUDENTHAL, R. and HEBBORN, P. Int. J. Biochem. **1,** 150 (1970).
28. BERTINO, J.R., BOOTH, B.A., BIEBER, A.L., CASHMORE, A. and SARTO-RELLI, A.C. J. Biol. Chem. **239,** 479 (1964).
29. TRAGER, W. J. Parasitol. **56(4),** 627 (1970).
30. GUTTERIDGE, W.E., JAFFE, J.J. and McCORMACK, J.J. JR., Biochim. Biophys. Acta **191,** 753 (1969).
31. BERTINO, J.R., PERKINS, J.P. and JOHNS, D.G. Biochemistry **4,** 839 (1965).
32. ZAKRZEWSKI, S.F., HAKALA, M.T. and NICHOL, C.A. Mol. Pharmacol. **2,** 423 (1966).
33. KAUFMAN, B.T. Biochem. Biophys. Res. Comm. **10,** 449 (1963).
34. REYES, P. and HUENNEKENS, F.M. Biochem. Biophys. Res. Comm. **28,** 833 (1967).
35. JOHNS, D.G., CAPIZZI, R.L., NAHAS, A., CASHMORE, A.R. and BERTINO, J.R. Biochem. Pharmacol. **19,** 1528 (1970).
36. McCULLOUGH, J.L. and BERTINO, J.R. Biochem. Pharmacol. **20,** 561 (1971).
37. JACKSON, G.J. and SIDDIQUI, W.A. J. Parasitol. 727 (1965).
38. MOULDER, J.W. The Biochemistry of Intracellular Parasitism. Univ. of Chicago Press, Chicago (1962).
39. VOORHEIS, H.P. Trans. Roy. Soc. Trop. Med. Hyg. **65,** 241 (1971).
40. SMYTH, J.D. The Physiology of Trematodes. W.H. Freeman and Co., San Francisco (1966).
41. STOKSTAD, E.L.R. and KOCH, J. Physiolog. Rev. **47,** 83 (1967).
42. ROGERS, W.P. in Chemical Zoology, Vol. 3 (Eds. M. FLORKIN and B.T. SCHEER), p. 379. Academic Press, New York (1969).

43. ARONSON, C.E. and JAFFE, J.J. Biochem. Pharmacol. **15,** 1995 (1966).
44. TRIGG, P.I. and GUTTERIDGE, W.E. Parasitology **62,** 113 (1971).
45. LANGER, B.W. JR., PHISPHUMVIDHI, P., JIAMPERPOOM, D. and WEID-HORN, R.P. Mil. Med. **134,** 1039 (1969).
46. GUTTMAN, H.N. and WALLACE, F.G. in Biochemistry and Physiology of Protozoa, Vol. 3 (Ed. S.H. HUTNER), p. 459. Academic Press, New York (1964).
47. WALSH, C.J. and SHERMAN, I.W. J. Protozool. **15,** 763 (1968).
48. BUNGENER, W. and NIELSEN, G. Z. Tropenmed. Parasitol. **19,** 185 (1968).
49. JAFFE, J.J. Trans. N.Y. Acad. Sci. **29,** 1057 (1967).
50. SENFT, A.W. J. Parasitol. **56(4),** 314 (1970).
51. JAFFE, J.J. Nature **230,** 408 (1971).
52. JAFFE, J.J. and DOREMUS, H.M. J. Parasitol. **56,** 254 (1970).
53. HEATH, R.L. and HART, J.L. J. Parasitol. **56,** 340 (1970).
54. BUNGENER, W. and NIELSEN, G. Z. Tropenmed. Parasitol. **18,** 456 (1967).
55. CLEAVER, J.E. in Thymidine Metabolism and Cell Kinetics (Eds. A. Neuberger and E.L. TATUM), p.60. John Wiley and Sons, New York (1967).
56. LARSSON, A. and REICHARD, P.J. J. Biol. Chem. **241,** 2540 (1966).
57. WORLD HEALTH ORGANIZATION. Expert Committee on Onchocerciasis. WHO Tech. Rep. Series 87, Geneva (1954).
58. ROLLO, I.M. in The Pharmacological Basis of Therapeutics, 4th Edition (Eds. L.S. GOODMAN and A. GILMAN), p. 1144. MacMillan Co., New York (1970).
59. WILLS, E.D. and WORMALL, A. Biochem. J. **47,** 158 (1950).
60. TOWN, B.W., WILLS, E.D., WILSON, E.J. and WORMALL, A. Biochem. J. **47,** 149 (1950).

Explanatory paragraphs for tables
Table 1.

Dihydrofolate reductase activity in crude extracts of parasitic protozoa and helminths

Source of Enzyme	Developmental Status	Sex	Estimated[a] Specific Activity	Ref
Plasmodium spps.	Erythrocytic Stages	—	1 -3	9-11
Crithidia spps.	Culture Form	—	150	12
Trypanosoma spps. *(Stercoraria* and *Salivaria)*	Bloodstream Form	—	0.5-30	10,13,14
T.(S.) cruzi	Culture Form	—	90 -100	14
S. mansoni	Adult	Mixed	0.6	15
N. brasiliensis	Adult	Mixed	0.5	16
D. immitis	Adult	Male	6	15
D. immitis	Adult	Female	14	15
L. carinii	Adult	Male	9	15
L. carinii	Adult	Female	25	15
D. witei	Adult	Mixed	13	15
O. volvulus	Adult	Mixed	50	15
Rat Liver	Adult	—	1.5-2	15,16

[a] Units per mg protein, a unit of DHF reductase being defined as that quantity which catalyzes the reduction of 1 mμmole of DHF per minute.

Sources of enzyme were particle-free supernatant fractions of freed, disrupted plasmodia, crithidia, trypanosomes, S. mansoni, and N. brasiliensis; acetone powders of trypanosomes, S. mansoni, D. immitis and L. carinii; and lyophilized D. immitis, L. carinii, D. witei and O. volvulus, following their suspension in different buffers (Tris-HCl, phosphate, or citrate), 0.07-0.1M, pH range 5.8-7.5. The assays were carried out in one of these buffers in the presence of NADPH (70-250 μM), DHF (50-100 μM), and a sulfhydryl agent, 2-mercaptoethanol or N-acetylcysteine (10-40mM), final volume 1-3 ml, 20°-37°. The course of the reaction was followed by measuring the decrease in absorbance at 340 nm.

Table 2.
Apparent affinities (Km) of dihydrofolate (substrate) and NADPH (cofactor) for dihydrofolate reductases of parasitic protozoa and helminths

| Source of Enzyme | M x 10⁻⁶ | | |
	Km DHF	Km NADPH	Ref
Plasmodium spps.	2.6- 3.0	1.0- 5.0	9-11
Crithidia spps.	3 - 4	10	12
Trypanosoma spps. *(Stercoraria* and *Salivaria)*	3 -10	5 -14	10,13,14
S. mansoni	7	28	15
N. brasiliensis	5	8	16
D. immitis	4	15	15
L. carinii	5	20	15
D. witei	5	20	15
O. volvulus	5	—	15
Rat Liver	0.2	5	15,25

Apparent affinity constants (Km) were determined by plotting the reciprocal of reaction velocity at 37° against the reciprocal of different concentrations of DHF or NADPH, during a period of time when the reaction was linear.

Table 3.
Comparative sensitivity of dihydrofolate reductases from parasitic protozoa and helminths to folate analogs and certain antiparasitic drugs.

Source of Enzyme	Concentration (X 10^{-8}M) causing 50 % Inhibition									Ref
	MTX	NSC	PYR	TRI	BW 57-43	BW 60-212	H-Q	NO-Q	SUR	
P. berghei	0.07	ND	0.05	7.0	0.8	1.7	ND	ND	ND	9
P. knowlesi	0.1	ND	0.1	3.0	0.5	1.0	ND	ND	ND	11
T. rhodesiense	0.1	0.04	20	25	400	60	200	20	3000	14,15
T. cruzi	0.15	ND	100	1000	300	30	ND	ND	ND	14
C. fasciculata	0.8	ND	300	2000	4000	110	ND	ND	ND	12
C. oncopelti	0.6	ND	500	2500	6600	100	ND	ND	ND	12
S. mansoni	0.07	0.1	700	1000	ND	1000	500	40	3000	15
N. brasiliensis	0.15	ND	70	1000	1500	50	ND	ND	ND	16
D. immitis	0.2	0.03	1500	2000	400	1000	200	20	1000	15
L. carinii	1	0.1	1000	1000	ND	200	200	10	1000	15
D. witei	3	0.4	400	5500	ND	800	700	100	1500	15
O. volvulus	3	0.04	4000	4000	ND	200	100	100	200	15
Mouse Red Cell	ND	ND	100	>100,000	60	ND	ND	ND	ND	9
Rat Liver	0.2	0.3	140	33,000	14	46	60	60	7000	15,22

MTX: 2,4-diamino-N^{10}-methylfolic acid (methotrexate); NSC: aspartic acid, N-(p-[[(2,4-diamino-5-chloro-6-quinazolinyl)-methyl]amino] benzoyl]-, dihydrate, L-(CCNSC 529,861); PYR: 2,4-diamino-5-p-chlorophenyl-6-ethylpyrimidine (pyrimethamine); TRI: 2,4-diamino-5-(3',4',5'-trimethoxybenzyl) pyrimidine (trimethoprim); BW 57-43: 1-(p-butylphenyl)-1,2-dihydro-2,2-dimethyl-4,6-diamino-s-triazine; BW 60-212: 2,4-diamino-6-butylpyrido [2,3-d]pyrimidine; H-Q: 2,4-diamino-6-(3,4-dichlorobenzylamino) quinazoline; NO-Q: 2,4-diamino-6-[(3,4-dichlorobenzyl)-nitroamino]quinazoline; SUR: Suramin.

ND: Not done.

At least several levels of each compound were tested in the standard reaction systems after 2-10 minutes preincubation with the enzyme, NADPH, sulfhydryl agent and buffer. In most cases the reaction was started by the addition of DHF. In each case, the reaction series contained approximately the same number of enzyme units generally an amount sufficient to give a control rate of 0.010-0.020 absorbance unit per minute at 340 nm.

16

PHYSIOLOGICAL ADAPTABILITY OF MALARIA PARASITES

R.E. Howells, W. Peters and C.A. Homewood
Department of Parasitology,
Liverpool School of Tropical Medicine,
Pembroke Place, Liverpool L3 5QA.

Introduction

Comprehensive reviews of plasmodial metabolism have been given by McKee (1951), von Brand (1966) and Peters (1969, 1970). Various aspects of protozoan biochemistry have also been reviewed by a number of other authors (see Peters, 1969; Florkin & Scheer, 1967; Bryant, 1970; Garnham, 1966 and Danforth, 1967). The difficulties of biochemical studies on parasites were elaborated on by Bryant (1970) and the ultra-structural studies of Killby & Silverman (1968), Cook et al. (1969) and Ladda (1969) on plasmodia "freed" from the host erythrocytes by various techniques, convincingly demonstrated the difficulties of obtaining undamaged plasmodia, uncontaminated by host tissue, for biochemical studies. It is hardly surprising that the metabolism of even the most intensively studied form of the malarial parasite, the erythrocytic trophozoite, remains incompletely understood. McKee (1951) considered it unfortunate that our knowledge of the metabolism of the sporozoite and exoerythrocytic stages was not more extensive in view of their significance in malarial prophylaxis, but during the last two decades, with the exception of the large number of studies on the fine structure of the parasites, surprisingly little effort has been made to rectify this situation. Recent work in our laboratory suggests that changes occur in the metabolism of the rodent malaria parasite *Plasmodium berghei* during its life cycle. The evidence for these cyclical changes will be presented here and their possible significance in the acquisition of chloroquine resistance by this and other species of plasmodia will be discussed.

Metabolic changes during the life cycle of *Plasmodium berghei*

The erythrocytic forms of *Plasmodium berghei,* although preferentially parasites of reticulocytes develop readily in normocytes (Fig. 1) and appear to digest erythrocyte stroma via a cytostome (review by Peters, 1970). The age of the host red cell influences the amount of pigment formed by the parasite (Howells **et al.**, 1968) and the morphology of the mitochondrion-like organelle of the plasmodium. In normocyte parasites the "mitochondrion" is a simple, acristate membrane bounded sac, but in reticulocytes some degree of crista development is observed (Ladda, 1969; Howells, 1970). Additionally whorls of membranes may be found in the parasite cytoplasm and these have also been considered to possess a mitochondrion-like function (Rudzinska & Trager, 1959). A number of workers have demonstrated that malarial parasites, including *P. berghei,* utilise oxygen and that their respiration is sensitive to cyanide and, in the case of *P. knowlesi,* carbon monoxide (see Bryant, 1970). Electron microscope cytochemical techniques have revealed the presence of cyto-chrome oxidase activity on the membranes of the mitochondria and of the multi-lamellate whorls (Howells **et al.**, 1969a & b; Theakston **et al.**, 1969) and the biochemical studies of Scheibel & Miller (1969a & b) have demonstrated the enzyme in host-cell free preparations of *P. knowlesi*. Repeated cytochemical tests have failed to demonstrate the presence of succinic dehydrogenase in the parasites (Howells, 1970). Bowman **et al.** (1961), Bryant **et al.** (1964) demonstrated that lactate is the major end product of glucose catabolism by *P. berghei* freed from their host cell and there is a preponderance of evidence that the erythrocytic forms of *P. berghei* do not utilise the citric acid cycle (reviewed by Danforth, 1967). Malate dehydrogenase (MDH) has been attibuted to the parasite (Sherman, 1966; Carter, 1970; Trigg **et al.** 1970) and may be involved in the process of CO_2 fixation described in these parasites by Siu (1967) and Nagarajan (1968**b**).

The function of cytochrome oxidase and the oxygen requirement of the parasite remains enigmatic. Scheibel & Pflaum (1970) consider that cytochrome oxidase is not necessarily of physiological significance to *P. knowlesi,* but although the amount of glucose utilised by this plasmo-dium is constant for aerobic and anaerobic conditions, the amount of lactate produced aerobically is approximately half that produced anaero-bically (Scheibel & Miller 1969**b**).

Fulton & Spooner (1956) demonstrated that in addition to lactate, *P. berghei* produced acetate and formate by the catabolism of glucose. Scheibel & Pflaum (1971) have shown that during *P. knowlesi* fermenta-tions acetate and formate are found in physiologically significant amounts. The latter workers make the interesting suggestion that the oxidative conversion of pyruvate to acetate may be associated with an electron transport mediated synthesis of ATP.

Overall, however, it may be concluded that the erythrocytic stages of *P. berghei* possess acristate mitochondria, catabolise glucose largely via the Embden Meyerhoff glycolytic pathway and that lactate is a major end product of this catabolism. An obstacle to this generalisation are the results of Nagarajan (1968**b**), who described citric acid cycle activity in the parasites.

In addition to trophozoites, schizonts and gametocytes of *P. berghei* are found within erythrocytes. With regard to the schizont there is no evidence that its mode of glucose catabolism differs from that of the trophozoite.

The gametocytes of *P. berghei* when immature possess acristate mitochondria but in mature forms cristate mitochondria are found (Howells, 1970). Aikawa **et al.** (1969) have described the fine structure of the gametocytes of a number of species of avian, reptilian and mammalian species of plasmodia and Kass **et al.** (1971) have described the fine structure of the gametocytes of *P. falciparum.* There is little information on the metabolism of gametocytes of plasmodia. Beyer (1962) showed that the gametocytes of *Eimeria intestinalis* were the first stages of the life cycle to utilise succinic dehydrogenase, but even after cooling blood to room temperature succinic dehydrogenase could not be detected by light microscope cytochemical techniques within the gametocytes of *P. berghei.*

There have been no studies on the carbohydrate metabolism of the ookinete. A feature of interest in this stage of the parasite however is the crystalline inclusions described by Garnham **et al.** (1962) in the ookinetes of *P.c. bastianelli.* Similar structures are found in the ookinete of *Leucocytozoon simondi* (Desser & Iretiak, 1971). From cytochemical evidence the latter authors considered that the crystallines were composed of a proteolipid material in which the lipid is neutral and postulated that they serve as a food reserve for the parasite. In *L. simondi* the crystalloid material aggregates into a central core in the oocyst and later, fragments pass into each forming sporozoite (Desser & Wright, 1967). The particles are not found in normal maturing oocysts of *P. berghei* but crystalline material has been observed in abnormal oocysts of this species (Davies & Howells, 1971; Davies **et al.,** 1970) and in ookinetes and abnormal oocysts of an avian malaria (Terzakis, 1969). In plasmodia the particles have been considered of viral origin and the cause of the abnormality of the host oocyst. It is equally possible however that the crystalline material may also serve as a food reserve in the malaria parasite and that its presence in abnormal oocysts is due to its incomplete utilisation by degenerating parasites. Baker (1963) demonstrated that the particles within the ookinete of *P.c. bastianelli* were devoid of RNA.

The oocysts of *P. berghei*, with few exceptions, are found as extracellular bodies attached to the wall of the mosquito midgut, lying within the coelom and separated from the epithelial cells by the basement membranes. The cyst is surrounded by a fibrous wall (Fig. 2) and there is no evidence of a cytostome for the ingestion of host material by pinocytosis or phagocytosis. A prominent ultrastructural difference is found in the morphology of the mitochondria (see Howells, 1970). In the oocyst they appear as elongate, tubular cristate bodies distributed through the cytoplasm.

Cytochemical tests performed at both the level of the light microscope and electron microscope have demonstrated cytochrome oxidase activity within the oocyst (Howells, 1970) and at the electron microscope level the mitochondrial localisation of the enzyme is apparent. In contrast to the erythrocytic trophozoites the oocyst stage contains succinic dehydrogenase activity, as demonstrated at the light microscope level by the nitro-blue tetrazolium technique (Howells, 1970) and at the level of the electron microscope by the ferricyanide method of Kerpel-Fronius & Hajos (1968) (Figure 3). The localisation of the deposit has also been studied in the epithelial cells of the mosquito midgut epithelium - the intra-mitochondrial localisation within the epithelial cells serving as a positive control in this test. The oocyst stages also possess demonstrable NAD- and NADP-dependent isocitrate dehydrogenase (Figures 4 & 5), following the technique given by Pearse (1961) for the localisation of these enzymes. The oocyst therefore differs considerably from the erythrocytic parasites both in the fine structure of the mitochondria and in its cytochemical characteristics. Von Brand (1966) has stressed that the demonstration of the presence of individual enzymes of the citric acid cycle does not prove the presence of a functional cycle in a parasite. The obvious difficulties of obtaining sufficient numbers of free oocysts have to the present precluded pure biochemical studies on the oocyst but it is feasible that within the anopheline host *P. berghei* does switch to a metabolic pathway involving the citric acid cycle.

Ball & Chao (1963) have successfully dissected oocysts of *P. relictum* from the midgut of culicine mosquitoes and have grown the oocysts **in vitro**. The oocyst appears an essentially extracellular parasitic form of *Plasmodium* and the studies of Ball & Chao (1963) indicate that the mosquito stomach is not essential for its growth and development. Schneider (1968) however, failed to grow young oocyst of *P. gallinaceum* free of host tissue and suggested that this was due to the absence of actively growing host cells.

Garnham (1966) described the dependence of the sporozoites of *Plasmodium* on exogenous glucose. The sporozoites possess cristate mitochondria and cytochemically have been found to possess demonstrable succinic dehydrogenase and cytochrome oxidase (Howells, 1970 a & b). In the pre-erythrocytic schizont of *P. berghei* histochemical tests (Howells & Bafort, 1970) have demonstrated the presence of cytochrome oxidase but the apparent absence of succinic dehydrogenase and the mitochondria of the liver schizont morphologically resemble those of the blood stages in being acristate (Bafort & Howells, 1969; Bafort, 1970).

There is therefore evidence for the presence of a cyclical change in the metabolism of the rodent malarial parasite *P. berghei* during its life cycle and this change may be compared with those described for other parasitic protozoa such as the African trypanosomes (Vickerman, 1971), *Leishmania donovani* (Janovy, 1967; Simpson, 1968) and the coccidian *Eimeria intestinalis* (Beyer, 1963).

The relationship of the cyclical changes observed in *P. berghei* to other malarial parasites

In the avian malarial parasites cristate mitochondria are found within the intra-erythrocytic trophozoites (Aikawa **et al.**, 1967), the exo-erythrocytic schizont (Aikawa, 1968), the oocysts and sporozoites (Terzakis **et al.**, 1967). There is also a considerable amount of biochemical evidence for the presence of a citric acid cycle within the blood stages of these parasites (see Danforth, 1967 and Peters, 1970). There would therefore appear to be no metabolic switch similar to that postulated for *P. berghei* in the avian malarias. Sherman **et al.** (1969) considered that glycolysis and CO_2 fixation are the major pathways of carbohydrate metabolism in the avian malarias and that the evidence for a citric acid cycle may alternatively be explained by the presence of transamination pathways related to the process of CO_2 fixation within the parasites. Scheibel & Pflaum (1970) also considered that the incorporation of radioisotope from ^{14}C glucose or $NaH^{14}CO_3$ into glutamic acid, alanine and aspartic acid as well as citrate, malate and succinate (Ting & Sherman, 1966; Sherman & Ting, 1968; Sherman **et al.**, 1969) should not be taken as evidence that these parasites have a Krebs cycle since incorporation in terms of μ moles of substrate is very low.

With the exception of *P. knowlesi* biochemical information on the metabolism of simian and human malarias is fragmentary and has been reviewed by Peters (1970). There is insufficient evidence to determine whether the changes observed in *P. berghei* occur also these other species. The erythrocytic trophozoites of *P. falciparum* in some respects resemble those of the avian species (see Peters, 1970) and the gametocytes of this parasite (Kass **et al.,** 1971) possess mitochondria with typically protozoan cristae.

Physiological adaptation and the acquisition of resistance to chloroquine.

Howells **et al.** (1970) proposed a theory for the mechanism of chloroquine resistance in malarial parasites, using *P. berghei* as an experimental model. It was proposed that since chloroquine appears to exert a primary effect on the phagosomes or food vacuoles of the parasite, causing clumping of the pigment and autophagosome formation, an effect of chloroquine on the parasite would be the cessation of haemoglobin digestion and the starvation of the parasite of amino acids obtained from this source. The presumptive evidence for a citric acid cycle in the sporogonic stages of the parasite has been described above and it was proposed that a switch to the citric acid cycle in the resistant parasites could confer resistance by enabling the parasites to synthesise amino acids by the transamination of citric acid cycle intermediates. The presence of a demonstrable succinic dehydrogenase activity in the parasites using the nitro blue tetrazolium technique at light microscope level presented preliminary evidence to support this theory. A feature of chloroquine resistant (RC strain) *P. berghei* is their obligate parasitism of reticulocytes and the reduction in pigment present in the parasites (Peters, 1965). A more recently developed chloroquine resistant (NS) strain (Peters **et al.,** 1969) in the absence of chloroquine develops within normocytes and produces pigment but in the presence of drug develops in reticulocytes and appears to lose its pigment (see Howells **et al.,** 1970). The habitation of the mouse reticulocyte by *P. berghei* does not of itself confer a marked chloroquine resistance on the parasites although parasites within reticulocytes exhibit a lowered sensitivity to the drug. Some intrinsic factor of the parasite may be assumed to be responsible for the high level of resistance in these strains. Reticulocytes themselves possess active mitochondria and in view of the difficulty of resolving the distribution of formazan deposit following light microscope techniques for succinic dehydrogenase activity, and of the doubts which have been raised regarding the specificity of localisation following staining with tetrazolium salts (Pearse & Hess, 1961) both the electrophoretic separation and the fine structural localisation of the enzyme were attempted. Attempts at the electrophoretic separation of succinic dehydrogenase repeatedly proved unsuccessful. Fine structural localisation using thiocarbamyl nitro blue tetrazolium (Seligman **et al.,** 1967) also failed and in our hands this tetrazolium salt frequently gave unsatisfactory results even on mammalian tissues. The failings of this salt have also been commented on by Seligman **et al.** (1971). The ferricyanide technique for the electron microscope localisation of succinic dehydrogenase activity (Kerpel-Fronius & Hajos, 1968; Kalina **et al.,** 1969) has also been attempted many times and with limited success. The apparent obstacles to the successful demonstration are as follows: 1) intact, unfixed erythrocytes do not appear to be permeable to one or more of the constituents of the test medium. Red cells have previously been shown to be relatively impermeable to succinate (Passow, 1964). 2) Lysis of the red cell membrane and incubation of the free parasites has been of limited value since the incubation of freed parasites resulted in such considerable morphological damage that in most instances parasite tissue would not be recognized with any degree of certainty. 3) In many experiments no apparent activity was observed even in obviously mammalian mitochondria. This may be due to the liberation of haem from haemoglobin during the preparative procedure since Keilin & Hartree (1947) demonstrated that haematin inhibits succinic dehydrogenase. 4) The enzymes of mouse liver and heart muscle withstand brief (5 minutes) fixation in cold formalin but following similar pretreatment of infected blood no activity would be demonstrated, even at the light microscope level. 5) There is a possibility that the ferricyanide may be reduced by the mouse haemoglobin. Theakston & Fletcher (1971) used sodium nitrite to convert haemoglobin to methaemoglobin prior to testing for glucose-6-phosphate dehydrogenase in malaria infected parasites. Similar treatment of blood prior to testing for SDH localisation has proved unsuccessful.

From the experiments performed to date no consistent evidence for the localisation of the enzyme in parasitised reticulocytes has been obtained but in the few where some degree of success in preserving both morphology and activity has been achieved the enzyme was located within the reticulocyte mitochondria but not in the parasite (Figures 6 & 7). In contrast to the results obtained with blood, control experiments performed on mouse heart muscle proved consistently successful with the ferricyanide techniques. The precise intra-mitochondrial localisation of enzyme activity in heart muscle is illustrated in Figure 8.

Even the successful demonstration of SDH activity within the parasite would be inconclusive evidence of Krebs cycle activity since in the helminths fumarate reductase has been demonstrated to function in CO_2 fixation (Saz, 1970). Electrophoretic studies on NADP-dependent isocitrate dehydrogenase (Fig. 9) have revealed a single band form in extracts from uninfected reticulocytes and reticulocytes infected with chloroquine sensitive and chloroquine resistant parasites. There is no difference in the band position in these various samples. It is of interest that a stronger activity was observed in preparations of reticulocytes infected with chloroquine resistant parasites than from uninfected reticulocytes. In these experiments reticulocyte counts exceeded 90 % as revealed by intravital staining with brilliant cresyl blue. Langer **et al.,** (1967), Sherman (1966) and Carter (1970) have identified a variety of parasite isoenzymes by electrophoretic studies. Further work is in progress in our laboratory on resolving host and parasite enzymes by the application of a combination of histochemical staining and immunoprecipitation techniques to electrophoresed preparations.

We must conclude that we have thus far been unsuccessful in obtaining further evidence to support the theory for the presence of a Krebs cycle in chloroquine resistant *P. berghei.* Morphological evidence further demonstrates that the "mitochondria" of resistant parasites are of the same size as those of sensitive strains (Howells, in preparation), and that in both cyclically and non-cyclically transmissible resistant strains the mitochondria are predominantly acristate (Figs. 10 & 11). Even the reduction of pigment in chloroquine resistant parasites may not be directly related to resistance since sensitive parasites in reticulocytes show reduced amounts of pigment (Howells **et al.,** 1968) and in the chloroquine resistant NS strain pigment is still observed in some parasites even when maintained on a daily regimen of chloroquine (Figs. 11 & 12).

Warhurst & Chance (personal communication) have recently shown that cyclically-transmissible chloroquine resistant strains of *P. berghei* differ from chloroquine-sensitive strains in possessing a satellite to the main peak DNA, both cyclically and non-cyclically transmissible sensitive parasites lacking the satellite.

A more tangible difference between the chloroquine resistant and sensitive strains of *P. berghei* was observed in studies on the accumulation of the drug by the parasites. Demonstration of lower concentrations in resistant than in sensitive strains led Macomber **et al.** (1966) to suggest that in resistant strains the mechanism by which chloroquine is accumulated in sensitive parasites is impaired. Fitch (1969) postulated a decrease in the number, affinity or accessibility of chloroquine receptor sites in resistant form. Partition profile studies (Rollo, 1968 and 1969) have shown that the accumulation of basic antimalarials within the red cell is dependent upon differences between the pH of the plasma and the pH within the red cell. Following the demonstration by Allison & Young (1964) that chloroquine is concentrated in mammalian lysosomes, Homewood **et al.** (1971) have extended these observations to propose a physico-chemical basis for the accumulation of chloroquine within parasitised erythrocytes. These latter workers consider that a low pH within the parasite lysosomes (an assumption supported by the localisation of acid phosphatase within the pigment vesicle of *P. berghei* (Aikawa & Thompson, 1971) and the presence of acid proteases in the parasite (Cook **et al.,** 1961)) would be effective in inducing an accumulation of the drug within these organelles. Concentration of the drug, it is argued, would cause a rise in the pH of lysosome contents and this change may in some way cause the clumping of the pigment. Chloroquine in high concentrations inhibits certain mammalian proteolytic enzymes (Cowey & Whitehouse, 1966). Evidence to support this concept was presented by the **in vitro** incubation of parasites in media made slightly alkaline with bicarbonate. It was observed that raising the extracellular pH of the parasites correspondingly increased the amount of clumping of the pigment. Homewood **et al.** (1971) postulate that the loss of acid from the parasites would effectively confer resistance to chloroquine. Other than the reduced pigment formation in chloroquine resistant (RC strain) *P. berghei* (Peters **et al.,** 1965), however, there is no evidence of changes in the lysosome pH or in the lysosomal enzymes of chloroquine-resistant parasites.

There are a number of other theories for the site of chloroquine action. Skelton **et al.** (1968) demonstrated that *P. lophurae* possesses coenzymes Q8 and Q9 and that the DPNH-oxidase systems of this parasite, with which the ubiquinones are involved, were inhibited by chloroquine. In a later study (Skelton **et al.,** 1969) ubiquinone production by *P. knowlesi, P. cynomolgi, P. falciparum,* and *P. lophurae* was demonstrated.

Siu (1967) presented evidence for a CO_2-fixing pathway in *P. berghei* and demonstrated that chloroquine at a concentration of $1.9 \times 10^{-4}M$ inhibited the action of the parasite's phosphoenolpyruvic carboxykinase by 50 %.

Ciak & Hahn, (1966) considered that chloroquine acts by binding to the parasite DNA and several alternative sites of action have been suggested by various other workers (discussed by Peters, 1970).

Conclusions

In summary, evidence has been presented for a cyclical change in the metabolism of *P. berghei,* possibly involving the acquisition of citric acid cycle activity by the sporogonic stages. An earlier theory (Howells **et al.,** 1970) postulated that chloroquine resistance in plasmodia might be explained by the utilisation of the Krebs cycle by the intra-erythrocytic trophozoites. Supporting evidence for this theory has not been obtained.

References

AIKAWA, M., HUFF, C.G. and SPRINZ, H. (1967). J. Cell. Biol., **34,** 229.

AIKAWA, M., HUFF, C.G. and SPRINZ, H. (1968). Am. J. trop. Med. Hyg., **17,** 156.

AIKAWA, M., HUFF, C.G. and SPRINZ, H. (1969). J. Ultrastruct. Res., **26,** 316.

AIKAWA, M. and THOMPSON, P.E. (1971). J. Parasit., **57,** 603.

ALLISON, A.C. and YOUNG, M.R. (1964). Life Sciences, **3,** 1407.

BAFORT, J. (1971) Ann. Soc. belge Méd. trop., **L1/1.**

BAFORT, J. and HOWELLS, R.E. (1970). Trans. R. Soc. trop. Méd. Hyg., **64,** 467, 1.

BAKER, J.R. (1963). Trans. R. Soc. trop. Med. Hyg., **57,** 233.

BALL, G.H. and CHAO, J. (1963). Ann. Epiphyties, **14,** 205.

BEYER, T.V. (1963). Sbornik Rabot Akademiia Nauk SSSR Institut Tsitologii, **3,** 70.

BOWMAN, I.B.R., GRANT, P.T., KERMACK, W.O. and OGSTON, D. (1961). Biochem. J., **78,** 472.

BRAND, T. von (1966). Biochemistry of Parasites, pp. 1-429. Academic Press, New York and London.

BRYANT, C. (1970). In "Advances in Parasitology", ed. B. Dawes, **7,** 139.

BRYANT, C., VOLLER, A. and SMITH, M.J.H. (1964). Am. J. trop. Med. Hyg., **13,** 515.

CARTER, R. (1970). Trans. R. Soc. trop. Med. Hyg., **64,** 401.

CIAK, J. and HAHN, F.E. (1966). Science, N.Y., **151,** 347.

COOk, R.T., AIKAWA, M., ROCK, R.C., LITTLE, W. and PRINZ, H. (1969). Milit. Med., **134,** 866.

COWEY, F.R. and WHITEHOUSE, M.W. (1966). Biochem. Pharmacol., **15,** 1071.

DANFORTH, W.F. (1967). In "Research in Protozoology" (T.T. Chen, ed.) Vol. 1, pp. 201- 306. Pergamon Press, Oxford.

DAVIES, E.E. and HOWELLS, R.E. (1971). Trans. R. Soc. trop. Med. Hyg., **65,** 13.

DAVIES, E.E., HOWELLS, R.E. and VENTERS, D. (1971). Ann. trop. Med. Parasit., **65,** 403.

DESSER, S.S. and WRIGHT, K.A. (1968). Can. J. Zool., **46,** 303.

FITCH, C.D. (1969). Proc. Nat. Acad. Sci., U.S., **64,** 1151.

FLORKIN, M. and SCHEER, B.T. (1967). "Chemical Zoology" (ed.) volume 1, "Protozoa" (ed. G.W. Kidder). Academic Press, New York and London.

FULTON, J.D. and SPOONER, D.F. (1956). Expl. Parasit., **5,** 59-78.

GARNHAM, P.C.C. (1966). "Malaria parasites and other haemosporidia", pp. 1-1114. Blackwell Scientific Publications, Oxford.

GARNHAM, P.C.C., BIRD, R.G. and BAKER, J.R. (1962). Trans. R. Soc. trop. Med.

Hyg., **56**, 116.
HOMEWOOD, C.A., WARHURST, D.C., PETERS, W. and BAGGALEY, V.C. (1971).
Nature, Lond., (in press).
HOWELLS, R.E. (1970). Ann. trop. Med. Parasit., **64**, 181.
HOWELLS, R.E. (1970). Ann. trop. Med. Parasit., **64**, 223.
HOWELLS, R.E., PETERS, W. and THOMAS, E.A. (1968). Ann. trop. Med. Parasit.,
62, 267.
HOWELLS, R.E., THEAKSTON, R.D.G., FLETCHER, K.A. and PETERS, W. (1969).
Trans. R. Soc. trop. Med. Hyg., **63**, 6.
HOWELLS, R.E., PETERS, W. and FULLARD, J. (1969). Milit. Med., **134**, 1026.
HOWELLS, R.E., PETERS, W., HOMEWOOD, C.A. and WARHURST, D.C. (1970).
Nature, Lond., **228**, 625.
HOWELLS, R.E. and BAFORT, J. (1970). Ann. Soc. belge Méd. trop., **50**, 587.
KASS, L., WILLERSON, D. Jr., RIECKMANN, K.H., CARSON, P.E. and BECKER,
R.P. (1971). Am. J. trop. Med. Hyg., **20**, 187.
KALINA, M., WEAVERS, B. and PEARSE, A.G.E. (1969). Nature, Lond., **221**, 479.
KEILIN, D. and HARTREE, E.F. (1947). Biochem. J., **41**, 503.
KERPEL-FRONIUS, S. and HAJOS, F. (1968). Histochemie, **14**, 343.
KILLBY, V.A.A. and SILVERMAN, P.H. (1969). Am. J. trop. Med. Hyg., **18**, 836.
LADDA, R.L. (1969). Milit. Med., **134**, 825.
LANGER, B.W., Jr., PHISPHUMVIDHI, P. and FRIEDLANDER, Y. (1967). Expl.
Parasit., **20**, 68.
MACOMBER, P.B., O'BRIEN, R.L. and HAHN, F.E. (1966). Science, N.Y., **152**,
1374.
MANWELL, R.D. and LOEFFLER, C.A. (1961). J. Parasit., **47**, 285.
McKEE, R.W. (1951). In "Biochemistry and physiology of Protozoa". (A. Lwoff, ed.)
Vol 1, pp. 251-322. Academic Press, New York.
MOON, T.W. and HOCHACHKA, P.W. (1971). Biochem. J., **123**, 695.
NAGARAJAN, K. (1968a). Expl. Parasit., **22**, 19.
NAGARAJAN, K. (1968b). Expl. Parasit., **22**, 33.
PASSOW, H. (1964). In "The Red Blood Cell. A Comprehensive Treatise", edit.
Bishop, C. and Surgenor, D.M. Academic Press, New York and London, Chapter 3,
71-145.
PEARSE, A.G.E. (1961). "Histochemistry Theoretical and Applied", J. and A.
Churchill, Ltd., London. 2nd edition.
PEARSE, A.G.E. and HESS, R. (1961). Experientia (Basel), **17**, 136.
PETERS, W. (1965). Annls Soc. belge Med. trop., **45**, 365.
PETERS, W. (1969). Trop. Dis. Bull., **66**, 1.
PETERS, W. (1970). "Chemotherapy and Drug Resistance in Malaria" Academic
Press, New York and London.
PETERS, W., ROBINSON, B.L., RAMKARAN, A.E. and PORTUS, J.H. (1969).
Trans. R. Soc. trop. Med. Hyg., **63**, 5.
ROLLO, I.M. (1968). Fedn. Proc. Fedn. Am. Socs. exp. Biol., **27**, 537.
ROLLO, I.M. (1969). Molecular pharmacology of antimalarials. Abstracts 8th Int.
Congr. trop. Med. Malar., Teheran, 1376-7.
RUDZINSKA, M.A. and TRAGER, W. (1959). J. biophys. biochem. Cytol. **6**, 103.
SAZ, H.J. (1970). J. Parasit., **56**, 634.
SCHEIBEL, L.W. and MILLER, J. (1969a). J. Parasit., **55**, 825.
SCHEIBEL, L.W. and MILLER, J. (1969b). Milit. Med., **134**, 1074.

SCHEIBEL, L.W. and PFLAUM, W.K. (1970). Comp. Biochem. Physiol., **37,** 543.
SCHNEIDER, I. (1968). Expl. Parasit., **22,** 178.
SELIGMAN, A.M., UENO, H., MORIZONO, V., WASSERBRUG, H.L., KATZOFF, L. and HANKER, J.S. (1967). J. Histochem. Cytochem. **15,** 1.
SELIGMAN, A.M., NIR, I. and PLAPINGER, R.E. (1971). J. Histochem. Cytochem., **19,** 273.
SHERMAN, I.W. (1966). J. Protozool., **13,** 344.
SHERMAN, I.W., RUBLE, J.A. and TING, I.P. (1969). Expl Parasit., **25,** 181.
SIMPSON, L. (1968). J. Protozool., **15,** 201.
SIU, P.M.L. (1967). Comp. Biochem. Physiol., **23,** 785.
SKELTON, F.S., PARDINI, S., HEIDKER, J.C. and FOLKERS, K. (1965). J. Am. chem. Soc., **90,** 5334.
SKELTON, F.S., LUNAN, K.D., FOLKERS, K., SCHNELL, J.V., SIDDIQUI, W.A. and GERMAN, Q.M. (1969). Biochemistry, **8,** 1284.
TERZAKIS, J.A. (1969). Milit. Med., **134,** 916.
TERZAKIS, J.A., SPRINZ, H. and WARD, R.A. (1967). J. cell. Biol., **34,** 311.
THEAKSTON, R.D.G. and FLETCHER, K.A. (1971). Life Sciences, **10,** 701.
TING, I.P. and SHERMAN, I.W. (1966). Comp. Biochem. Physiol., **19,** 855.
TRIGG, P.I., BROWN, I.N., GUTTERIDGE, W.E., HOCKLEY, D.J. and WILLIAMSON, J. (1970). Trans. R. Soc. trop. Med. Hyg., **64,** 2.
VICKERMAN, K. (1971). In "Ecology and physiology of parasites", ed. A.M. Fullis. University of Toronto Press, pp. 58-91.

246

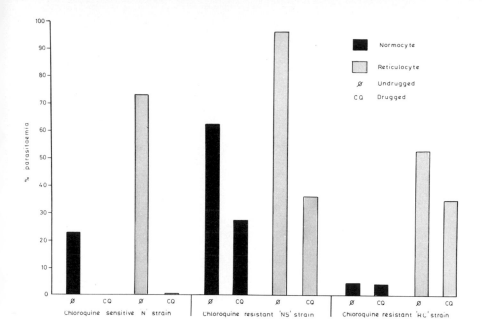

Fig. 1.
Effect of red cell age on the virulence and sensitivity to chloroquine of *P. b. berghei* in the laboratory mouse.

Histogram illustrates the parasitaemia attained in random-bred male white mice (Tuck, TFW) of 20 gram body weight by day +4 following initial infection with approximately 1×10^6 parasites on day 0. Reticulocyte mice were injected intraperitoneally with 0.2 ml of a 0.2 % solution of phenylhydrazine in sterile saline, daily from day -6 to day -1 prior to infection. On day 0 the blood of such mice showed a reticulocytosis of approximately 90 %, as revealed by intravital staining with brilliant cresyl blue. Normocyte mice were not dosed with phenylhydrazine and showed a reticulocytosis of 2 - 3 % on day 0.

Chloroquine drugged mice were injected i.p. with 8 mg/kg chloroquine diphosphate daily from day 0 to day +3 inclusive.

Undrugged mice were not dosed with chloroquine. See Howells **et al.** (1970) for details of strain.

N strain -   the drug sensitive Keyberg 173 strain
NS strain -   the cyclically transmissible chloroquine-resistant strain derived from the N strain
RC strain -   the non-cyclically transmissible chloroquine-resistant strain of *P. berghei* also derived from the N strain (see Peters, 1970).

Note that in the three strains examined the parasitaemia developed more rapidly in reticulocytes than in normocytes. Note also that in the N strain, although a parasitaemia failed to develop in drugged "normocyte" mice in drugged "reticulocyte" mice a parasitaemia of 0.4 % developed.

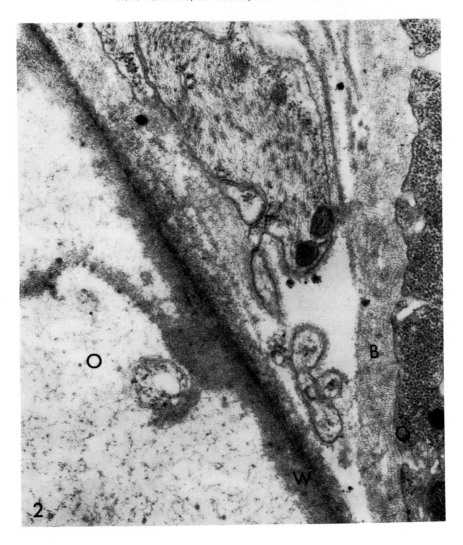

Fig. 2.
Electronmicrograph of the oocyst wall of the N67 strain of *P. berghei* and of the basement membrane of the midgut epithelium of *Anopheles stephensi.* (X 20,000)
The oocyst wall (W) is composed of a fine fibrous material with no ovious periodicity. Note that the basement membrane (B) of the mosquito epithelial cells exhibits a regular periodicity of banding at Å. O = oocyst.

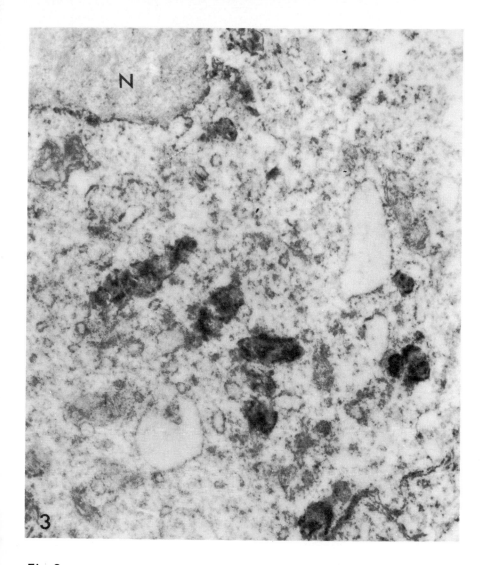

Fig. 3.
Electronmicrograph of the oocyst of the N67 strain of *P. berghei* following incubation in the test medium of Kerpel-Fronius & Hajos (1968) for the demonstration of succinic dehydrogenase activity. Note the discrete deposits of copper ferrocyanide associated with membranous structures considered to be the mitochondria within the cytoplasm of the oocyst. The poor morphological preservation of the specimen is the normal appearance of unfixed oocysts following incubation in the test medium and post-osmication in osmium tetroxide. N = nucleus.
(X 31,500).

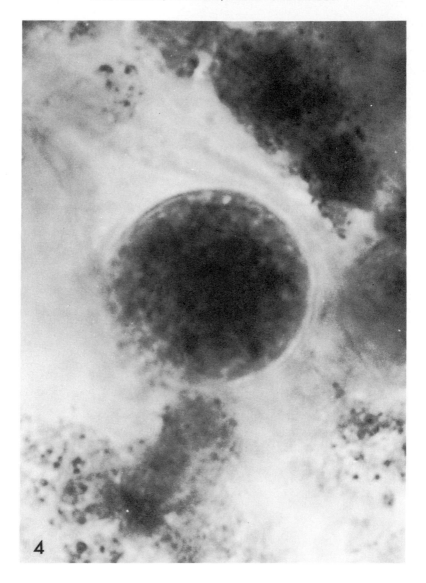

Fig. 4.
Localisation of NAD-dependent isocitrate dehydrogenase in the oocyst of
the N67 strain of *P. berghei* following the NBT technique given by Pearse
(1961). No formazan deposits were observed in control oocysts incubated
in medium lacking sodium isocitrate. (X 2,200).

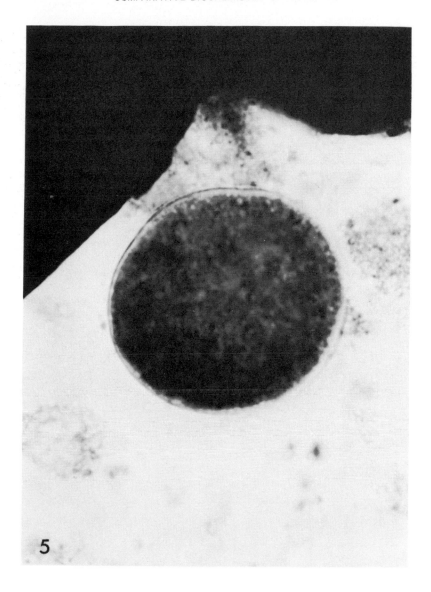

Fig. 5.
Localisation of NADP-dependent isocitrate dehydrogenase in the oocyst
of the N67 strain of *P. berghei* following incubation in the test medium
of the NBT technique given by Pearse (1961). No deposits were observed
in oocysts incubated in medium lacking isocitrate. (X 2,500).

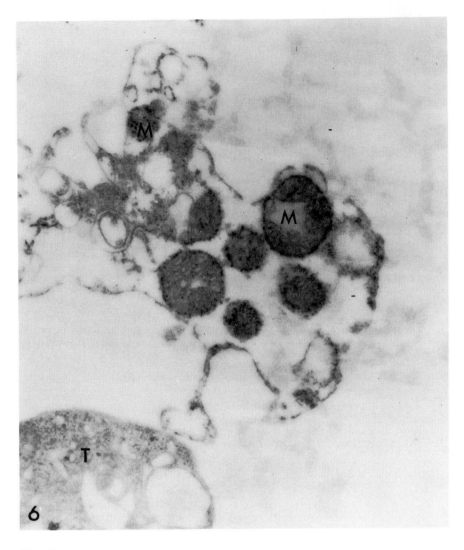

Fig. 6.
Electronmicrograph of a trophozoite of chloroquine-resistant RC strain *P. berghei* following incubation in the test medium of Kerpel-Fronius & Hajos (1968) for the demonstration of succinic dehydrogenase.
Note the absence of deposits from the parasite (T) tissues and the presence of copper ferrocyanide in the mammalian mitochondria (M). Some deposits are localised on the erythrocyte membranes. Unstained section of araldite embedded tissue. (X 31,100).

252

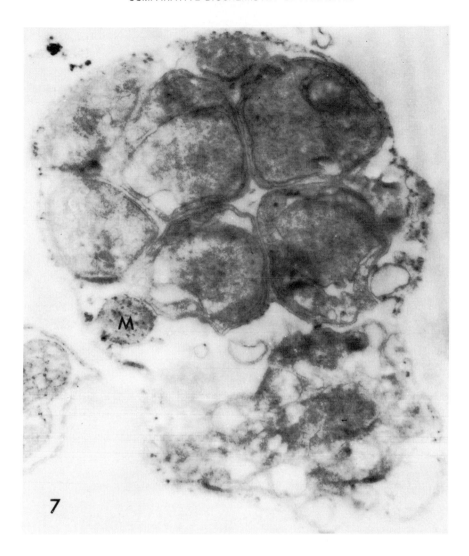

Fig. 7.
Preparation as in Fig. 6. Note deposit in reticulocyte mitochondria and absence from both the schizont and trophozoite within the cell. Refer to text for reservations of interpretation. M = mitochondria. (X 24,900).

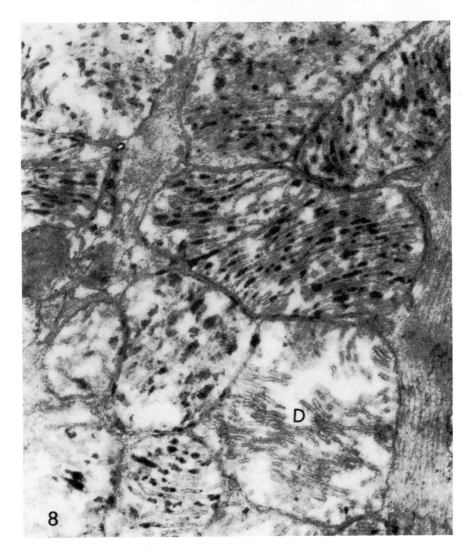

Fig. 8.
Localisation of succinic dehydrogenase activity within the heart muscle mitochondria of the laboratory mouse following incubation in the test medium of Kerpel-Fronius & Hajos (1968).
Note discrete localisation of deposit within the intra-cristal space of the mitochondrion, also the apparent absence of deposits from the damaged mitochondria (D). (X 52,500).

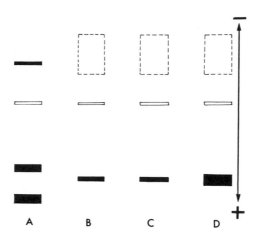

Fig. 9.
Diagram of starch gel zymogram of isocitrate dehydrogenase (NADP)
showing forms of the enzyme obtained from supernatants of uninfected
mouse liver (A) and reticulocytes (C), reticulocytes infected with chloroquine-sensitive, N strain *P. berghei* (B) infected with chloroquine-resistant
RC strain *P. berghei* (D).
Electrophoresis was run at 4°C for 17 hours at 200 V with a citrate-
phosphate buffer system, pH 7.0. Gel stained by the technique of Moon
& Hochackka (1971). Note that only a single enzyme form was obtained
from the uninfected reticulocytes and the N and RC infected reticulocyte
preparations.
Enzyme bands shown as solid black bands. Position of haemoglobin
delineated by broken lines - - - - - -.

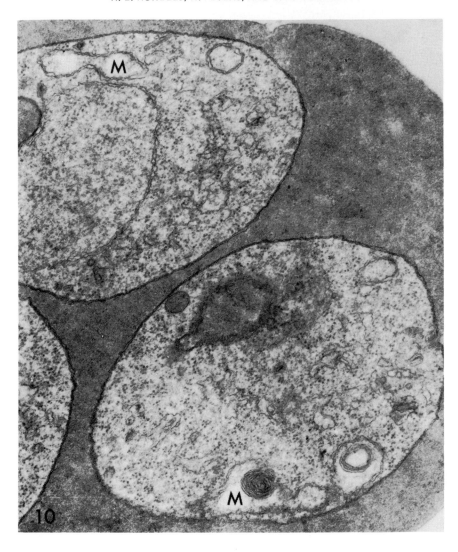

Fig. 10.
Electronmicrograph of trophozoites of the chloroquine-resistant RC strain of *P. berghei.* Note that the mitochondria (M) are acristate. Whorled membranes are apparent within one of the mitochondria. (X 31,000).

256

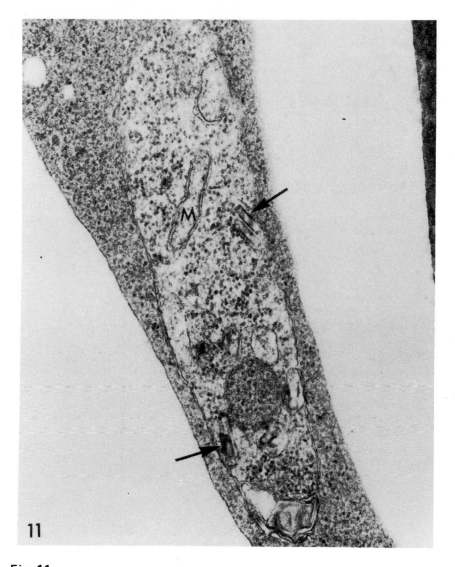

Fig. 11.
Electronmicrograph of trophozoites of the chloroquine-resistant NS strain of *P. berghei.* The infected blood was obtained from a mouse which had been dosed daily for 4 days with 50 mg/kg chloroquine.
Note that the "mitochondria" are acristate (M) and that grains of pigment (arrows) are presented in the parasite cytoplasm (cf. Fig. 10). (X 49,000).

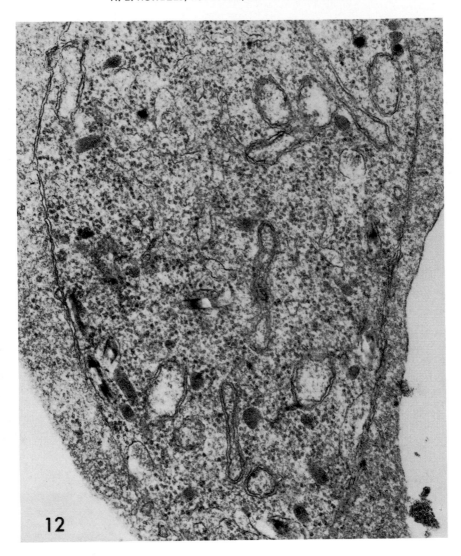

Fig. 12.
Electronmicrograph of a gametocyte of the chloroquine-resistant NS strain of *P. berghei.* Note acristate mitochondria and scattered pigment grains. (X 54,200).

17

CYTOCHEMICAL AND BIOCHEMICAL STUDIES
OF THE INTESTINAL CELLS OF *ASCARIS SUUM*

M. Borgers and H. Van den Bossche

Research Laboratoria, Janssen Pharmaceutica
Beerse, Belgium.

The very high content of ACPase[a] in **Ascaris** intestinal epithelium has been reported previously [1]. It appeared from these studies that about 80 % of the total activity was confined to the brush border, and that the remaining activity was localized in the numerous lysosomes, distributed over the whole cytoplasm but most abundant in the apical region.

Alkaline phosphatase (AlPase), on the other hand, was not detected cytochemically or biochemically.

In order to get more precise information about the specificity of phosphatases, a series of substrates have been used and tested over the pH range 5 to 9. Moreover, the effect of activators and inhibitors has been investigated.

The biochemical results are summarized in fig. 1 to 7. The pH dependence of different substrate hydrolysis by a cytoplasmic extract of **Ascaris** intestine is shown in fig. 1.

In order to determine the amount of ACPase activity bound to cytoplasmic particles, fractionation studies were performed. Free and total ACPase activities were measured on the light mitochondrial fraction (fig. 2). Electron microscopic examination indicated that this fraction was almost exclusively composed of fragments of microvilli and lysosomal particles (fig. 7).

To substantiate these results further, measurements of free and total acitivities were made of a second acid hydrolase, namely acid ribonuclease (fig. 3).

[a] ACPase = acid phosphatase.

These biochemical data support the assumption that the bulk of ACPase in *Ascaris* intestine is not bound to lysosomal particles, thus confirming its already reported cytochemical localization on the brush border[1].
Inorganic pyrophosphatase (iPPase) activity was investigated on an isolated brush border fraction and showed a peak activity at pH 2.8 (fig. 4). Its pH optimum (different from ACPase) and its response to activators (as Mn++) suggest that we deal with a second acid hydrolase, confined to the brush border.

Specific ortho- and pyrophosphates (G-6-P, AMP, ADP, IDP, TPP and ATP) tested at acid and neutral pH, gave a rather comparable distribution pattern. Although variations in intensity of reactivity towards these substrates and some differences in sensitivity towards activators and inhibitors were noted, clearcut answers with regard to specific acitivity could not be provided. We assume therefore that most of the activity towards these substrates could be attributed to non-specific ACPase.

However, a difference in response towards fluoride inhibition has been established for G-6-Pase. Its specific activity is, nevertheless, very low (fig. 5).

At alkaline pH, activity towards ADP and ATP is observed. The activity appeared to be strongly enhanced by Mg++ (fig. 6).

The second part of this paper is concerned with the topographical distribution at the ultrastructural level of these phosphatases in *Ascaris* intestine and compared with the localization of these enzyme activities in mammalian small intestine. A semi-quantitative evaluation of the activities, estimated by light and electron microscopy is shown in table 1.

The most striking observation is the close resemblance between ACPase distribution in *Ascaris* and AlPase in mouse small intestine.

Acid phosphatase is localized on the brush border, in lysosomal bodies and very seldom in the Golgi apparatus (fig. 8). Complex cytolysomes, frequently localized in the supranuclear region, stain strongly for ACPase (fig. 9).

The fact that the exact physiologic significance of the brush border enzyme is poorly understood in mammals, makes it difficult to speculate upon the divergence in phosphatase content between mammals and parasites. However, if one admits that the non-specific phosphatases are related in a fundamental way to breakdown and (or) transport of foodstuffs, the high content of brush border ACPase in *Ascaris* could be related to extracellular digestion.

Whether this digestion is needed to facilitate intake and transport or whether it is a build-in defense mechanism to hydrolyse foreign material remains pure speculative.

The labelling of the lateral plasma membranes for ACPase activity occurred very infrequently and was confined to damaged area or to area where cells were extruded in the lumen (fig. 10). Whether this acitivity has a functional significance in the process of cellular extrusion is still an open question. On the other hand, it is not unlikely that this localization is an artifact of enzyme diffusion due to leakage of enzymes from the brush border into the intercellular space through altered terminal bars at spots where cells are extruded.

From the cytochemical experiments, using different substrates, (ATP, ADP, TPP and G-6-P) at pH 5 to pH 7.2, it appeared that nearly the same sites (brush border and lysosomes) are stained as with β-glycerophosphate (fig. 11 and 12).

Addition of ACPase inhibitors in the different incubating media resulted in a complete loss of activity, so that from the cytochemical point of view there is no evidence whatsoever for the existence of specific phosphatase activity in the pH range from 5 to 7.2. The presence of specific G-6-Pase as found biochemically could not be confirmed by cytochemical means. Its presence in tracer amounts only and its well known sensitivity to cytochemical treatment does probably not allow its cytochemical detection. This is in contrast again with the rather high level of G-6-Pase in mouse small intestine where the endoplasmic reticulum and the nuclear membranes of the absorbing cells are uniformly positive.

Other striking differences compared to mammalian species are 'the absence of nucleoside diphosphatases in endoplasmic reticulum and nuclear envelope, the absence of TPPase in the Golgi apparatus and the absence of ATPase on the lateral cell membranes.

The only enzyme activity which could be localized in the alkaline pH range was ADPase activity in the lammelar elements of the Golgi apparatus and only in a few cases on the brush border (fig. 13 and 14).

Non-specific alkaline phosphatase was not found in *Ascaris* intestinal cells (fig. 15).

The tracer activity or even the total lack of some specific phosphatases might reflect the low metabolic rate of the *Ascaris* intestinal epithelial cells, as compared with mammalian species.

References:

1. BORGERS, M., VAN DEN BOSSCHE, H. and SCHAPER, J.:
   J. Histochem. Cytochem., **18,** 519 (1970).

Substrates	Brush border		Lateral plasma membranes		Lysosomes		Endoplasmic reticulum-nuclear envelope		Golgi apparatus		Mitochondria	
	M	A	M	A	M	A	M	A	M	A	M	A
β -GP pH 9	++	⊙	-	-	++	⊙	-	-	++	⊙	-	-
β -GP pH 7	+	+	-	-	+	+	-	-	-	-	-	-
β -GP pH 5	-	⊕	-	⊕	++	⊕	-	-	+	+	-	+
ATP pH 7.2	+	+	+	⊙	+	+	-	-	-	-	+	⊙
ATP pH 8.8	++	-	+	-	+	-	+	-	+	-	-	-
ADP pH 7.2	+	+	-	-	-	-	+	-	+	-	-	-
ADP pH 8.8	++	⊕	-	-	+	-	-	-	++	⊕	-	-
TPP pH 7.2	+	+	-	-	-	-	++	⊙	++	-	-	-
TPP pH 8.8	++	-	-	-	+	+	+	-	+	-	-	-
AMP pH 7.2	+	+	-	-	-	+	+	-	-	-	-	-
G-6-P pH 6.7	+	++	-	-	-	++	++	⊙	-	-	-	-

The activities are graded as follows : - no activity;   + = weak to moderate;   ++ = strong activity.

Table 1.

Semi-quantitative evaluation of phosphatase activities in different cell organelles of *Ascaris* intestinal (A) and mouse small intestinal (M) cells.

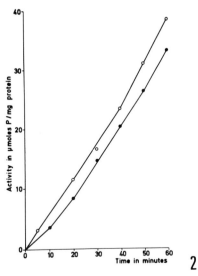

Fig. 1.
pH dependence of disodium phenylphosphate (o), β-glycerophosphate (x), glucose-6-phosphate (△), and 5'-AMP (●) hydrolysis by a cytoplasmic extract of *Ascaris* intestine.
A, sodium acetate-acetic acid buffer; B, sodium cacodylate-HCl buffer; C, tris (hydroxymethyl) aminomethane buffer; D, glycine-NaOH buffer.

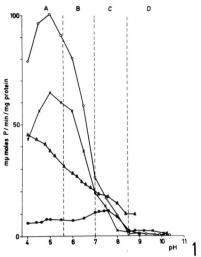

Fig. 2.
Free and total acid phosphatase activity in the light mitochondrial fraction of *Ascaris* intestine. ● — ●, free activity; o — o, total activity.

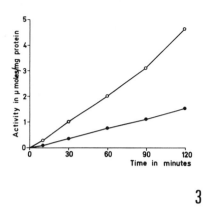

3

Fig. 3.
Free and total acid ribonuclease activity in the light mitochondrial fraction of *Ascaris* intestine. ● — ●, free activity, ○ — ○, total activity.

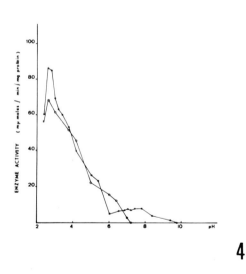

4

Fig. 4.
Effect of pH and Mn++ on the inorganic pyrophosphatase activity in isolated brush borders of *Ascaris* intestine. ● — ●, + 5 mM MnCl$_2$; ○ — ○, - MnCl$_2$.

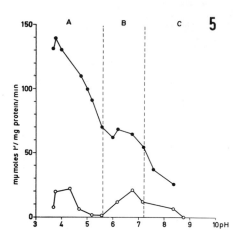

Fig. 5.
pH dependence of glucose-6-phosphate hydrolysis by a particulate fraction of *Ascaris* intestine and effect of 0.01 M NaF upon the hydrolysis.
● — ●, without NaF; ○ — ○, with NaF.

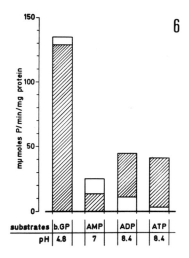

Fig. 6.
Effect of Mg++ upon hydrolysis of ortho- and pyrophosphate substrates. ▨ , + 5 mM Mg++, ▢ , - Mg++.

266

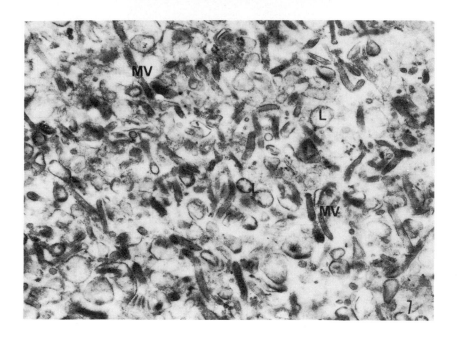

Fig. 7.
Electron microscopic survey of a light mitochondrial fraction. Microvilli
fragment (MV) and lysosome-like structures (L) are predominant
(X 16.000).

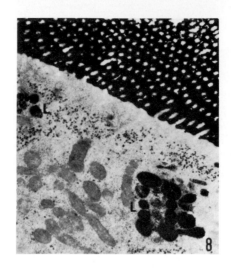

Fig. 8.
ACPase brush border and apical lysosomes (L) are heavily stained.
(β-glycerophosphate at pH 5) (X 10.030).

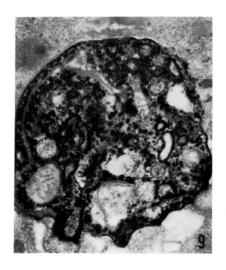

Fig. 9.
ACPase (β-glycerophosphate at pH 5). A complex cytolysome loaded
with reaction product (X 21.700)

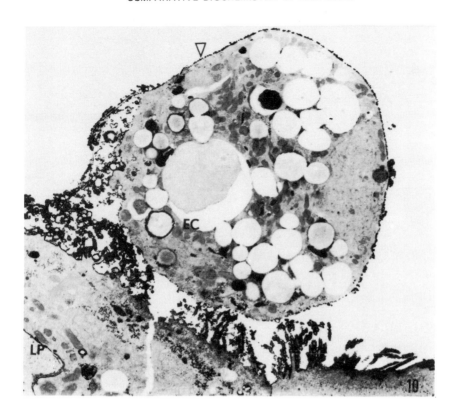

Fig. 10.

ACPase (β-glycerophosphate at pH 5). The plasma membrane (arrow) and lysosomes of an extruded cell (EC) are stained for enzymatic activity. A portion of the lateral membrane (LP) is also reactive (X 5.600).

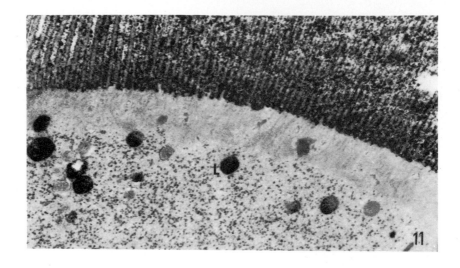

Fig. 11.
Reactivity towards G-6-P at pH 6.7. The lead phosphate precipitate is localized at identical sites as seen in fig. 8. L, apical lysosomes (X 13.700).

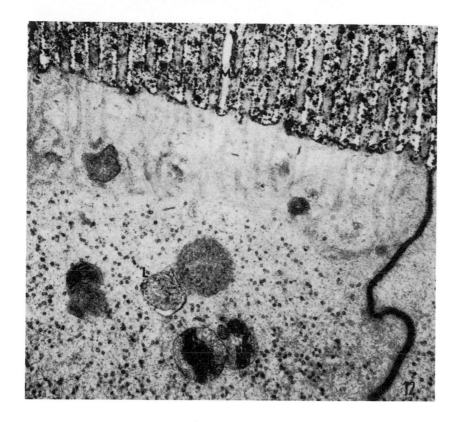

Fig. 12.

Reactivity towards AMP at pH 7.2. A moderate amount of precipitate is shown on the microvilli (MV). The lysosomes (L) present in the apical cytoplasm are almost completely devoid of reaction product (X 23.800).

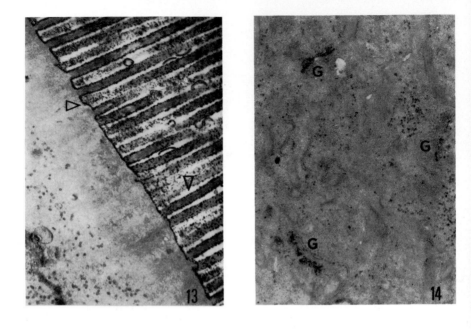

Fig. 13.
Reactivity against ADP at pH 8.8. The extraneous compartment of the plasma membrane and of the microvilli (arrow) is moderately stained (X 17.500).

Fig. 14.
Reactivity against ADP at pH 8.8. Some Golgi saccules (G) are filled with lead phosphate precipitate (X 11.130).

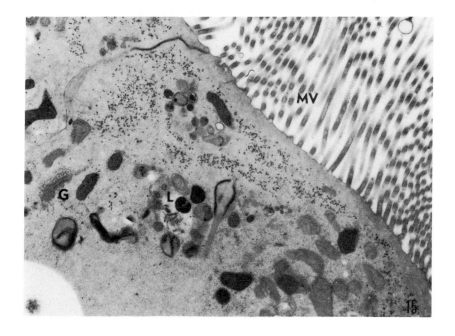

Fig. 15.

Alkaline phosphatase (β-glycerophosphate at pH 9). Neither brush border (MV) nor cytoplasmic organelles, as lysosomes (L), show the slightest activity (X 10.030).

18

FACTORS INFLUENCING THE MOVEMENT OF MATERIALS ACROSS THE INTESTINE OF *ASCARIS**

Calvin G. Beames, Jr. and Gary A. King
Department of Physiological Sciences
Oklahoma State University
Stillwater, Oklahoma 74074, U.S.A.

Adult *Ascaris* are free in the upper small intestine of their vertebrate host and feed on the intestinal contents which are largely fluid and rich in partially digested nutrients. The worm's cuticle is permeable to water, certain ions and some chemicals; but its intestine is undoubtedly the principal route for the absorption of nutrients (Fairbairn, 1957 and 1960; Lee, 1965; and Brand, 1966).

The intestine of *Ascaris* is a straight tube which is often divided into three regions, the anterior region, the mid region or mid-gut, and the posterior region. The wall of the intestine consists of a single layer of cells supported on its pseudocoelomic surface by a basement membrane (basal lamells). Microvilli cover the luminal surface of these cells and suggest an absorptive function. Studies of secretory activity (Lee, 1962) and determinations of the distribution of digestive enzymes (Carpenter, 1952) suggest that the anterior region is primarily secretory while the mid and posterior regions are involved mainly with absorption.

Until recently, very little attention has been given to the requirements or mechanisms for movement of materials across the intestine; and the data available still remains fragmentary. In 1969 Harpur determined that the organic acid end-products of the carbohydrate catabolism of *Ascaris* left the worm by the feces. His observations also suggested that there was selective reabsorption of organic acids which were excreted into the lumen of the intestine. Sanhueza, et al. (1968) measured the movement of several sugars across *Ascaris* mid-gut **in vitro.** They found that glucose

* The senior author's current investigations are supported in part by the NIH, U.S. Public Health Service Grant AI-06047.

and fructose moved across the intestine, the rate of entry of glucose being controlled partially by a process sensitive to phloridizin and dependent upon the presence of sodium ions in the lumen. With the condition they employed little or no absorption of galactose or 3-o-methylglucose was observed. They showed that specific processes for glucose transport existed on the luminal surface of the epithelial cells but hexoses were not moved across the intestine against a concentration gradient. In other studies, Castro and Fairbairn (1969) compared cuticular and intestinal absorption of glucose by *Ascaris.* They reported that isolated ribbons of the intestine absorbed glucose freely against a concentration gradient. With the conditions they employed 3-o-methylglucose was not accumulated against a concentration gradient.

Our initial interest in the movement of materials across the intestine of *Ascaris* was stimulated by the limited ability of the worm's tissues to synthesize long chain acids from acetate (Beames et al. 1967). Our results showed that acetate incorporation into lipids would account for only 12 % of the total lipid in the eggs passed each day by an adult female worm. It seemed reasonable that the majority of the lipids in the eggs were being obtained from the worm's diet and so a series of experiments were designed to test this possibility.

In our first experiments everted sacs of isolated *Ascaris* intestine were prepared and incubated in 5 ml of a palmitic-1-C^{14} acid-albumin solution for one hour at 37 C in a test tube apparatus similar to the one described by Crane and Wilson (1958). Chemical and radiochemical purity of the palmitic acid was determined by gas-liquid chromatography. The method described by Johnston (1958) was followed to prepare the fatty acid-albumin solution. An appropriate gas mixture was bubbled gently through the solution during the incubation period. Movement of fatty acid across the intestine was measured by standard radioactive tracer techniques. At the end of the incubation, the sac preparation was removed, washed 3 times in separate beakers of saline and blotted dry. Fluid (hemolymph) in the sac was collected and saved. The gut was sliced longitudinally, washed 3 times in saline and blotted dry. Total lipid was extracted separately from the hemolymph and gut and a 1 ml aliquot of fatty acid-albumin solution by following the procedure of Folch et al. (1951).

The results are presented in Table I. Movement of palmitic-1-C^{14} acid across the everted gut of *Ascaris* was measured under aerobic and anaerobic conditions and in the presence and absence of glucose and sodium iodoacetate. At the end of the experiment, under anaerobic conditions with added glucose, approximately 87 % of the total radio-

276

activity was in the luminal solution and 6 % was in the pseudocoelomic solution (hemolymph). When glucose was omitted from the anaerobic system, there was less radioactivity removed from the luminal solution. Similar results were observed when iodoacetate was added to the anaerobic system and when the system was aerobic. The percent radio-activity in the mid-gut was only slightly higher for the anaerobic system supplied with exogenous glucose than it was for the other experimental systems. The highest percent radioactivity in the pseudocoelomic solution was observed with the anaerobic system plus glucose.

Omitting glucose, adding iodoacetate or incubating the everted sac preparation in the presence of oxygen, had limited effect upon the movement of fatty acid into the mid-gut. On the other hand, the movement of fatty acid into the pseudocoelomic fluid did seem to be effected by factors which influenced the metabolic activity of the intestinal cells.

The results suggest that fatty acid moved into the cells via a passive process, but energy from the catabolism of carbohydrate was required for the movement of fatty acid into the pseudocoelomic fluid. In this regard the process was similar to that described for the movement of long chain fatty acids across the vertebrate intestine (Isselbacher, 1965).

Under normal conditions fatty acids in the luminal fluid of the intestine of *Ascaris* undoubtedly are present in complex with bile salts as micelles. To determine the effect of bile salt upon the movement of fatty acids across the worm's intestine we used sodium glycochenodeoxycholate and followed the procedure described by Hofmann (1963) to prepare micellar solutions. Everted sac preparations did not lend themselves to quantitative determinations, so the measurements were carried out with sacs of the mid-gut of *Ascaris* prepared and incubated as described by Beames (1971). The results are presented in Table II and the rates of movement of fatty acids complexed with albumin are included for comparison. The rate of movement was increased when the fatty acids were complexed with bile salt as micelles. In all instances the addition of glucose to the experimental system facilitated the rate of movement.

Experiments were carried out to determine the effect of the concentration of fatty acids in the luminal solution upon their rate of movement across the intestine. The results are presented in Figure 1. Increasing the concentration of fatty acid in the luminal solution markedly increased the rate at which the fatty acid moved from the luminal solution to the pseudocoelomic fluid.

Our experience with the effect of exogenous glucose and various gas phases upon the movement of long chain fatty acids prompted us to investigate the effect of these factors upon the movement of 3-o-methyl-gucose and other sugars across the intestine of *Ascaris.* The results were reported recently (Beames 1971). Movement of 3-o-methylglucose from the luminal solution to the pseudocoelomic fluid did occur when glucose was present and 95 % N_2 - 5 % CO_2 was the gas phase. The sugar was moved against a concentration gradient and the movement was direc-tional. When the gas phase was 95 % O_2 - 5 % CO_2 or 99+ % N_2, there was a marked reduction in the movement of the sugar. Similar measure-ments were obtained with fructose. With galactose very little movement occurred under any of the experimental conditions.

In summary, it appears that significant amounts of fatty acid do move across the intestine of *Ascaris.* The rate of movement is a function of the concentration of fatty acid in the worm's intestine. Movement is facilitated when the fatty acid is complexed with a bile salt and when glucose is present in the incubation medium.

Results reported by Sanhueza et al. (1968), Castro and Fairbairn (1969), and Beames (1971) suggest that certain sugars are actively transported by the intestine of *Ascaris*. The mechanism appears similar in many respects to what has been described for mammalian intestine. It differs from mammalian intestine, however, in that galactose is not transported.

Observations regarding certain factors which influence the movement of fatty acids and sugars across the intestine suggest that **in vitro** processes require exogenous glucose and CO_2 and are sensitive to high partial pressures of oxygen. These observations seem to agree with what is known of the worm's carbohydrate metabolism. This metabolism has been discussed in detail by Dr. Saz in this symposium and recently in an excellent review (Saz, 1971). The requirement for exogenous glucose appears to result from the limited endogenous carbohydrate present in the intestinal epithelium and the fact that it is rapidly depleted **in vitro.** The effect of oxygen upon the movement of materials across the intestine may be as follows. Under anaerobic conditions (which is the normal environment for adult *Ascaris*) a system of flavoproteins in the mitochondria of the worm's tissue shuttle electrons from reduced pyri-dine nucleotide to fumaric acid. The product is succinic acid. In this process ATP is formed from ADP and inorganic phosphate. When the tissue is exposed to oxygen, the shuttle of electrons to fumaric acid is reversed because a flavin oxidase present in the mitochondria shuttles electrons to molecular oxygen and forms hydrogen peroxide. Because there is little catalase in the tissues of *Ascaris,* (Laser, 1944) the hydrogen perioxide produced could accumulate to toxic levels. Further-more, the action of the flavin oxidase interferes with the catabolism of carbohydrates and the formation of nucleoside triphosphates. Hence, exposing the intestine to a high partial pressure of oxygen should, and apparently does, inhibit physiological activities requiring metabolic energy, such as the movement of long-chain fatty acids and sugars across the intestine. The requirement of CO_2 for the movement of materials can be explained by considering the terminal steps of glycolytic pathway of the tissues of *Ascaris*. These steps differ from those of mammalian tissues, bacteria, and yeast. In the worm, phosphoenolpyruvic acid is not converted to pyruvic acid. Instead, it is condensed with CO_2 to form oxaloacetic acid (Bueding and Saz, 1968). Under experimental conditions where CO_2 is removed from the system, carbohydrate metabolism would be inhibited. This, in turn, should, and apparently does, inhibit physiological processes that require energy derived from metabolic reactions.

In closing, I wish to emphasize that these data represent the **in vitro** situations. The observations are valuable, but care must be taken in using them to draw conclusions regarding the physiological potentials and limitations of the worm **in vivo.**

References

BEAMES, C.G., Jr., HARRIS, B.G. and HOPPER, F.A., Jr. 1967. Comp. Biochem. Physiol. **20,** 509.

BEAMES, C.G., Jr. 1971. J. Parasit. **57,** 97.

BUEDING, E. and SAZ, H.J. 1968. Comp. Biochem. Physiol. **24,** 511.

CARPENTER, M.F.P. 1952. The Digestive Enzymes of *Ascaris lumbricoides* var. suis: Their Properties and Distribution in the Alimentary Canal. Dissertation. Michigan Univ. Microfilms, Publ. no. 3729. Ann Arbor, Michigan. 183.

CASTRO, G.A. and FAIRBAIRN, D. 1969. J. Parasit. **55,** 13.

CRANE, R.K. and WILSON, T.H. 1958. J. Appl. Physiol. **12,** 145.

FAIRBAIRN, D. 1957. Exptl. Parasit. **6,** 491.

FAIRBAIRN, D. 1960. In Nematology (J.N. Sasser and W.R. Jenkins eds.). The University of North Carolina Press. Chapel Hill. 267.

FOLCH, J., ASCOLI, I., LEES, M., MEATH, J.A. and LEBARON, F.N. 1951. J. Biol. Chem. **191,** 833.

HARPUR, R.P. 1969. Comp. Biochem. Physiol. **28,** 865.

HOFMANN, A.F. 1963. Biochem. J. **89,** 57.

ISSELBACHER, K.J. 1965. Fed. Proc. **24,** 16.

JOHNSTON, J.M. 1958. Proc. Soc. Exp. Biol. Med. **98,** 836.

LASER, H. 1944. Biochem. J. **38,** 333.

LEE, D.L. 1962. Parasit. **52,** 241.

LEE, D.L. 1965. The Physiology of Nematodes. Oliver and Boyd. Edinburgh, 154 p.

SANHEUZA, P., PALMA, H.O.R., PARSONS, D.S., SALINAS, A. and OBERHAUSER, E. 1968. Nature. **219,** 1062.

SAZ, H.J. 1971. Am. Zool. **11,** 125.

Von BRAND, T. 1966. Biochemistry of Parasites. Academic Press, New York, 429 p.

Table 1.

Total activity recovered from luminal solution, mid-gut and pseudocoelomic solution after palmitic acid-1-C^{14}-albumin solution was placed on the luminal side of *Ascaris* mid-gut.

System	Percent of total activity after incubation base on initial activity ($\pm$ sd)		
	Luminal	Mid-gut	Pseudocoelomic
Anaerobic (95 % N_2 - 5 % CO_2)			
plus glucose	87.0 ± 2.8 (5)	6.9 ± 2.2 (5)	6.0 ± 1.8 (5)
minus glucose	92.3 ± 1.9 (5)	4.3 ± 1.1 (5)	3.4 ± 1.8 (5)
plus glucose and 0.1 M Iodoacetate	90.3 ± 2.2 (4)	5.2 ± 2.4 (4)	4.4 ± 2.14 (4)
Aerobic (95 % Air - 5 % CO_2)			
plus glucose	90.9 ± 1.0 (5)	6.5 ± 1.0 (5)	2.5 ± 0.6 (5)

The luminal solution contained 0.12 mM Na palmitate-1-C^{14} (102,167 dpm/m mole) in *Ascaris* saline containing 0.5 % Bovine albumin, pH 8.0. Solution on the pseudocoelomic side of the gut contained 0.12 mM Na palmitate in hemolymph. Each solution contained 0.04 M glucose unless otherwide indicated. Temperature 37 C. Time 60 minutes. The number in ( ) indicates the number of determinations.

Table 2.
Effect of Bile Salt upon the Movement of Various Fatty Acids Across *Ascaris* Mid-Gut (mμ moles/cm^2 ± sd).

Fatty Acid (0.20 mM)	Glucose in Medium (0.04 mM)	Fatty Acid Complexed With	
		Albumin	Glycochenode-oxycholate
Palmitic Acid	+	1.54 ± 0.54 (7)	8.19 ± 1.32 (8) 3
	−	0.84 ± 0.34 (3)	7.20 (2)
Oleic Acid	+	2.00 ± 0.33 (5)	9.16 ± 1.92 (6)
	−	0.97 (1)	5.68 (2)
Linoleic Acid	+	1.93 ± 0.67 (3)	5.74 ± 1.68 (6)
	−	0.34 (1)	2.06 (2)

Solution on the luminal side of the intestine contained 0.20 mM fatty-1-C^{14} acid (322,325 dpm/μmole) in 0.5 % Bovin albumin in saline, pH 8.0 or in 6.0 mM sodium glycochenodeoxycholate solution, pH 7.6. Solution on the pseudocoelomic side of the gut contained 0.20 mM fatty acid in hemolymph. Each solution contained glucose where indicated. Gas phase 95 % N$_2$ - 5 % CO$_2$. Temperature 37 C. Time 60 minutes. The number in ( ) indicates the number of determinations.

Fig. 1.

Each point represents an average of 6 determinations. The 0.1, 0.2 and 0.4 mM concentrations of fatty acid were complexed with sodium glycocheno-deoxycholate at concentrations of 4, 6 and 12 mM respectively. Glucose was added to the solutions to give an initial concentration of 0.4 mM. Gas phase 95 % N_2 - 5 % CO_2. Temperature 37 C. Time 1 hour.

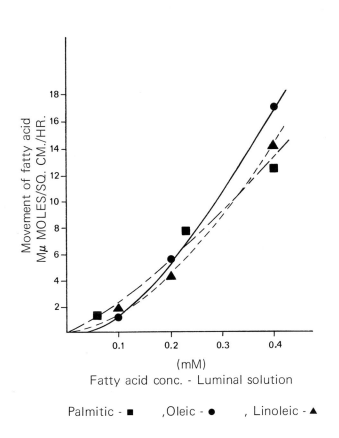

Palmitic - ■ ,Oleic - ● , Linoleic - ▲

19

ACID PHOSPHATASES IN THE INTESTINAL CELLS
OF TWO NEMATODE LARVAE:
ANISAKIS SP. AND *TRICHINELLA SPIRALIS*

E.J. Ruitenberg

Laboratory of Pathology,
National Institute of Public Health
Postbus 1, Bilthoven, The Netherlands

1. Introduction

In order to compare a number of enzyme histochemical properties of two nematode larvae: *Anisakis* sp. and *Trichinella spiralis* various enzyme systems were studied. Special attention was given to the alimentary tract of both larvae, since Borgers **et al.** (1970) noted an interesting histochemical pattern in the intestine of adult *Ascaris suum.* In this parasite acid phosphatase acitivity was observed in the brush border, while alkaline phosphatase was absent. The scope of our investigation was a) to study the histochemical distribution pattern of various phosphatases and leucine aminopeptidase in *Anisakis* sp. and *T. spiralis* larvae and b) to study the isoenzyme pattern of acid phosphatases in both larvae.

2. Histochemical properties of Anisakis sp. larvae

Anisakis sp. larvae may cause the so called herring worm disease (anisakiasis) in man (Van Thiel **et al.,** 1960; Van Thiel, 1962; Ruitenberg, 1970). This parasite belongs to the family Heterocheilidae Railliet and Henry. 1915 (Yamaguti, 1961). Davey (1971) described the genus in detail.

Anisakis sp. larvae lay normally encysted in the abdominal cavity of the herring. Sometimes, however, they are found in the musculature. When the herring is gibbed, most larvae are removed from the abdominal cavity. Those remaining and the ones penetrated into the musculature might reach the consumer. In man, who should be regarded as an accicental host, the penetrated larva may cause acute abdominal colics.

However, the disease can be efficiently prevented when the fish is treated adequately before it reaches the consumer (Ruitenberg, 1970).

Anisakis sp. larvae measure about 20 mm in length with a diameter of 0.4 mm. The larvae are transparent with exception of the white ventriculus ("stomach"). On the oral end a boring tooth is present, while a caudal spine (mucron) is located at the terminal portion of the larva.

The histologic features of the larva are: cuticle; musculature; lateral lines; alimentary canal and excretory organ. Since enzyme histochemical studies revealed differences in the musculature it appeared suitable to divide this tissue in 3 portions, i.e. a marginal rim, a peripheral and a central zone. The lateral lines are arranged in a longitudinal fashion on both sides of the larva. The alimentary canal is composed of the esophagus, the ventriculus ("stomach") and the intestinal tract. The epithelial cells of the alimentary canal are cylindric in shape with a basal nucleus. In the intestinal tract the luminar aspect has a brush border. The epithelial cells are supported by a basement membrane, which is referred to as a basal lamella. The excretory organ which has probably a secretory function (Ruitenberg and Loendersloot, in press) is located in the cranial part of the larva with an excretory pore situated ventrally of the boring tooth in front of the head.

In our experiments *Anisakis* larvae were used, which were provided by the Institute for Fishery Products (TNO), IJmuiden, The Netherlands (Director, Ir. J. van Mameren).

The larvae were collected from the abdominal cavity of freshly brought in herring. The larvae were placed in seawater diluted with tapwater (1:1) and sent to our Institute. Here the most active larvae were selected for the experimental work. No data could be obtained on the period elapsed between the time the herring was caught and the time the larvae arrived at our Institute. Since the transport - with a possible starvation of the larvae - might have affected the activity of the larvae an uncontrolled variable was introduced. Since no attempts were made to hatch the larvae **in vitro,** this variable had to be accepted.

For histochemical studies cryostat section (10 μthick) were cut and stained with hematoxylin and eosin (H and E), the periodic acid-Schiff (PAS) stain for among others neutral mucopolysaccharides and glycoproteins with and without diastase digestion. The following enzymatic reactions were studied: alkaline phosphatase (Burstone, 1958); acid phosphatase (Barka's method) (Barka and Anderson, 1965); 5-nucleotidase (Wachstein and Meisel, 1957) and adenosine triphosphatase (Wachstein and Meisel, 1957).

Previously changes in the enzyme histochemical properties of *Anisakis* larvae, penetrated into the rabbit stomach, were described (Ruitenberg and Loendersloot, 1971). In this study alkaline and acid phosphatase were stained with a combined method. The alkaline phosphatase was seen as a blue precipitate, the acid phosphatase activity as a red precipitate.

For the photographic demonstration of the (enzyme) histochemical reactions in this paper sections of larvae penetrated into the rabbit stomach were used. This was done since better pictures could be obtained from the penetrated larvae and no differences in (enzyme) histochemical reaction could be observed between fresh and penetrated larvae.

2.1. Results

Although starvation (transport of the larvae in modified seawater) might have influenced the activity of the enzymes, no differences in the enzyme histochemical properties of the larvae of various shipments were found.

Intestinal tract

Since no differences were observed in the histochemical reactions of the epithelial cells in the various parts of the intestinal tract, the results will be described together (table 1).

Brush border

All phosphatases except alkaline phosphatase were present and displayed a moderate to strong activity (Fig. 1).

Epithelial cells

Intracellularly a strong PAS activity was present (Fig. 2). By means of the diastase digestion technique this PAS-positive material proved to be glycogen. Of the phosphatases only adenosine triphosphatase showed some activity.

2.2. Discussion

In contrast to the findings in mammalian tissues is the absence of the alkaline phosphatase activity in the brush border (Jervis, 1963). Alkaline phosphatase activity is necessary for the resorption of nutrients from the intestinal lumen. For this purpose, however, the *Anisakis* sp. larva possesses 3 other phosphatases (acid phosphatase, 5-nucleotidase, adenosine triphosphatase) which show a very high activity in the brush border. Although no ultrastructural studies of the intestinal cells were performed, light microscopic observations and the results of the enzyme histochemical studies made it clear that the intestinal cells were covered with a brush border. This is in accordance with the description of the ultrastructural features of other nematodes as described by Jamuar (1966) for *Nippostrongylus brasiliensis,* by Jenkins and Erasmus (1969) for *Metastrongylus* sp. and by Lee (1969) for *Ancylostoma caninum.*

It is conceivable that the absence of alkaline phosphatase and the presence of acid phosphatase is due to the pH in the intestinal lumen. Also Jamuar (1966) observed acid phosphatase activity in the brush border of the intestinal cells of *Nippostrongylus brasiliensis.*

The presence of acid phosphatase, the absence of alkaline phosphatase and the presence of glycogen in the *Anisakis* sp. larvae are in accordance with the situation in the intestinal cells of adult *Ascaris suum* (Borgers **et al.**, 1970).

3. Histochemical properties of T. spiralis larvae

In order to compare the results obtained with *Anisakis* sp. larvae similar studies were conducted on *T. spiralis* larvae present in the tongue and diafragm of rats 1, 3, 12 and 24 months after infection with 1000 *T. spiralis* larvae.

A precise correlation of the enzymes with the morphological structures could only be made at 24 months after infection. For histochemical studies cryostat sections (10μ thick) of tongue and diaphragm were cut and stained with H and E and PAS. The following enzymatic reactions were studied: alkaline phosphatase (Burstone, 1958), acid phosphatase (Barka's method) (Barka and Anderson, 1965), and adenosine triphosphatase (Wachstein and Meisel, 1957).

T. spiralis larvae possess the following tissues: cuticle, musculature, dorsal, ventral and lateral lines, alimentary canal and genital primordium. The alimentary canal is composed of: esophagus (with stichosome cells around the caudal portion), mid-gut and hind-gut.

3.1. Results

The distribution pattern of the various phosphatases studied are presented in table II.

Intestinal tract

Both alkaline phosphatase and acid phosphatase are absent in the intestinal tract. Adenosine triphosphatase, however, is present, especially in the mid- and hind-gut.

Intracellularly a strong PAS activity was present. This material proved to be glycogen.

3.2. Discussion

In contrast to the observations made in *Anisakis* sp. larvae and in other nematodes acid phosphatase activity is lacking in the gut cells. In accordance with the observations made in other nematodes is the absence of alkaline phosphatase activity and the presence of glycogen.

The question then arose which enzyme system(s) would be available for the *T. spiralis* larvae for the transport of nutrients from the intestinal lumen to the intestinal cells and for the digestion of these nutrients.

4. Leucine aminopeptidase activity in Anisakis sp. and T. spiralis larvae

For this purpose the activity of leucine aminopeptidase (Pearse, 1961) was studied both in *Anisakis* sp. and *T. spiralis* larvae.

4.1. Results

In table III the distribution pattern of leucine aminopeptidase in the alimentary canal of *Anisakis* sp. and *T. spiralis* is presented.

In *Anisakis* sp. larvae a strong leucine aminopeptidase activity is present in the brush border of the intestinal cells.

In *T. spiralis* larvae a moderate activity was observed in the cells of the intestinal tract.

4.2. Discussion

In *Anisakis* sp. larvae leucine aminopeptidase, was observed exclusively in the brush border of the intestinal cells indicating a role in the transport of nutrients from the intestinal lumen.

In *T. spiralis* larvae leucine aminopeptidase was observed intracellularly in the cells of the intestinal tract. Due to the limited mangification of the light microscope it was not possible to observe whether this enzyme was also present in the brush border. However, for the digestion of nutrients *T. spiralis* is equipped with this enzyme.

From the therapeutic point of view this observation is interesting. Among the enzymes studied which are directly related with the digestion of nutrients only leucine aminopeptidase was present in *T. spiralis* larvae. It is conceivable that a specific therapy for trichinosis could be achieved by the treatment of infected individuals by means of an anti-leucine aminopeptidase serum. Further studies in order to examine this possibility are now underway in our laboratory.

5. Isoenzyme pattern of acid phosphatases in Anisakis sp. and T. spiralis larvae.

In order to obtain further information on the acid phosphatases of both *Anisakis* sp. and *T. spiralis* larvae a pilot study of the isoenzyme pattern by means of the polyacrylamide gel electrophoresis method was performed. For the source of the larvae see under sections 2 and 3. The *T. spiralis* larvae used were obtained from rats 6 weeks after infection with 1000 larvae. Since it was impossible to obtain isolated structures from the 2 parasites these studies were conducted with material from whole larvae. It should be borne in mind that in *Anisakis* sp. larvae a strong activity is present in the brush border of the intestinal cells, while some activity was observed in the musculature and a weak activity in the lateral lines. In *T. spiralis* larvae a strong activity of acid phosphatase could be demonstrated in the following structures: ventral, dorsal and lateral lines, stichosome cells and the genital primordium. Larval extracts were prepared according to Edwards et al. (1971). Electrophoresis of the extracts on acrylamide gel was performed largely according to Akroyd (1968). After electrophoresis the gel slab was incubated in the appropriate substrate mixture. For the demonstration of acid phosphatase activity Barka's method was employed (Barka and Anderson, 1965).

5.1. Results

In the extract of *Anisakis* sp. larvae a single band of acid phosphatase activity could be detected (Fig. 3).
In the extract of *T. spiralis* larvae two bands of enzyme activity could be demonstrated.

5.2. Discussion

From the last results it may be concluded that acid phosphatase in *Anisakis* sp. larvae do not possess isoenzymes. This is at variance with the situation in *Nippostrongylus brasiliensis* (Edwards et al., 1971). In this parasitic nematode a marked pattern was observed. *T. spiralis* larvae, however, possess two isoenzymes.

Conclusion

Our studies indicate that in the alimentary canal of *Anisakis* sp. larvae acid phosphatase is only present in the brush border of the intestinal cells. In *T. spiralis* larvae acid phosphatase activity could not be detected in the gut. Both nematode larvae lack alkaline phosphatase activity. Glycogen could be demonstrated in the intestinal cells of both larvae.

The isoenzyme pattern of acid phosphatase in *Anisakis* sp. larvae consisted only of one band, while in *T. spiralis* larvae two bands could be demonstrated.

Finally, leucine aminopeptidase might be an important enzyme for the transport and digestion of nutrients in both *Anisakis* sp. and *T. spiralis* larvae.

References

AKROYD, P. (1968). In chromatographic and electrophoretic techniques. Vol. II, 2nd edition. Ed. Smith, I. Heinemann, London.

BARKA, T. and ANDERSON, P.J. (1965). Histochemistry. Theory, practice and bibliography. Harper and Row, New York, Evanston and London.

BORGERS, M., VAN DEN BOSSCHE, H. and SCHAPER, J. (1970). J. Histochem. Cytochem. **18,** 519.

BURSTONE, M.S. (1958). J. Histochem. Cytochem. **6,** 322.

DAVEY, J.T. (1971). J. Helminth. **45, 51.**

EDWARDS, A.J., BURT, J.S. and OGILVIE, B.M. (1971). Parasitology **62,** 339.

JAMUAR, M.P. (1966). J. Parasit. **52,** 1116.

JENKINS, T. and ERASMUS, D.A. (1969). Parasitology **59,** 335.

JERVIS, H.R. (1963). J. Histochem. Cytochem. **11,** 692.

LEE, C.C. (1969). Exp. Parasit. **24,** 336.

PEARSE, A.G.E. (1961). Histochemistry. Theoretical and applied. (2nd edition). J. and A. Churchill, Ltd., London, U.K.

RUITENBERG, E.J., (1970). Anisakiasis. Pathogenesis, diagnosis and prevention. Thesis Utrecht.

RUITENBERG, E.J. and LOENDERSLOOT, H.J. (1971). Tijdschr. Diergeneeskunde **96,** 247.

RUITENBERG, E.J. and LOENDERSLOOT, H.J. (in press). Histochemical properties of the excretory organ of *Anisakis* sp. larvae. J. Parasitology.

THIEL, P.H. van, KUIPERS, F.C. and ROSKAM, R.Th. (1960). Trop. geogr. Med. **12,** 97.

THIEL, P.H. van (1962). Parasitology **52,** 16.

WACHSTEIN, M. and MEISEL, E. (1957). Am. J. clin. Path. **27,** 13.

YAMAGUTI, S. (1961). Systema helminthum. Vol. III. The nematodes of vertebrates. Interscience Publishers, Inc., New York.

Table 1.

Enzyme histochemistry of *Anisakis* sp. larvae

| | Phosphatases | | | |
	AIP	AcP	5-Nucl.	ATPase
Cuticle	−	−	±	±
Musculature				
marginal	−	++	+	+
peripheral	−	−	−	−
central	−	−	−	−
Lateral lines	−	±	−	±
Alimentary canal				
brush border	−	++++	++++	++
intracellular	−	−	−	±
cellular membrane	−	−	±	±
basal lamella	−	−	+	+
Excretory organ	−	−	++	+++

AIP: alkaline phosphatase; AcP: acid phosphatase; 5-Nucl.: 5-nucleotidase;
ATPase: adenosine triposphatase.

Table II.

Enzyme histochemistry of *T. spiralis* larvae

	Phosphatases		
	AlP	AcP	ATPase
Cuticle	—	—	—
Musculature			
peripheral	—	—	—
central	—	—	—
Lines			
dorsal, ventral	—	+++	—
lateral	—	+++	—
Alimentary canal			
esophagus	—	—	+
mid gut	—	—	+++
hind gut	—	—	+++
Stichosome cells	—	++	++
Genital primordium	—	+++	—

AlP: alkaline phosphatase; AcP: acid phosphatase; ATPase: adenosine triphosphatase.

Table III.

Leucine aminopeptidase in the alimentary canal of two Nematode larvae

	Anisakis **sp.**	*T. spiralis*
Esophagus	−	−
Gut		
brush border	++++	?
intracellular	−	++

Fig. 1.

Cross section of an *Anisakis* sp. larva, showing strong acid phosphatase activity in the brush border of the epithelial cells of the gut. The larva is located in the stomach of a rabbit. Cryostat section. Alkaline and acid phosphatase (X 144).

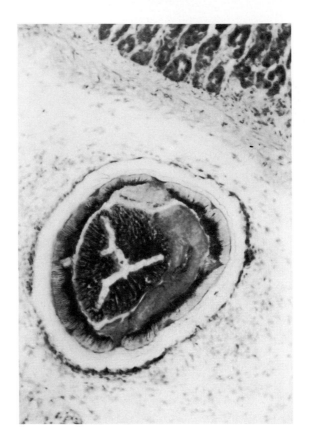

Fig. 2.
Cross section of an *Anisakis* sp. larva, showing PAS-positive material in the epithelial cells of the gut. The larva is located in the stomach of a rabbit. Cryostat section, PAS-stain (X 144).

Fig. 3.
One band of acid phosphatase activity from homogenate of *Anisakis* sp. larvae. Polyacrylamide gel technique.

20

STUDIES ON ACID HYDROLASES AND ON CATALASE OF THE TRYPANOSOMATID *CRITHIDIA LUCILIAE*

Y. Eeckhout*

Laboratoire de Chimie physiologique,
Université de Louvain, Belgium

The results reported in the present paper were obtained in the course of an investigation on the properties of lysosomal and peroxisomal marker enzymes in homogenates of the trypanosomatid *Crithidia luciliae*. Lysosomes, **i.e.** cytoplasmic organelles containing several acid hydrolases that display structure-linked latency (de Duve, 1959), have to our knowledge never been demonstrated in cell-free extracts of kinetoplastid flagellates. The semi-quantitative observations made by Seed, Byram and Gam (1967) and Molloy and Ormerod (1971) only suggest the existence of particle-bound acid phosphatase in extracts of blood trypanosomes of the *brucei* group. Several cytochemical data (see the recent reviews by Jadin, 1970 and Brooker, 1971) indicate that in many trypanosomatids acid phosphatase is present in the flagellar pocket or reservoir, in some Golgi saccules, in the spongiome surrounding the contractile vacuole and in some cytoplasmic granules. However, it must be recognized that cytochemical staining methods provide only qualitative results for a single acid hydrolase, namely phosphatase. Therefore, quantitative studies on the acid hydrolases are necessary to our knowledge of lysosomes and digestion in trypanosomatid flagellates. This was the object of the exploratory investigations outlined in the first part of this paper. A preliminary account of some of these experiments has been presented (Eeckhout, 1970).

The starting point of the second part of our work was the observation that young crithidiae displayed much higher (up to 100 times) catalase activity than old ones. The evolution of catalase activity with age of cells

* Chargé de Recherches du Fonds National de la Recherche Scientifique.

and the influence of compounds such as cycloheximide, hemin and hydroxyurea on this evolution disclosed, as will be seen, some interesting relations between catalase biosynthesis, hemin uptake and cell division.

Materials and methods

All the experiments were carried out on a strain of **Crithidia luciliae** kindly provided by Dr. M. Steinert (University of Brussels). Unless stated otherwise, the flagellate was cultivated axenically in Boné's medium (Boné and Steinert, 1956) wherein phosphate buffer was replaced by Tris buffer. Culture conditions were similar to those described by Steinert (1969). The cells were collected at the end of the logarithmic phase or at the beginning of the stationary phase by low speed centrifugation of the culture. The cells were washed twice with 0.15 M NaCl-0.010 M Tris (pH 7.5). Washed crithidiae were homogenized with an Ultra-Turrax homogenizer (20.000 r.p.m., two to four times 20 sec.); the suspension medium was either 0.25 M sucrose or 0.15 M NaCl, both ice-cold and buffered at pH 7.5 with 10 mM Tris. Cell-free extracts were prepared by low-speed centrifugation of the homogenate at $0°$. A soluble fraction was separated from a particulate fraction by centrifuging the cell-free extract at 3.10^6 **g**-min at $0°$.

The enzyme assays were performed at $25°$. Most of the techniques used were based on those described by de Duve **et al.** (1955) and Baudhuin **et al.** (1964). Determinations of β-fructofuranosidase were performed by the method of Dahlqvist (1964). Acid proteinase was assayed by measuring the acid-soluble radioactivity liberated from denatured ^{125}I-labelled hemoglobin. It was verified that the measured activities were proportional to the enzyme concentration and constant with incubation time. One unit of phosphatase and β-fructofuranosidase activity corresponds to the amount of enzyme that decomposes 1μmole of substrate/min. The unit of catalase activity is defined by Baudhuin **et al.** (1964). Total protein was determined by the method of Lowry **et al.** (1951). Cells were counted in a hemacytometer. The technique described by Miller and Palade (1964) was used for cytochemical localization of acid phosphatase with the optical microscope. A more complete description of enzyme assays and other techniques will be reported elsewhere.

Results

Subcellular localization of the acid hydrolases.

The four hydrolases investigated (β-glycerophosphatase, β-fructofuranosidase, proteinase and deoxyribonuclease) displayed optimum activity in the acid pH range. Less than 10 % of the acid phosphatase activity was found to be latent in cell-free extracts and more than 60 % of the enzyme was recovered in the soluble fraction. These results could not be ascribed simply to damaging effects of the homogenizer because it was observed (Fig. 1) that even intact cells displayed high free phosphatase activity; the same was found for acid β-fructofuranosidase (Fig. 2). Thus the major part of both hydrolases is accessible to the substrates in intact cells.

When the substrate concentration was lowered, the latency of acid phosphatase (Fig. 3) and β-fructofuranosidase (Fig. 4) in intact cells increased. This indicates that both enzymes are localized in a compartment of the cell that is accessible to the substrates but limited by a diffusion barrier.

Cytochemical detection of acid phosphatase (Fig. 5) was carried out on intact unfixed cells under conditions similar to the biochemical assay. Most cells displayed a single deposit of lead sulfide which appeared white under phase contrast in the part of the cell called flagellar pocket or reservoir. This observation confirms previous work done by other authors (see Jadin, 1970 and Brooker, 1971) on several trypanosomatid species. Moreover, our biochemical data indicate that at least 60 % of the total acid phosphatase is located in the flagellar pocket and its extensions. The alternative possibility that the enzyme is localized at the cell surface and that the deposition of lead phosphate in the flagellar pocket is the result of an artifact is rendered unlikely by the stainings observed when inorganic phosphate was substituted for the substrate in the Gomori medium (Fig. 5(3)): lead phosphate did not precipitate in the flagellar pocket but appeared principally at the cell surface.

Also worth mentioning here are some recent results (see Table 1) concerning two other acid hydrolases: proteinase and deoxyribonuclease. 60 to 70 % of both enzymes sedimented in a cell-free extract and Triton X-100 caused considerable activation and solubilization of the particle-bound acid proteinase. These preliminary observations suggest that, in contrast to acid phosphatase and β-fructofuranosidase, significant portions of proteinase and deoxyribonuclease are associated within subcellular particles, possibly lysosomes. More detailed studies are required to confirm the lysosomal nature and function of these particles.

On the other hand, more than 90 % of the catalase activity was found in the soluble fraction although, in contrast to acid phosphatase and β-fructofuranosidase, most of its activity was latent in intact cells. These data indicate that the major part of catalase is present in the soluble part of homogenates. However, they do not exclude the existence of catalase containing peroxisomes in the intact cell, as suggested by recent biochemical and cytochemical results obtained by Muse **et al.** (1970) on *Trypanosoma conorhini* and *Crithidia fasciculata.* It is indeed possible that the peroxisomes of *C. luciliae* either contain only a small fraction of the total catalase activity, or are damaged by the homogenization procedure.

Variation of enzyme activities with age of cells.

As shown by Fig. 6, the amounts of acid phosphatase and β-fructofuranosidase did not vary appreciably with age of cells. In contrast, catalase activity per cell increased markedly during the lag phase and decreased subsequently during the log phase. No catalase activity could be detected in the culture medium but increasing quantities of both acid hydrolases appeared in the extracellular medium. (Fig. 7). The release of acid hydrolases was probably not the consequence of cell lysis for the following reasons: (1) Rate of release per cell did not increase with age of culture. (2) It is unlikely that 30 % of the cells would have been lysed after 30 hours without morphological alterations and without simultaneous release of catalase. Therefore the release of acid hydrolases in the culture medium is most probably the consequence of a physiological process.

Studies of catalase.

The experiments reported here were carried out in order to determine some parameters of catalase variation with age of cells. It is clear from Fig. 8 that cycloheximide inhibited the increase of catalase acitivity. The effect of this inhibitor of protein synthesis, although not specific for catalase (cell division was also blocked), strongly suggests that catalase increase is the consequence of enzyme synthesis.

It is well established (see Lwoff, 1951) that practically all trypanosomatids are incapable of synthesizing hemin which is known to be a part of the catalase molecule. As therefore expected, it was found (Fig. 9) that no catalase synthesis took place when hemin was omitted from the culture medium. It is also evident from Fig. 9 that *C. luciliae* is able to use the hemin part of hemoglobin for synthesizing catalase.

In the presence of hydroxyurea (Fig. 10), which reversibly inhibits division of *C. luciliae* (Steinert, 1969), the increase of catalase activity at the initial rate lasted much longer. As shown in Table 2, the total catalase activities calculated per mg of cell protein were also increased by hydroxyurea. Moreover the drug had a striking effect when added during the log phase (Fig. 11): catalase activity which was decreasing increased again while cell division was inhibited. These results suggest that inhibition of cell division stimulated the net increase of catalase activity per cell either by stimulating catalase synthesis or by inhibiting catalase degradation. Preliminary experiments indicating that hydroxyurea inhibited catalase degradation are in favour of the second hypothesis.

Conclusions

The first part of the work presented above is dealing with some properties of the acid hydrolases of *C. luciliae.*

Acid β-glycerophosphatase- the most popular marker enzyme for lysosomes- did not display the classical lysosomal properties: sedimentability and latency. But assays carried out on intact cells showed that these "abnormal" properties could not be due solely to the damaging effects of the Ultra-Turrax homogenizer. Indeed it was observed that intact cells, incubated in such conditions as to respect the integrity of cell structure, displayed more than 60 % free phosphatase activity. Similar results were obtained with β-fructofuranosidase. Cytochemical studies on acid phosphatase, carried out under conditions similar to those of the biochemical assays on intact cells, indicate that the flagellar pocket or reservoir is the site of localization of the enzyme. Other authors (see Brooker, 1971 and Creemers and Jadin, 1970) have previously detected acid phosphatase in the reservoir of several trypanosomatids but quantitative data were lacking. The occurence of a β-fructofuranosidase (measured at pH 7.6) in the cell membrane has been suggested by Cosgrove (1963) from his quantitative observations on sucrose and raffinose utilization by *C. luciliae*. Our enzymatic assays reveal that at least 60 % of acid phosphatase and β-fructofuranosidase activity are located in the reservoir of *C. luciliae.* The influence of substrate concentration on the activity of both enzymes in intact cells points to the presence of a diffusion barrier between the reservoir and the extracellular environment. It is possible that the desmosome- like structures linking the flagellum to the membrane of the anterior part of the reservoir (Anderson and Ellis, 1965; Brooker, 1970; Creemers and Jadin, 1970) represent the structural equivalent of this diffusion barrier. Important amounts of phosphatase and β-fructofuranosidase were found to be released in the culture medium. The kinetics of this release and the fact that catalase did not

301

appear simultaneously in the medium allow us to exclude cell lysis as the only cause of enzyme escape. If, as suggested by Vickerman (1969) and Brooker (1971), lysosomal enzymes are also secreted by bloodstream trypanosomes, they may contribute to the antigeniticity and/or the pathogenicity of the parasite. The exact physiological function of the reservoir remains an open question although recent electronmicroscopic data (see Preston, 1969 and Brooker, 1971) strongly suggest that this permanent invagination of the cell membrane is, together with the neighbouring cytostome, a site of pinocytosis. As proposed by Preston (1969), the acid hydrolases released in the flagellar pocket may initiate digestion before pinocytosis. On the other hand, the reservoir is the site of discharge of the contractile vacuole and it could also be a site of cellular defaecation (Brooker, 1971). As in *Tetrahymena pyriformis* (Müller, 1970), there are perhaps two populations of acid hydrolases in *C. luciliae:* the first, comprising phosphatase and β-fructofunanosidase would be mostly secreted through the flagellar pocket in the extracellular environment, while the second, comprising proteinase and deoxyribonuclease, would remain mostly inside the vacuolar apparatus as suggested by our preliminary results.

The second part of our investigation has been devoted to the evolution of catalase activity during growth and ageing of *C. luciliae* in culture. The rapid and striking increase of catalase activity during the lag phase was inhibited by cycloheximide and thus appeared to be the result of enzyme synthesis, the hemin part being taken from the medium. The subsequent decrease of activity was inhibited by agents such as hydroxyurea that inhibited cell division.

Ryley (1955) reported that *Strigomonas oncopelti,* which lacks the requirement for exogenous hemin, was catalase-negative. Wertlieb and Guttman (1963) observed that this flagellate had about the same catalase activity as the other monoxenous insect trypanosomatids when it was grown in hemin-containing medium.

It is tempting to speculate about the relation between catalase and cell division and about the function of catalase, but further experiments are needed to find out, for example, if catalase could be a storage form of hemin to be used for the synthesis of other hemoproteins (cytochromes?) required for cell division, and if the decline of catalase activity has anything to do with ageing of cells.

Acknowledgments

The author wish to thank Drs. Christian de Duve, Pierre Jacques and Gilbert Vaes for their helpful criticism and suggestions during the course of this study and the preparation of the manuscript. The very able assistance of Miss Josée Wille is also gratefully acknowledged. This work was supported in part by an EMBO postdoctoral fellowship.

References

ANDERSON, W.A. and ELLIS, R.A. (1965). J. Protozool., **12,** 483.
BAUDHUIN, P. BEAUFAY, H., RAHMAN-LI, Y., SELLINGER, O.Z., WATTIAUX, R., JACQUES, P. and de DUVE, C. (1964). Biochem. J., **92,** 179.
BONE, G.J. and STEINERT, M. (1956). Nature, **178,** 308.
BROOKER, B.E. (1970). Z. Zellforsch., **105,** 155.
BROOKER, B.E. (1971). Z. Zellforsch., **116,** 532.
COSGROVE, W.B. (1963). Exper. Parasitol., **13,** 173.
CREEMERS, J. and JADIN, J.M. (1970). Septième Congrès International de Microscopie Electronique, Société Française de Microscopie Electronique, Paris, p. 391.
DAHLQVIST, A. (1964). Anal. Biochem., **7,** 18.
de DUVE, C. (1959). In Subcellular Particles, p. **128,** ed. Hayashi, T. New York: Ronald Press.
de DUVE, C., PRESSMAN, B.C., GIANETTO, R., WATTIAUX, R. and APPELMANS, F. (1955). Biochem. J., **60,** 604.
EECKHOUT, Y. (1970). Archiv. Internat. Physiol. Biochim., **78,** 993.
JADIN, J.M. (1970). Thèse, Faculté des Sciences, Université de Paris.
LOWRY, O.H., ROSEBROUGH, N.J., FARR, A.L. and RANDALL, R.J. (1951). J. Biol. Chem., **193,** 265.
LWOFF, M. (1951). In Biochemistry and Physiology of Protozoa, vol. 1, p. 129, ed. Lwoff, A. New York: Academic Press.
MILLER, F. and PALADE, G.E. (1964). J. Cell. Biol., **23,** 519.
MOLLOY, J.O. and ORMEROD, W.E. (1971). J. Protozool., **18,** 157.
MULLER, M. (1970). J. Protozool., **17,** suppl. 13.
MUSE, K.E., BAYNE, R.A. and ROBERTS, J.F. (1970). J. Parasitol., **56,** 244.
PRESTON, T.M. (1969). J. Protozool., **16,** 320.
RYLEY, J.F. (1955). Biochem. J., **59,** 353.
SEED, J.R., BYRAM, J. and GAM, A.A. (1967). J. Protozool., **14,** 117.
STEINERT, M. (1969). FEBS Letters, **5,** 291.
VICKERMAN, K. (1969). J. Protozool., **16,** 54.
WERTLIEB, D.M. and GUTTMAN, H.N. (1963). J. Protozool., **10,** 109.

Table 1.
Influence of Triton X-100 on sedimentability of acid hydrolases[1]

Enzyme	Sedimentable activity (% of total)	
	Control	Triton X-100 added
Phosphatase	38 (102)	30 (107)
Proteinase	73 (132)	12 (107)
Deoxyribonuclease	67 (93)	35 (−)
Protein	39 (91)	30 (91)

(1) A small volume of 10 % Triton X-100 was added to a cell-free extract to give a final concentration of 0.1 %; the same volume of water was added to another portion of the same extract, to serve as control. Samples of both preparations were centrifuged at 3.10^6 **g** - min. The cell-free extracts and both fractions were assayed for their total enzyme activities. Recoveries (activities of sediment + soluble fraction in % of activity of the extract) are given in parentheses. Acid proteinase displayed a marked latency in the control sediment: activity determined in the absence of Triton X-100 was 43 % of total activity measured in the presence of 0.1 % Triton X-100; Triton X-100 had no effect on the activity of the soluble fraction.

Table 2.
Influece of hydroxyurea on catalase activity [1]

Age of culture (hours)	Controls		Hydroxyurea	
			50 μg/ml	200 μg/ml
24	107	110	142	213
48	47	48	126	274

(1) Values (milliunits/mg protein) were calculated from the experiment illustrated on Fig. 10.

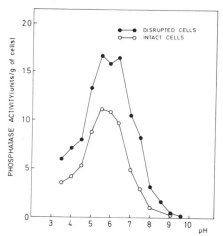

Fig. 1.
Phosphatase activity of intact and disrupted cells at various pH values. 0.25 ml of a suspension of cells (50 mg wet weight per ml, in 0.15 M NaCl - 0.010 M Tris pH 7) were incubated 10 min. at 25° at the indicated pH values in a final volume of 0.5 ml containing 0.05 M β-glycerophosphate, 0.1 M Tris and 0.1 M acetate. Cells were completely disrupted by three successive cycles of freezing (in a mixture of propan-2-ol and solid CO_2) and thawing.

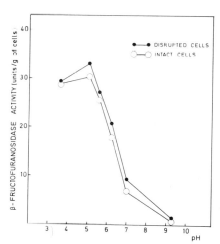

Fig. 2.
β-fructofuranosidase activity of intact and disrupted cells at various pH values. 0.1 ml of a suspension of cells (10 mg wet weight per ml) were incubated 10 min. at 25° at the indicated pH values in a final volume of 0.2 ml containing 12.5 mM sucrose, 0.1 M Tris and 0.1 M acetate. Suspension medium and cell disruption: see legend to Fig. 1.

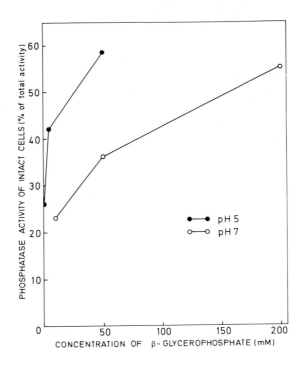

Fig. 3.
Influence of substrate concentration on acid phosphatase activity of intact cells. Total activity measured in the presence of Triton X-100 at a final concentration of 0.1 % (w/v) (experiment at pH 5) or after three cycles of freezing and thawing (experiment at pH 7).

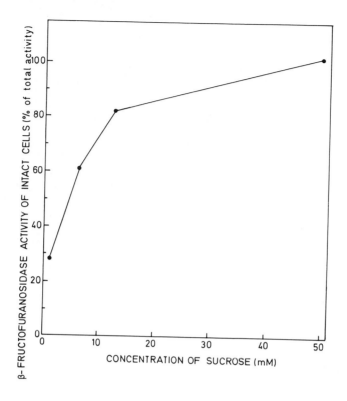

Fig. 4.
Influence of substrate concentration on β-fructofuranosidase activity of intact cells at pH 7. Total activity was measured on cells disrupted by three successive cycles of freezing and thawing.

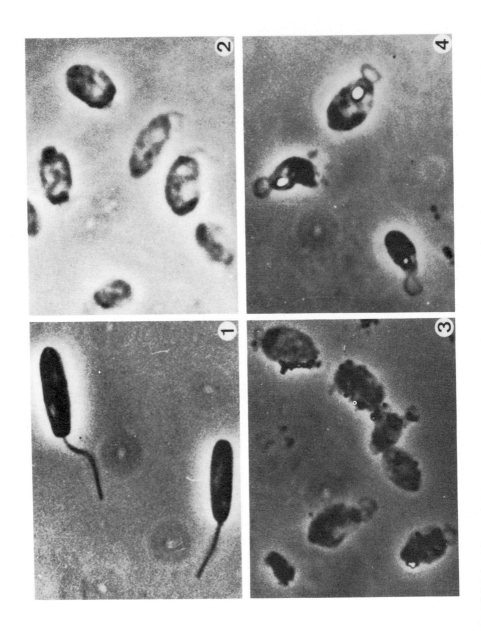

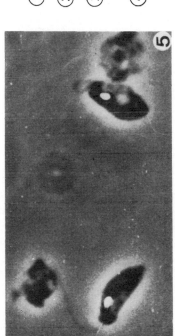

① LIVING CELLS

② GOMORI WITHOUT SUBSTRATE

③ GOMORI WITHOUT SUBSTRATE
   + PHOSPHATE

④ ⑤ GOMORI WITH SUBSTRATE

10μ

Fig. 5.
Cytochemical localization of acid phosphatase. Intact cells were incubated 10 min. at 25° either (4 and 5) in Gomori medium containing 3 mM Pb(NO₃)₂, 10 mM β-glycerophosphate, 0.15 M NaCl and 0.05 M acetate buffer, pH 5 or (2) in the same medium without β-glycerophosphate, or (3) in medium 2 supplemented with 0.18 mM phosphate.

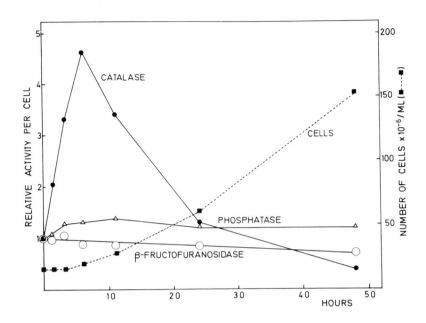

Fig. 6.
Variation of enzyme activities with age of cells. Culture was inoculated at zero time with 16.10^6 cells/ml; glucose was replaced by glycerol (20 mM) in boné's medium in order to measure β-fructofuranosidase activity in the medium. Samples of the culture were taken under sterile conditions at various times. An aliquot of each sample was centrifuged, the cell-free supernatant was poured off and the pellet of cells was resuspended in fresh ice-cold culture medium containing Triton X-100 0.5 % (w/v). The results shown were obtained on the resuspended cells. At zero time, activity (milliunits/10^6 cells) was 0.104 for catalase, 3.28 for acid phosphatase and 3.58 for acid β-fructofuranosidase.

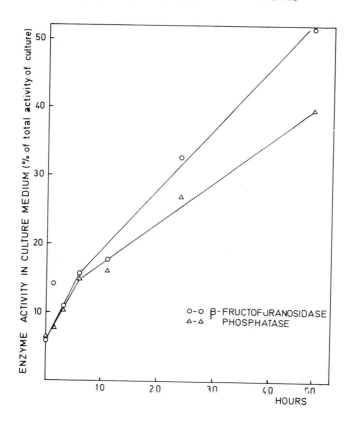

Fig. 7.

Variation of enzyme activities in the medium. Same experiment as that of Fig. 6. Enzyme activities were determined on the whole culture (cells + medium), on the medium and on the centrifuged cells. Results are expressed as activities of the medium in % of the activities in the whole culture (cells + medium). All recoveries were between 94 and 110 %.

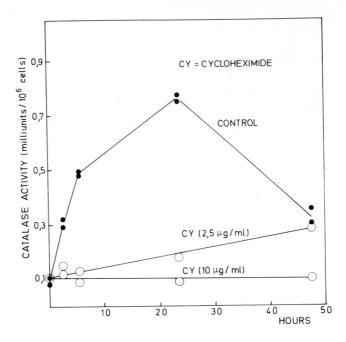

Fig. 8.
Influence of cycloheximide on catalase synthesis. Culture inoculated at zero time with 10^7 cells/ml. Hemin was replaced by hemoglobin (0.1 mg/ml) in the culture medium.

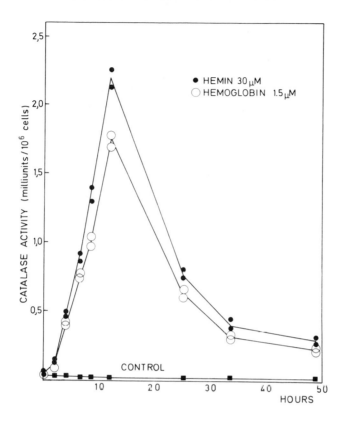

Fig. 9.
Influence of hemin and hemoglobin on catalase synthesis. Cultures inoculated at zero time with 9.10^6 cells/ml. Controls: Boné's medium without hemin. Number of cells $\times 10^{-6}$/ml after 49 hours: 33 (controls), 249 (hemin), 186 (hemoglobin).

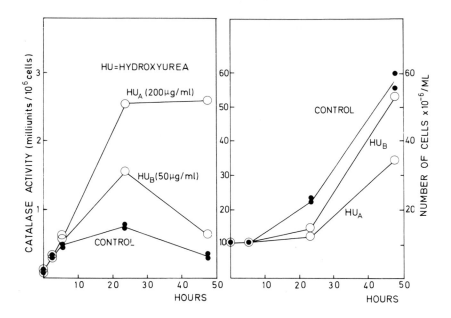

Fig. 10.
Influence of hydroxyurea on catalase synthesis. Hemin was replaced by hemoglobin (0.1 mg/ml) in the culture medium.

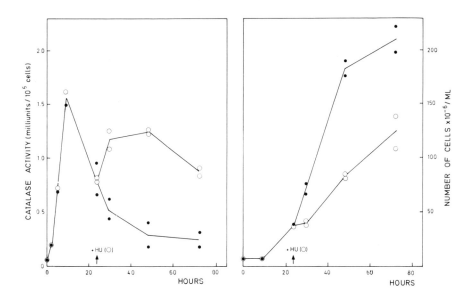

Fig. 11.
Effect of hydroxyurea (HU) added during the log phase. Final concentration of hydroxyurea: 0.2 mg/ml. Hemin was replaced by hemoglobin (0.1 mg/ml) in the culture medium.

21

STRUCTURAL AND BIOCHEMICAL CHANGES IN NIPPOSTRONGYLUS BRASILIENSIS DURING DEVELOPMENT OF IMMUNITY TO THIS NEMATODE IN RATS.

D.L. Lee*

Houghton Poultry Research Station,
Houghton, Huntingdon.

Until very recently most work on immunity to parasitic animals has been on the effect of the parasite on the host and very little attention has been paid to changes in the parasite brought about by the immune response of the host.

The immune response of rats to the nematode *Nippostrongylus brasiliensis* has been widely used as a model system for studying immunity to nematodes (see Ogilvie & Jones, 1971) as rats develop a strong active immunity to this nematode in a relatively short time.

The life cycle is briefly as follows: eggs passed in the faeces of the rat hatch in a faecal/soil mixture and the larvae feed upon bacteria in the mixture. The third stage larva migrates to the surface, attaches the loosened cuticle of the second stage larva to the substratum and uses this sheath, which is usually open at the front end, as a refuge. This third stage larva, which is the infective stage, penetrates the skin of the host, migrates to the lungs, where it moults to the fourth stage, and then migrates via the trachea and oesophagus to the small intestine where it lives between the villi. Here it moults to the adult stage. Most worms reach the intestine by the third day after infection. The adult worms, which measure 2-6 mm. in length, begin to pass eggs about 6 days after infection and this continues for several days, with a gradual and then rapid reduction in numbers from the 9th or 10th day. Most of the nematodes are expelled from the 12th day to the 15th day after infection, although a small number of worms, usually males, do survive for a longer time. These rats then have a strong resistance to further infections.

* Present address: Department of Pure and Applied Zoology, The University of Leeds, Leeds.

Ogilvie & Hockley (1968) and Lee (1969a) have studied the structure of *N. brasiliensis* during a primary infection and noted marked changes in the structure of these worms as the host became immune. Ogilvie & Hockley (1968) showed that the intestinal cells of female worms taken from rats 6 to 8 days after infection were similar in appearance to those described by Jamuar (1966). Granular endoplasmic reticulum filled most of the cell; the mitochondria were arranged around the periphery of the intestine; large dense sphaerocrystals and a few small vacuoles were also present. Lee (1969a) confirmed this for both male and female worms taken 5 and 10 days after infection and gave a much more complete description of the intestine (Fig. 1).

Ogilvie & Hockley (1968) showed that from about the 14th day after infection the intestinal cells of female worms were extensively vacuolated, the endoplasmic reticulum was swollen, the mitochondria were slightly altered in appearance but the microvilli appeared undamaged and no immune precipitate was observed in the lumen. Lee (1969a) showed that in 12 day old worms (both sexes) there was a decrease in amount of granular endoplasmic reticulum a decrease in number of secretory granules, and large irregular areas containing a fine filamentous material had appeared. In 13 day old worms the granular endoplasmic reticulum was collected into discrete areas of the cell, fewer mitochondria were present and there were very few secretory granules. The lumen of the intestine contained a mucus-like material dispersed throughout the lumen but not congregated against the microvilli. In the intestinal cells of 15 to 16 day old worms the cytoplasm was much less dense, there was a big increase in the number and size of lipid droplets, a big reduction in the amount of granular endoplasmic present, and large numbers of vacuolated areas had appeared. The lumen of the intestine of many individuals was full of cell debris, presumably of host origin and this may mean that the worms have a reduced ability to digest their food. The changes begun in the 15 to 16 day old worms were more marked in 19 day old worms (Fig. 2).

Lee (1969a) also studied changes in the so-called excretory glands (now shown to be a form of digestive gland (Lee 1969b, 1970) and called sub-ventral or exo-digestive glands) the body wall and the gonads of male and female worms. There was no apparent change in the body wall and the exo-digestive glands until the 15th day after infection when large numbers of big lipid droplets were found in the cytoplasm (Fig. 4). The glands appeared to be functioning normally. Noticeable changes had taken place in the male reproductive tract of 13 day old worms. There was resorption and breakdown of spermatozoa by the walls of the seminal vesicle and the anterior vas deferens. The spermatozoa inside these cells were surrounded by multi-laminate membranes which were probably lysosomes and were in varying stages of disruption (Fig. 3). This also occured in 15 day and 19 day old worms but large numbers of lipid droplets were also present in the epithelial cells of the reproductive tract and in the spermatocytes. In the female reproductive tract of 15 day and 19 day old worms large numbers of lipid droplets had appeared in the oocytes and many lysosome-like bodies were present. The amount of lipid in the epithelial cells had also increased. Few eggs were present in the uterus.

These changes in *N. brasiliensis* can be correlated with the onset and development of immunity in the host. Ageing is not responsible for these changes (Ogilvie & Hockley, 1968). Most changes occured in the cells of the intestine but changes also occured in other tissues of the body. One noteable change is the apparent increase in the amount of lipid in the 15 day old worms. This could be an actual increase in the amount of lipid or a change in the nature of the lipids once the immune reaction had begun to affect the parasite. Lee (in press) has studied the lipids of male and female *N. brasiliensis* 7 and 15 days after infection to see if these changes could be confirmed by means of biochemical techniques. There is an increase in total lipid from 10.9 % $\pm$ 2 at day 7 to 15.5 % $\pm$ 2 at day 15 in females (42 % increase) and from 8.4 % $\pm$ 1 to 14.6 % $\pm$ 2 in males (74 % increase). The biggest changes from 7 to 15 days are in the neutral lipids (Table 1).

These changes in structure and lipid composition are almost certainly secondary changes brought about by a primary effect of the host on the parasite. Lee (1969a) suggested that one effect of the immune response may be to interfere with the normal behaviour patterns of the nematode. If the immune response over-rides the thigmokinetic behaviour of the nematodes they will be driven out of the mucosa into the lumen of the intestine and this will result in a movement along the intestine as they try to find a more suitable environment. Brambell (1965) has shown that this migration does occur. Once away from the mucosa the worms will then be unable to feed effectively and they will also have moved from an aerobic to an anaerobic environment. This would result in interference with lipid metabolism and would explain the build-up of triglycerides in the worms.

Chandler (1937) suggested that the immune response of the host may affect the nutrition of *N. brasiliensis;* anti-enzymes could be developed by the host and these would inhibit the activity of the enzymes by means of which this nematode digests and assimilates the hosts tissues.

Lee (1969b; 1970) has shown that the oesophageal glands and the exo-digestive glands of *N. brasiliensis* contain a number of enzymes, some of which could be used in the digestion of host tissues and may act as antigens. One of the enzymes found by Lee (1970) in the exo-digestive glands was cholinesterase. The role of this enzyme in a mainly histolytic secretion is not known but Lee (1970) suggested that it may play a role in maintaining the position of the nematode in the mucosa. This has been supported by Ogilvie & Jones (1971). Cholinesterase is found in the venom of some snakes and in the salivary secretion of some biting invertebrates, such as leeches, and seems to play a role in paralysing the prey or inhibiting the pain receptors at the site of the bite. *N. brasiliensis,* which causes ulceration of the hosts mucosa, could be thrown out by local spasm of the mucosa as a result of this ulceration. If, however, one role of the cholinesterase is to inhibit the nerve endings of the mucosa in this region, or to affect the neuro-muscular transmission in that region of the intestine, then this will stop such local spasms and may also slow down peristalsis. Symons (1966) has shown that the passage of food through the upper jejunum is slower in rats infected with this nematode. Acetylcholinesterase is present in large amounts in *N. brasiliensis* (Sanderson, 1969) and it is thought to be released by the worms (Sanderson & Ogilvie, 1971).

Edwards, Burt & Ogilvie (1971) have studied this, and other enzymes, in normal, immune damaged (14 days after infection) and in worms grown in reinfected rats (adapted worms) and have found that isoenzymes of acetylcholinesterase showed changes in damaged and adapted worms and that these could be related to the effects of immunity.

Using vertical flat-slab acrylamide gel electrophoresis they found seven bands staining for esterase in all worms examined (Fig. 5). Three of these in normal worms and five in immune damaged worms (Bands A-C) were identified as acetylcholinesterases.

Compared with normal (8 day) worms, there was an increase in bands B and C in 10, 12 and 14 day old worms. Worms removed from reinfected rats showed rather similar esterase bands to those of damaged worms but band A, which was present in normal worms and not in damaged worms, was greatly increased in adapted worms. Edwards **et al.** (1971) suggested that antibodies act on the acetylcholinesterase isoenzymes of *N. brasiliensis.* In both immune damaged and in adapted worms, isoenzymes B and C are increased and B^1 and B^2 appear (Fig. 5). In damaged worms, isoenzyme A disappears at about the time expulsion of the worm occurs but worms grown in reinfected hosts produce much more isoenzyme A than normal worms (Fig. 5). It is suggested that these adapted worms are more resistant to the action of antibodies against their acetylcholineste-rases, take much longer to become damaged and so have an enhanced resistance to the expulsion mechanism.

These results have shown that a lot of information on the effect of the immune response on parasitic worms can be obtained by careful study of the structure of the worm, examining the behaviour of the worm and by studying the biochemistry and physiology of the worm at different periods of time after infection. It will be interesting to see if similar results are forthcoming on other nematodes which cause a well marked immunity in their hosts or which appear to evade the immune response of the host.

References

BRAMBELL, M.R. (1965). Parasitology **55,** 313.

CHANDLER, A.C. (1937). American Journal of Hygiene **26,** 309.

EDWARDS, A.J., BURT, J.S. & OGILVIE, B.M. (1971). Parasitology **62,** 339.

JAMUAR, M.P. (1966). Journal of Parasitology **52,** 1116.

LEE, D.L. (1969a). Parasitology **59,** 29.

LEE, D.L. (1969b). Symposia of the British Society for Parasitology **7,** 3.

LEE, D.L. (1970). Tissue & Cell **2,** 225.

LEE, D.L. (in press). Changes in adult *Nippostrongylus brasiliensis* during the development of immunity to this nematode in rats. 2. Total lipids and neutral lipids. Parasitology **63.**

OGILVIE, B.M. & HOCKLEY, D.J. (1968). Journal of Parasitology **54,** 1073.

OGILVIE, B.M. & JONES, V.E. (1971) Experimental Parasitology **20,** 138.

SANDERSON, B.E. (1969). Comparative Biochemistry and Physiology **29,** 1207.

SANDERSON, B.E. & OGILVIE, B.M. (1971). Parasitology, **62,** 367.

SYMONS, L.E.A. (1966). Experimental Parasitology **18,** 12.

Table 1.
Total lipids and neutral lipids of male and female *Nippostrongylus brasiliensis* during the course of a primary infection in rats.

	Age of nematodes (days)			
	7 ♀*	7 ♂*	15 ♀+	15 ♂+
Total lipid as % of dry wt.	$10.9 \pm 2\%$	$8.4 \pm 1\%$	$15.5 \pm 2\%$	$14.6 \pm 2\%$
Some neutral lipid components expressed as % of total neutral lipid.				
Triglyceride	13	0.5	24	10
Free fatty acids	41	68	32	37
Cholesterol	16	16	21	28
Mono- and diglycirides	20	14.4	21	15

* Average of five experiments.

+ Average of six experiments.

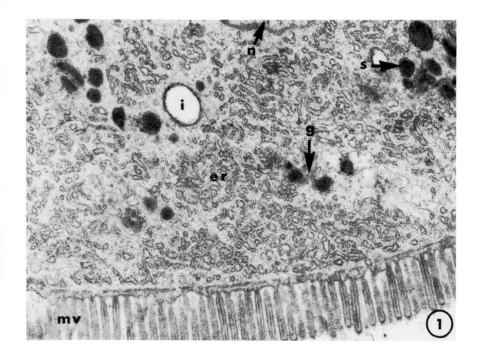

Fig. 1.
Electron micrograph of a section through the intestine of a 10-day-old female *Nippostrongylus brasiliensis.* (X 22.000). (From Lee, 1969a: reproduced with the permission of Cambridge University Press). **er,** endo plasmic reticulum; **g,** Golgi apparatus; **i,** crystalline inclusion; **mr,** microvilli; **n,** nucleus; **s,** secretory granule.

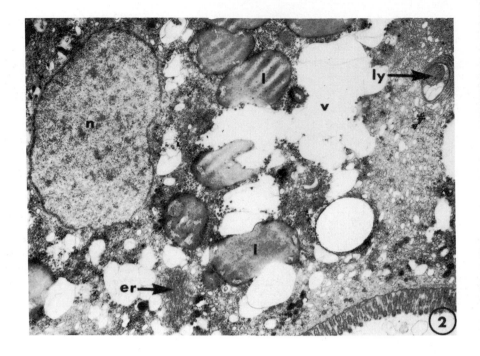

Fig. 2.
Electron micrograph of a section through the intestine of a 19-day-old female *Nippostrongylus brasiliensis*. (X 10.000). (From Lee, 1969a: reproduced with the permission of Cambridge University Press). **er,** endoplasmic reticulum; **l,** lipid; **ly,** lysosome; **n,** nucleus; **v,** vacuole.

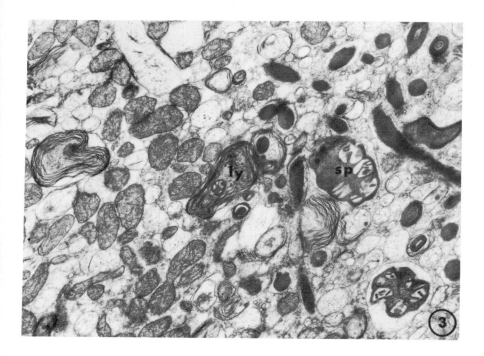

Fig. 3.
Section through the seminal vesicle of a 13 day-old-male to show spermatozoa in the walls of the seminal vesicle. The spermatozoa are in varying stages of disruption within lysosome-like structures (X 20.000). (From Lee, 1969a: reproduced with the permission of Cambridge University Press). **ly**, lysosome; **sp**, spermatozoon.

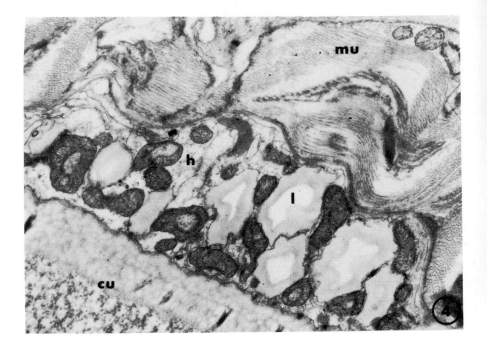

Fig. 4.
Electron micrograph of a section through the body wall of a 16-day-old male to show the accumulation of lipid droplets in the hypodermis (X 20.000). (From Lee, 1969a: reproduced with the permission of Cambridge University Press). **cu**, cuticle; **h**, hypodermis; **l**, lipid; **mu**, muscle.

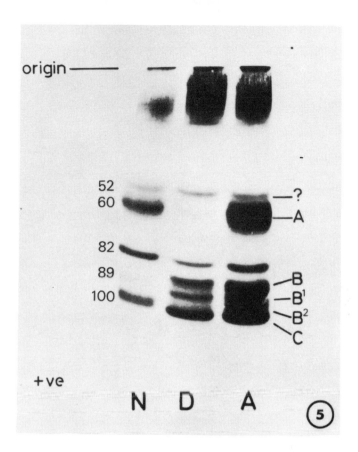

Fig. 5.
Comparison of the esterase pattern present in normal (N) damaged (D) and adapted (Ad) *Nippostrongylus brasiliensis.* Note that acetylcholinesterases B and C increase and B^1 and B^2 appear in D and Ad worms, acetylcholinesterase A is reduced in damaged (D) and greatly increased in adapted (AD) worms. Unlabelled bands are non-specific esterases. (From Edwards **et al.** 1971; reproduced with the permission of the author and Cambridge University Press).

22

ADAPTATIONS IN OXIDATIVE METABOLISM DURING THE TRANSFORMATION OF *TRYPANOSOMA RHODESIENSE* FROM BLOODSTREAM INTO CULTURE FORM.

I.B.R. Bowman, H.K. Srivastava* and I.W. Flynn

Department of Biochemistry
University of Edinburgh Medical School
Teviot place
Edinburgh, EH8 9AG, U.K.

Types of infection by *T. rhodesiense* in mammalian blood may vary from those of an acute character with rapidly increasing parasitaemia and early death of the host animal to chronic types with many remissions and relapses. The variation can be related to the degree of pleomorphism of the strain used: monomorphic strains consisting of long slender (LS) trypomastigotes cause an acute infection whereas pleomorphic strains cause the more chronic type of infection. In a relapsing infection long slender forms predominate as the parasitaemia increases and in the remission phase they are replaced by a high percentage of short stumpy (SS) cells. A third morphological cell type found in pleomorphic strains, the intermediate short stumpy (ISS) form, may represent a developmental stage of the SS form. Loss of the pleomorphic character as occurs in syringe passaged, rodent adapted strains is accompanied by a loss of infectivity to other mammals and to the tsetse fly and these LS forms can not be cultured. From this it is concluded that the SS form is an essential transitional stage between the blood form and epimastigote. In addition to considerable differences in morphology a number of biochemical features distinguish LS & SS trypomastigotes from culture epimastigotes. In summary the LS form lacks a mitochondrion, a tricarboxylic acid cycle and a cytochrome system [1, 2]. Oxygen uptake is mediated by L-glycerol-3-phosphate (α GP) oxidase [3] located in discrete extramitochondrial particles [4]. In contrast the culture epimastigote has a highly developed mitochondrion [5], possesses a functional TCA cycle [6] and a cytochrome mediated electron transport system [7]. It is the SS form

---

* Present address: Department of genetics, Haryana agricultural university, Hissar, India.

which initiates the metabolic transformation or respiratory switch in Vickerman's terminology[8] between these disparate LS and culture forms. Vickerman [8] has shown that the SS congener has devloped a mitochondrial tubule containing cristae and NADH-tetrazolium reductase activity which may be related to the acquired ability of these stages to use α-oxoglutarate as an energy source [8, 9, 10]. These results suggested that a metabolic switch to intramitochondrial oxidation had occurred. Our evidence[11] suggests that this switch is incomplete in the blood-stream stages and this report describes the sequence of metabolic changes in the transitions LS to SS to epimastigote forms.

Substrate oxidation by the blood stream forms

Table 1 shows that the most striking feature of the metabolism of the SS forms is the ability to oxidise α-oxoglutarate in addition to glucose and glycerol; only the latter two substrates are oxidised by the LS form. This finding of α-oxoglutarate supported oxidation is not surprising as it had been shown previously [8, 9, 10] to maintain the motility of the SS forms selectively. None of the other substrates tested was oxidised by whole cells. However, it can be seen that in lysed cell preparations pyruvate is oxidised almost as well as α-oxoglutarate suggesting that pyruvate is impermeable to the intact trypanosome membrane. Pyruvate and α-oxo-glutarate oxidative decarboxylases have been synthesised by the SS stage and distinguish clearly the SS or ISS from LS forms. It is likely that the activities of other enzymes of the TCA cycle are limiting in the SS form as the intermediates of this cycle are poorly oxidised.

The products of metabolism of glucose, α-oxoglutarate and pyruvate are shown in Table 2. The carbon balances with glucose and α-oxoglutarate were obtained with intact cells, pyruvate with lysed cells. It is significant that only 60 % of the glucose carbon used is metabolised to pyruvate in the SS forms - the LS form produces ⟨ 90 % as pyruvate. 10 % of carbon is found in CO_2, 7 % in succinate and 9 % in acetate none of which is produced to any significant extent by LS stages[12]. Glycerol (8 %) is produced from αGP by a phosphatase which is extremely active in trypanosomes. It can be seen that α-oxoglutarate is simply decarboxyla-ted to succinate and CO_2. The small percentage of pyruvate may result from the further slow metabolism of succinate. In lysed preparations pyruvate is decarboxylated to acetate and CO_2. It can be concluded that the SS stages develop a chondriome and with it some of the enzymes of the TCA cycle in particular the oxidative decarboxylases, but succinoxi-dase seems to be limiting due either to a lack of the flavoprotein dehydrogenase or to a lack of a cytochrome electron transport system. Cytochromes are not detectable in the SS forms therefore NADH generated in oxidation processes is reoxidised by α GP oxidase. It is possible that an autoxidisable flavoprotein is present as an alternative oxidase as it can be shown that in conditions in which α GP oxidase is inactive, where DHAP as electron acceptor is removed by gel filtration, there is still a residual oxygen uptake in the presence of NADH (5-15 % of the original activity) which is rotenone and amytal sensitive.

The metabolic transformation of blood forms to mitochondrial mediated metabolism is initiated at the SS stage but is not completed until transfer to culture or the insect vector.

Metabolic Transformation in Culture

SS forms were isolated under sterile conditions from rat blood and transferred into the blood-agar biphasic system with Earle's saline overlay of Tobie, Mehlman and von Brand [13]. The rates of oxygen uptake by lysed cells at various stages of development in the bloodstream and of transformation in culture were followed polarographically at 26° using α-oxoglutarate, NADH, succinate, L-proline and α-glycerophosphate.

The metabolic changes occurring during transformation of SS into culture epimastigote form were studied and the results presented in Figures 1 and 2. In Figure 1 it can be seen that LS forms do not oxidise succinate or proline and the rates of oxygen uptake with these substrates is minimal in cell free preparations of trypomastigotes in which 75 % were in the SS or ISS form. The rate of oxidation of α GP by bloodstream forms (LS and SS) was highest of all substrates tested (1.3 μmol O_2/mg protein/h). However, within seven days of transfer into culture, oxygen uptake supported by succinate and proline approaches the maximum rate of the established (∞) culture form and the rate of proline oxidation is twice that of succinate and α GP. There is no significant increase in the oxidation rate of α GP in the culture forms. This could indicate the persistence of α-glycerophosphate oxidase, though this is not the only possibility as this oxidative process becomes cyanide sensitive in the culture form.

In Figure 2 it can be seen that there is little increase in the oxidation rate of NADH with increasing proportions of SS forms but within 24 h of culturing this rate increases to 0.5 μmol O_2/mg protein/h and stays constant for 3-4 days then approaches a maximum of 1.3 μmol O_2/mg protein/h at 7 days. α-oxoglutarate is not oxidised by LS forms but as has already been noted, this substrate is oxidised by SS forms to about

50 % of the maximum oxidation rate. The difference in the rates of oxidation of α-oxoglutarate quoted in Table 1 and Figure 2 are due to the different assay temparatures, 37° and 26° respectively. Maximum rates of oxidation of α-oxoglutarate and NADH are obtained within 7 days of transfer to culture. The progress curve of proline oxidation is redrawn in Figure 2 for comparative purposes and shows again that its rate of oxidation is twice that of other substrates tested.

With the exception of α GP it can be stated that there are marked increases in the rates of oxidation of the marker substrates within 3-4 days of transfer of trypomastigotes into culture and these rates reach the values found in established culture epimastigotes within 7 days at which time all trypanosomes have transformed morphologically into culture forms.

Cyanide Sensitivity of Developing Culture Forms

Table 3 shows the effect of cyanide (3 mM) on the oxygen uptake supported by the test substrates. There is a gradual increase in inhibition by cyanide in the bloodstream stages and early culture stages and after 3 days in culture the oxidation of NADH, succinate, α-oxoglutarate and proline is wholly inhibited by cyanide. The concentration of cyanide at 3 mM is excessively high compared to those concentrations (10^{-5} - 10^{-6} M) required to cause complete inhibition of cytochrome aa$_3$ in mammalian mitochondria. Lower concentrations (10^{-5} M) inhibited oxygen uptake by only 50 % (Figure 3). It should be noted that α GP oxidation has become cyanide sensitive suggesting that the cytoplasmic α GP oxidase has been superseded by a cytochrome dependent α GP oxidase system. It is concluded that the cytochrome oxidase of the culture form is not of the aa$_3$ type but some other cytochrome oxidase less sensitive to cyanide inhibition and that this is synthesised within 3 days in culture.

Cytochromes of Established Culture Forms

In the light of the poor inhibition by cyanide spectral analysis of the epimastigotes was carried out to identify the terminal oxidase. Interpretation of the spectra was complicated by contamination of the trypanosome preparation by haemoglobin and haemiglobin from the blood agar biphasic culture medium [13]. *T. rhodesiense* EATRO 173 is easily established in culture and has been sub-cultured for over one year. The culture organisms were harvested and washed nine times in an attempt to

332

remove the last trace of haemiglobin and the cells were then sonicated. The difference spectrum (Figure 4, lower curve) of dithionite reduced versus oxidised (aerated) preparations shows an α band at 555-556nm, β band 524-526 nm and Soret band at 430 nm showing the presence of cytochromes b and c, but although there is a small peak at 600 nm the absence of a peak at 444 nm shows that cytochrome aa_3 is absent. The upper curve (Figure 4) is the CO-difference spectrum giving peaks at 570, 540 and 418 nm which is consistent with the presence of cytochrome o. The absence of peaks at 590, 550 and 430 nm is a further indication of the lack of cytochrome aa_3. It should be emphasized that a contaminant CO complex with haemoglobin would give the same spectrum. The CO difference spectrum also showed a small peak at 625-628 nm. Since the methaemoglobin of the culture medium may be present intracellularly and therefore would not be removed by washing of the cells, a mitochondrial preparation was made from cells sonicated in 0.3M sucrose, 24 mM tris, 1 mM EGTA, pH 7.4 for 3 min at 3 Amp. The mixture was centrifuged at 1250 g for 10 min and the supernatant fluid centrifuged at 15,000 g for 10 min. The resulting pellet was washed 3 times by repeated centrifugation and finally suspended in 100 mM phosphate buffer, pH 7.4. The difference spectra of this mitochondrial fraction are given in Figure 5. The lower curve is the reduced (dithionite) versus oxidised (aeration) spectrum in which the α, β and γ peaks are rather sharper than with unfractionated lysates. The middle spectrum is obtained when cyanide is added to the reduced sample. No change is observed in the positions of the maxima but the peak heights are considerably potentiated. These spectra indicate the presence of cytochromes b and c and again the absence of cytochrome aa_3. The small peak at about 600 nm shown in Figure 4 is also to be found in the mitochondrial preparations. The CO difference spectrum with peaks at 570 nm, 540 nm and 418 nm identifies cytochrome o. In those experiments with broken cells or mitochondrial fractions any contaminating methaemoglobin would be reduced by dithionite to haemoglobin and so give an identical spectrum with CO as cytochrome o. This possibility can be avoided by metabolic reduction of the cytochrome system in washed intact cells. Glucose (10 mM) was used as reducing substrate. The reduced minus oxidised spectrum is shown in Figure 6 (lower curve) along with the CO difference spectrum obtained by saturating a metabolically reduced cell suspension with CO and setting this against a metabolically reduced reference suspension. The results are essentially the same as before showing the presence of cytochromes b, c and o and the absence of aa_3.

Difference Spectra of *T. rhodesiense* **Pittam Strain.**

This established culture form was grown in bulk in the blood broth medium of Pittam [14] and an acetone dried powder prepared. The dithionite reduced minus ferricyanide oxidised spectrum of this material is shown in Figure 7. Peaks at 555-556, 526 and 430 nm indicate cytochromes b and c. The small peak at 608-610 nm is not indicative of aa_3 since there is no peak at 444 nm. Furthermore the CO-difference spectrum (upper curve) shows no band at 590 nm of aa_3; instead there is a marked peak at 630 nm of some other CO binding pigment, perhaps cytochrome d. The other bands in the CO spectrum are, as in the case of freshly prepared *T. rhodesiense* EATRO 173, most likely due to cyto-chrome o. There appears to be little CO binding pigment tentatively identified as cytochrome d in the EATRO 173 strain, as the extinction at 630 nm is very small and not present in all preparations.

Pyridine Haemocromes

Pyridine haemochrome of the acid acetone extract of *T. rhodesiense* Pittam gave absorption bands at 556, 526 and 418 nm consistent with presence of haem b. The reduced pyridine haemochrome of the acid acetone insoluble residue had absorption peaks at 553, 520 and 415 nm similar to the values quoted by Hill[15, 16] for haem c derived from cytochrome c_{555} isolated from *C. fasciculata* or *T. rhodesiense.* A similar cytochrome c_{555} has recently been isolated from *C. oncopelti* [17]. It is concluded that the strains of *T. rhodesiense* examined here contain this atypical cytochrome c_{555}.

General conclusions

The epimastigote form of *T. rhodesiense* has an atypical cytochrome system, consisting of cytochromes b, c_{555} and o, and this system is synthesised within 3-4 days after introduction of the organisms to culture. Depending perhaps on strain differences or on culture conditions, a second CO-binding pigment with some of the properties of cytochrome d may be present. Multiple cytochrome oxidases have been reported in *C. fasciculata* [15] and in *C. oncopelti* [18].

Concomitant with the development of this cytochrome system in the early stages of culture *T. rhodesiense* EATRO 173, there is a marked development of enzyme systems for the oxidation of NADH, succinate and proline, and a potentiation of the oxidative metabolism of α-oxoglutarate. The high rate of proline oxidation in the established culture (insect mid-gut) form parallels the dependence of tsetse fly tissue on proline as an energy source[19]. Whereas the bloodstream trypomastigotes of *T. rhodesiense* show a strict requirement for carbohydrate, it is possible that the epimastigotes rely upon the oxidation of proline and other amino-acids as an energy source, as does the insect host.

This work was supported by grants from the Trypanosomiasis Panel of the Overseas Development Administration, U.K.

References

1. RYLEY, J.F. (1956). Biochem. J., **62,** 215.
2. FULTON, J.D., and SPOONER, D.F. (1959). Exp. Parasit., **8,** 137.
3. GRANT, P.T. and SARGENT, J.S. (1960). Biochem. J., **76,** 229.
4. BAYNE, R.A., MUSE, K.E. and ROBERTS, J.F. (1969). Comp. Biochem. Physiol., **30,** 1049.
5. VICKERMAN, K. (1970). The African Trypanosomiases, edited by Mulligan, H.W. (George Allen and Unwin Ltd., London) p. 60.
6. BRAND, T. von, TOBIE, E.J. and MEHLMAN, B. (1950). J. Cell Comp. Physiol., **35,** 273.
7. RYLEY, J.F. (1962). Biochem. J., **85,** 211.
8. VICKERMAN, K. (1965). Nature, **208,** 762.
9. BALIS, J. (1964). Rev. Elev. Med. Vet. Pays. Trop., **17,** 361.
10. RYLEY, J.F. (1966). Proc. Intern. Congr. Parasitol., Rome, (Pergamon Press, Oxford) p. 41.
11. BOWMAN, I.B.R., FLYNN, I.W. and FAIRLAMB, A.H. (1970). J. Parasit., **56,** 402.
12. GRANT, P.T. and FULTON, J.D. (1957). Biochem. J., **66,** 242.
13. TOBIE, E.J., von BRAND, T. and MEHLMAN, B. (1950). J. Parasit., **36,** 48.
14. PITTAM, M.D. in DIXON, H. and WILLIAMSON, J. (1970). Comp. Biochem. Physiol., **33,** 127.
15. HILL, G.C. and WHITE, D.C. (1968). J. Bacteriol., **95,** 2151.
16. HILL, G.C., GUTTERIDGE, W.E. and MATHEWSON, N.W. (1971). Biochem. Biophys. Acta (in press).
17. PETTIGREW, G. and MEYER, T. (1971). Biochem. J. (in press).
18. SRIVASTAVA, H.K. (1971). FEBS Letters (in press).
19. BURSELL, E. (1966). Comp. Biochem. Physiol., **19,** 809.

Table 1.

Substrate utilisation by *T. rhodesiense* **EATRO 173 SS**

Substrates (25 μmol) were incubated with whole cells (1.1 mg protein) in Krebs saline (3 ml) in conventional Warburg respirometers with KOH in centre wells. Cell lysates (5.7 mg protein) were suspended in a reaction mixture (3 ml) containing 3 mM EDTA, 25 mM nicotinamide, 5 mM KCl, 5 mM MgCl$_2$, 66 mM potassium phosphate buffer, pH 7.4, 30 mg bovine plasma albumin, substrates (25 μmol) and cofactors ADP and NAD (5 μmol). Rates of oxygen uptake were measured manometrically at 37° over the first 30 minutes.

	μmol O$_2$/mg Protein/h	
Substrate	**Whole Cells**	**Lysate**
Glucose	8.3 (12)*	2.00 (6)
α-oxoglutarate	3.95 (4)	1.00 (11)
Glycerol	7.95 (3)	—
L-α-glycerophosphate	—	2.40 (3)
Pyruvate	0.00 (5)	0.78 (9)
Succinate	0.00 (4)	0.23 (5)
Citrate	0.00 (2)	0.23 (3)
Isocitrate	0.00 (2)	0.28 (2)
Fumarate	0.00 (2)	0.08 (4)
Malate	0.00 (2)	0.11 (3)

* Figures in parenthesis denote numbers of determinations.

Table 2.
Metabolic products of *T. rhodesiense* EATRO 173 SS

Whole cells (1.1 mg protein) were incubated with glucose (25 μmol) or α-oxoglutarate (25 μmol) in saline (3 ml). Samples were taken at zero time and after 35 minutes, deprotenised with $HClO_4$ (0.5 ml, 0.33 M) and neutralised with K_2HPO_4 (0.6 M) prior to analysis. Lysates (5.7 mg protein) were incubated with pyruvate (25 μmol) in the fortified reaction mixture given in Table 1. Results are expressed as a percentage of substrate uitlised.

| | Substrate | | |
Product	Glucose	α-oxoglutarate	Pyruvate
Pyruvate	60	3	—
CO_2	10	20	31
Succinate	7	75	—
Glycerol	8	—	—
Acetate	9	—	62
Citrate	⟨ 1	—	—
Hexose phosphate	⟨ 1	—	—
Phosphoenolpyruvate	⟨ 1	—	—
	94 %	98 %	95 %

Table 3.

The effect of cyanide on the oxidative metabolism of lysates of *T. rhodesiense*

Conditions were the same as those described in Figure 1. After a steady rate of oxygen uptake was recorded with each of the substrates, KCN (0.1 ml, 30 mM) was injected in the chamber.

	% sensitivity to KCN (3×10^{-3}M)			
	α-oxoglu-tarate	Proline	Succinate	NADH
Bloodstream forms				
LS	0.0	0.0	0.0	2.5
SS (75 %)	25.0	10.0	3.8	15.0
Culture forms				
1-day	31.0	19.3	19.4	26.0
3-day	100.0	100.0	100.0	100.0
Established	100.0	100.0	100.0	100.0

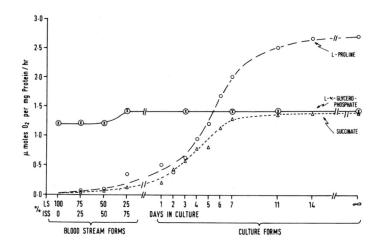

Fig. 1.
Water lysed cells were incubated in the fortified reaction mixture (3 ml) given in Table 1, at 26° in a polarograph chamber fitted with a Clark oxygen electrode. After a steady endogenous rate of oxygen uptake was obtained substrates (5 μmol) were injected by microsyringe, ⊗ , L-glycerol-3-phosphate; ○, L-proline; △, succinate.

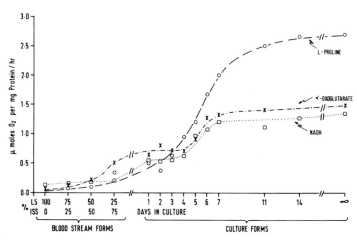

Fig. 2.
Conditions as in Figure 1. X, α-oxoglutarate; □, NADH; ○, L-proline.

339

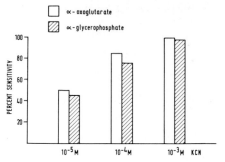

Fig. 3.
Conditions as in Figure 1. α-glycerophosphate or α-oxoglutone (5 μmol) was injected into the chamber and after a steady rate of oxygen uptake was recorded KCN freshly prepared and neutralised to pH 7.4 was added to give the concentrations shown.

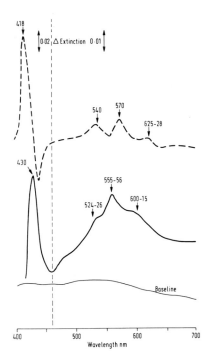

Fig. 4.
Difference spectra of sonicated preparations (5.0 mg protein/ml) of *T. rhodesiense* EATRO 173, in 0.1 M phosphate buffer, pH 7.4. The lower solid line represents the difference spectrum of respiratory pigments reduced with dithionite minus pigments oxidised by vigorous aeration at 2°. The upper dashed line is the difference spectrum of a preparation reduced with dithionite and saturated with CO compared with a preparation reduced by dithionite.

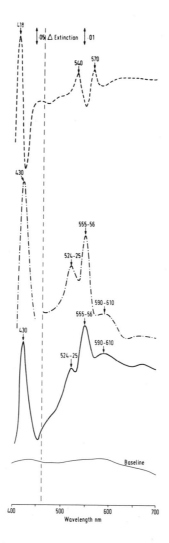

Fig. 5.
Difference spectra of a mitochondrial preparation of *T. rhodesiense* EATRO 173 epimastigotes suspended in 0.1 M phosphate buffer, pH 7.4 at a concentration of 2.5 mg protein per ml. Lower solid curve: dithionite reduced minus oxidised aerated spectrum. Middle broken curve: the reduced sample treated with cyanide minus oxidised spectrum. Upper dashed line: the difference spectrum of mitochondria reduced with dithionite and saturated with CO minus a dithionite reduced sample.

341

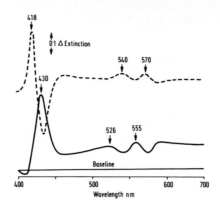

Fig. 6.
Difference spectra of suspensions of intact cells of *T. rhodesiense* EATRO 173 epimastigotes. The lower solid line represents the spectrum of cells metabolically reduced with glucose (10 mM) minus cells aerated without substrate. The upper dashed line is the spectrum of metabolically reduced and CO treated cells minus cells in which the pigments were reduced by glucose metabolism.

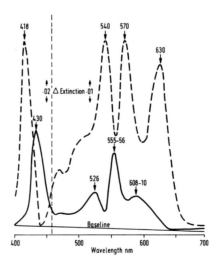

Fig. 7.
Difference spectra of acetone dried *T. rhodesiense* Pittam strain suspended in 0.1 M phosphate buffer, pH 7.4, at a concentration of 5 mg protein per ml. Lower solid line represents the spectrum of a suspension with the pigments reduced by dithionite versus a sample oxidised by aeration. The upper dashed line indicates the spectrum of respiratory pigments after reduction with dithionite and saturation with CO minus pigments reduced by dithionite.

23

EFFECTS OF BONGKREKIC ACID ON MALARIA PARASITES
(PLASMODIUM LOPHURAE)
DEVELOPING EXTRACELLULARLY IN VITRO

W. Trager
The Rockefeller University
New York, New York 10021

Twenty years ago I observed a marked effect of adenosine triphosphate (ATP) and pyruvate on the prolonged extracellular survival and development **in vitro** of a bird malaria parasite *Plasmodium lophurae* [10]. The effect was first noted in experiments of 2 days' duration but it was later also shown after only 1 day **in vitro** [12]. By starting cultures with young uninucleate trophozoites freed from their host erythrocytes by immune hemolysis one could see that omission of either ATP or pyruvate resulted in fewer multinucleate forms and more degenerate ones. ATP could not be replaced by adenosine-5'-monophosphate (AMP) but it could be replaced by adenosine diphosphate with phospho-enolpyruvate. Furthermore, parasites incubated 16 hr in the absence of added ATP and then supplied with ^{14}C-labeled proline incorporated only about half as much label as parasites that had been provided originally with ATP (at 2 or 4 mM) [13].

The function of the exogenous ATP is not clear. We know that malaria parasites require preformed purines [1, 2, 14, 15], but the culture medium already contains adenosine (0.04 mM) and adenosine monophosphate (mixed 2', 3' isomers) at 1.4 mM. The interesting possibility exists that external ATP is needed for the functioning of a transport mechanism across the outer of the 2 membranes surrounding the parasite [8]. This outer membrane is probably derived originally from the red cell membrane, as indicated in the beautiful electron micrographs of Ladda et al. [7] showing entrance of a merozoite into a red cell. This membrane however soon becomes so closely integrated with the parasite membrane that when the parasites are removed from their host cells they retain both membranes [5, 8, 9] . If the outer membrane is originally an inside-out red cell membrane its ATPase would be on its outer surface where, in nature, it would have access to the ample ATP supplies of the erythrocyte. With the parasite removed from the host cell ATP would have to be supplied in the medium.

These hypothetical considerations suggested investigation of ATPase inhibitors, of which the antibiotic bongkrekic acid was found to be effective.

Materials and Methods

The maintenance of synchronous infections of *P. lophurae* in ducklings, the preparation of free parasites by immune hemolysis, and their development **in vitro** in a red cell extract medium have all been previously described in detail [see 11, 13].

Assessment of the extent of development of the extracellular parasites after 1 day **in vitro** was based on two different kinds of criteria, one morphological, the other biochemical. In Giemsa-stained films 200 parasites were counted and classified as normal with 1 nucleus, normal with 2 nuclei, normal with over 2 nuclei, and degenerate. The counts for each experiment were done with the actual slide number covered by another coded number attached by an assistant, thus reducing subjective bias. Since experiments were begun with 90 % or more of the parasites uninucleate trophozoites, the extent of increase in proportion of multi-nucleate forms provided a good criterion of growth.

In addition each culture received, usually after 16 hr of incubation, a suitable amount of a ^{14}C-labeled precursor. Three to 4 hr later, after sampling for the stained films, the parasites were quickly washed with a glucose buffer and then precipitated with trichloracetic acid. This precipitate was washed, solubilized, and its radioactivity determined by scintillation counting.

The bongkrekic acid was a gift kindly sent by Dr. W. Berends of the Technical University at Delft, The Netherlands. It was received as a solution of 100 mg in 4 ml 2 N NH$_4$OH. Small amounts were diluted 10-fold with redistilled water, sterilized by passage through a Swinnex millipore filter, and stored at $2°$. Atractyloside, purchased from Calbiochem., was dissolved at 20 mg in 5 ml redistilled water and similarly sterilized. The solution was stored frozen at $-20°$.

Results and Discussion

Atractyloside and bongkrekic acid are two chemically very different materials that specifically inhibit, though by different mechanisms, the adenine nucleotide translocase of mitochondria [3, 4, 6, 16] . No effect of atractyloside on the extracellular development of *P. lophurae* as judged morphologically has been noted (Table 1). Bongkrekic acid, on the other hand, does have such an effect, and an even greater effect on the incorporation of proline (Table 2). The proportion of degenerate parasites was not increased (as it was if ATP was omitted), but development to multinucleate forms was always inhibited in flasks with bongkrekic acid at concentrations like those effective for mitochondria. It should be noted that the addition of NH_4OH at a concentration corresponding to that calculated to have been introduced with the antibiotic had no effect (Table 2, Exp. C).

Bongkrekic acid was also effective if the basal medium was a full-strength rather than a 0.3 strength red cell extract (Table 3). A full-strength extract contains duck hemoglobin at 14 %. This experiment illustrates the deleterious effect of omission of ATP and pyruvate (less striking than generally seen with a more dilute red cell extract   ) and the inhibiting effect of bongkrekic acid at 35 as well as at 70 μg/ml.

Bongkrekic acid had little or no effect on incorporation of methionine as compared to its effect on proline incorporation (Table 4). This experiment also shows control flasks that received NH_4 acetate to supply NH_4^+ at the same concentration at which it was introduced with the solution of bongkrekic acid. Methionine incorporation was about the same in all 3 sets, whereas proline incorporation was sharply reduced in the flasks with bongkrekic acid. This is in keeping with the fact that omission of ATP from the medium reduces proline incorporation [13] but does not affect methionine incorporation.

Increasing the concentration of ATP partially counteracted the inhibiting effect of bongkrekic acid (Table 5). This was evident in the extent of proline incorporation and even more in the proportion of multinucleate parasites. AMP at a similar high concentration had if anything a deleterious effect. It is interesting that the effect of bongkrekic acid on mitochondria was counteracted by ATP at about the same levels (10 mM) as were effective with *P. lophurae* [3] .

This phenomenon has yet to be investigated in more detail. In particular the effects of ADP and of other ATPase inhibitors, as oligomycin, should be studied. Ouabain had no effect, but this was not surprising since the culture medium is high in both potassium and protein. It will also be important to try to localize an ATPase activity of parasite membranes cytochemically at the electron microscope level. Meanwhile we can entertain the hypothesis that malaria parasites, like mitochondria, require exogenous ATP for the translocation of adenine nucleotides, or for energy-linked transport, or for both activities.

References

1. BUNGENER, W. and NIELSEN, G. (1968). Z. Tropenmed. Parasitol. **19,** 185.
2. BUNGENER, W. and NIELSEN, G. (1969). Z. Tropenmed. Parasitol. **20,** 67.
3. HENDERSON, P.J.F. and LARDY, H.A. (1970). J. Biol. Chem. **245,** 1319.
4. KEMP, A., Jr., OUT, T.A., GUIOT, H.F.L. and SOUVERIJN, J.H.M. (1970). Biochem. Biophys. Acta **223,** 460.
5. KILLBY, V.A. and SILVERMAN, P.H. (1969). Am. J. Trop. Med. Hyg. **18,** 836.
6. KLINGENBERG, M., GREBE, K. and HOLDT, H.W. (1970). Biochem. Biophys. Res. Comm. **39,** 344.
7. LADDA, R., AIKAWA, M. and SPRINZ, H. (1969). J. Parasitol. **55,** 633.
8. RUDZINSKA, M.A. (1969). Int. Rev. Cytol. **25,** 161.
9. RUDZINSKA, M.A. and TRAGER, W. (1961). J. Protozool. **8,** 307.
10. TRAGER, W. (1950). J. Exptl. Med. **92,** 349.
11. TRAGER, W. (1966). Trans. N.Y. Acad. Sci., Sec. II, **28,** 1094.
12. TRAGER, W. (1967). J. Protozool. **14,** 110.
13. TRAGER, W. (1971). J. Protozool. **18,** August issue.
14. VAN DYKE, K., TREMBLAY, G.C., LANTZ, C.H. and SZUSTKIEWICZ, C. (1970). Am. J. Trop. Med. Hyg. **19,** 202.
15. WALSH, C.J. and SHERMAN, I.W. (1968). J. Protozool. **15,** 763.
16. WIEDEMANN, M.J., ERDELT, H. and KLINGENBERG, M. (1970). Biochem. Biophys. Res. Comm. **39,** 360.

Table 1.
Development of *P. lophurae* extracellularly **in vitro**. Effects of ATP-pyruvate and atractyloside in 0.3 strength duck erythrocyte extract.

Exp.	Flasks	ATP mM	Pyruvate mM	Atractyloside μg/ml	Atractyloside mM	Av. % parasites >IN	Av. % parasites Deg.
A	1- 3	0	0	0		5	26
	4- 6	4	10	0		15	5
	7- 9	4	10	100	0.12	13	8
	10-12	4	10	20	0.024	12	5
B	1- 4	0	0	0		25	14
	5- 8	4	10	0		36	5
	10-12	4	10	150	0.18	37	3

Table 2.
Effects of Bongkrekic acid (BK) on extracellular development of *P. lophurae* and on incorporation of Proline-U-^{14}C. 0.3 strength red cell extract. Isotope (2.5 μc) added at 16 hr, samples taken 4 hr later. (BK prepared from stock solution in 2N NH_4OH).

Exp.	Flasks	BK μg/ml (mM)	Parasites per flask x 10^6	Av. % parasites ⟩IN	Deg.	Av. cpm/cult. per hr. expos.
A	1- 3	0	140	35	6	970
	10-12	70 (0.14)		24	1	320
B	1- 3	0	140	33	5	1000
	7- 9	70 (0.14)		28	7	222
	10-12	70* (0.14)		20	6	295
C	3,4	0	160	22	8	563
	5,6	0‡		25	4	504
	10-12	70 (0.14)		12	4	230

* Added from a separately prepared dilution.

$\ddagger$ These flasks contained NH_4OH at 0.005N.

Table 3.

Effects of Bongkrekic acid (BK) and omission of ATP and pyruvate on extracellular development of *P. lophurae* and incorporation of Proline-U-^{14}C. Full strength red cell extract. 100 x 10^6 parasites per flask. Isotope (2.5 μc) added at 16 hr, samples taken 4 hr later.

Flasks	ATP mM	Pyruvate mM	BK μg/ml	Av. % parasites >IN	Av. % parasites Deg.	Av. cpm/cult. per hr. expos.
1- 3	0	0	0	28	6	437
4- 6	2	5	0	40	4	465
7- 9	2	5	35	24	2	277
10-12	2	5	70	29	6	207

Table 4.

Extracellular development of *P. lophurae* and incorporation of Methionine-methyl-^{14}C (Me) and Proline-U-^{14}C (Pr) as affected by Bongkrekic acid (BK). 0.3 strength red cell extract. 100 x 10^6 parasites per flask. Isotopes (each at 2.5 μc per flask) added at 16 hr, samples taken 4 hr later.

Flasks	BK μg/ml	Av. % parasites >IN	Av. % parasites Deg.	Av. cpm/cult. per hr. expos. Me	Av. cpm/cult. per hr. expos. Pr
1,2				318	
3,4	0	25	4		425
5,6				292	
7,8	0*	31	4		400
9,10				300	
11,12	35	18	2		243

* Flasks 5-8 contained ammonium acetate at 0.0025N.

Table 5.
Partial prevention by ATP of effects of Bongkrekic acid (BK) on extracellular *P. lophurae.* 0.3 strength red cell extract, pyruvate present throughout at 5 mM. Proline-U-^{14}C (2.5 μc) added at 16 hr, samples taken 4 hr later.

Exp.	Flasks	ATP mM	AMP mM	BK mM	Av. % parasites with ⟩ IN	Av. cpm/cult. per hr. expos.
A*	1- 3	2	0	0	24	810
	4- 6	2	0	0.07	10	345
	7- 9	6	0	0.07	17	347
	10-12	10	0	0.07	25	413
B*	1- 3	2	0	0	28	533
	4- 6	2	0	0.07	16	218
	7- 9	2	10	0.07	6	189
	10-12	12	0	0.07	22	281

* Each flask received, for Exp. A 160 x 10^6 parasites, for Exp. B 200 x 10^6.

24

CARBOHYDRATE METABOLISM IN *ENTAMOEBA HISTOLYTICA*

R.E. Reeves
Department of Biochemistry
Louisiana State University School of Medicine
New Orleans, La. U.S.A. 70112

Growing cultures of **Entamoeba histolytica** require glucose or a glucose polymer which can be converted to glucose. No other substance appears to fulfill their energy requirements. Glucosamine, mannose, and 2-deoxyglucose are readily phosphorylated by the amebal glucokinase[1], but they do not support growth when substituted for glucose in the medium. When these sugars are presented in addition to glucose only the deoxysugar exerts an inhibitory effect upon culture growth, presumbably, because only it is accepted by the mechanism for active glucose transport into the cells.

Glucose-6-P may be converted to amebal glycogen in the usual manner or it may be metabolized by way of the glycolytic pathway. The dehydrogenase required to transform it to 6-phosphogluconate is absent from the armory of amebal enzymes. Not only is the dehydrogenase not found but, when glucose labelled at position 1 by ^{14}C was metabolized by resting cells CO_2 was not preferentially liberated from this position. This finding by Bragg and Reeves[2] speaks against the operation of a pentose phosphate shunt in these organisms.

The only unusual feature noted in amebal glucokinase was its great sensitivity to inhibition by AMP, $K_i = 10^{-6}M$. This feature distinguishes it from other investigated glucokinases. It may be speculated that a low adenylate energy charge causes amebal metabolism to switch from glucose to endogeneous glycogen reserves which may amount to as much as 50 percent of the dry weight of the cell.

No unusual biochemical properties were found associated with the amebal glucose-6-P isomerase. However, electrophoretic properties (see

351

Fig. 1) distinguished enzyme from the human pathogen, the low temperature strains which are non-pathogenic in man, and closely related species of amebae. The electrophoretic properties of the isomerase provide the best method for distinguishing among the closely related strains and species of ameba [3].

Amebal phosphofructokinase is a particulate enzyme sedimenting at 100,000 x **g**. This enzyme lacks the regulatory features commonly associated with its counterpart from mammalian sources. Its acitivity was not inhibited by ATP (in the presence of an excess of magnesium ion) and no conditions were found under which AMP or cyclic AMP were stimulatory. The only other phosphofructokinase which, to my knowledge, is so devoid of regulatory features is that from the slime mold *Dictyostelium discoideum* reported by Bauman and Wright [4].

The glycolytic enzymes catalyzing the steps between fructose-1,6-di-P and P-enolpyruvate have been identified by a number of investigators since the pioneering report by Hilker and White [5]. Work in our laboratory has confirmed their presence and that their activities are in excess of that necessary to account for the rate of glucose metabolism.

I would now like to digress from the subject of the individual glycolytic enzymes to take note of some recent findings on the rate of glucose metabolism and the end products arising therefrom. Montalvo et al. [6] report work with resting cells of *E. histolytica* under anaerobic conditions. The average findings, from six experiments, are listed in Table 1. In these experiments of one hour duration the average consumption of glucose was 0.7 μmole/min/ml packed cells. In experiments of shorter duration rates to 1 μmole/min were encountered. The identified products of glucose metabolism were CO_2, acetate, ethanol, and hydrogen. These accounted for 96 percent of the glucose carbon utilized.

We have recently encountered an aerobic facet of amebal metabolism. This surprised us since some degree of anaerobiosus is required for growth and multiplication of amebae in culture. Freshly harvested amebae consume oxygen and carbohydrate from their endogeneous glycogen reserves. After a period of preincubation their respiration is stimulated by exogeneous glucose. (Fig. 2). Glucose carbon is converted to the same end products, but the acetate/ethanol ratio increases from 0.37 in the anaerobic cells to about 2.2. Hydrogen gas production ceases under aerobic conditions.

Having checked out the enzymes catalyzing amebal glycolysis from glucose to P-enolpyruvate a new and challenging situation arose. Clearly glucose is metabolized to CO_2, acetate, and ethanol, but we were unable to identify the next enzyme of the glycolytic pathway, pyruvate kinase (I).

352

$$PEP + ADP \rightleftharpoons pyruvate + ATP \qquad (I)$$

The only other enzyme then known to catalyze the direct interconversion of PEP and pyruvate was PEP synthase which Cooper and Kornberg[7] had identified from *Escherichia coli*. Its equilibrium was heavily weighted in the reverse direction from that of pyruvate kinase, as is indicated by the direction of the larger arrows in equations I and II.

$$PEP + P_i + AMP \rightleftharpoons pyruvate + ATP \qquad (II)$$

One day I interrupted my search for amebal pyruvate kinase and proceeded to give a lecture on glycolysis to some graduate students. The topic that day included the Cooper-Kornberg enzyme (II) To dramatize the very different equilibrium between it and pyruvate kinase I wrote a third equation (III) noting that if such a reaction should occur the free

$$PEP + PP_i + AMP \rightleftharpoons pyruvate + P_i + ATP \qquad (III)$$

energies of PEP and PP_i, being greater than that of ATP, would cause the equilibrium be shifted again to the right.

As soon as equation III was on the blackboard I realized what might be missing from my search for pyruvate kinase in amebae. When the class ended I rushed back to the laboratory and within minutes had evidence that amebal "pyruvate kinase" requires inorganic pyrophosphate. The enzyme is now called pyruvate phosphate dikinase. It had been discovered by Hatch and Slack[8] in Australia, but the issue of Biochemical Journal bearing their announcement did not arrive in New Orleans until some months later. Evans and Wood[9] had also independently discovered the same enzyme.

The stoichiometry of the pyruvate phosphate dikinase reaction[10] raised a serious question, "Where might amebae get PP_i in the quantitiy required to convert PEP to pyruvate?" Setting aside worries about this problem a study of the amebal PEP carboxylase was initiated. There was no difficulty in demonstrating that an amebal homogenate could actively carboxylate PEP to oxaloacetate, but the nucleotide requirement of the reaction could not be established. None of the nucleotide diphosphates acted as acceptor for the phosphate group from PEP. Finally it emerged that the acceptor was orthophosphate (P_i). The amebal enzyme was PEP carboxytransphosphorylase which catalyzes reaction IV.

$$CO_2 + PEP + P_i \rightleftharpoons OAA + PP_i \qquad (IV)$$

The enzyme had been discovered by Siu and Wood[11], but had not been encountered in any organism save the propionibacteria. Like the bacterial enzyme the amebal enzyme was stimulated by cobaltous ion. The stoichiometry was shown to be in accord with that required by equation IV[12].

This reaction, it was then noted, provides an answer to the PP_i problem, but it raises still another question, "What happens to the oxaloacetate in an organism lacking a functioning citric acid cycle and which is not a producer of succinate?"

At the time the preceeding work was underway our attention was diverted to the amebal alcohol dehydrogenase. Since ethanol is a major end product of glucose metabolism it seemed obvious that the large amount of reduced NAD formed at the glyceraldehyde-P dehydrogenase step must be reoxidized at the alcohol dehydrogenase step. To our surprise amebal homogenates gave no evidence of an NAD-linked alcohol dehydrogenase. They did possess abundant activity which was strictly linked to NADP. The homogenate from one ml of packed cells was capable of converting acetaldehyde to ethanol at a rate of 25 μmoles/min. when NADPH was supplied. With NADH the rate was zero [13].

At this point the picture of amebal carbohydrate metabolism began to come into focus. Earlier, unpublished work on the two malate dehydrogenases had revealed that the enzyme which catalyzes the reduction of oxaloacetate to malate was present in quantity and was tightly linked to NAD. The other enzyme, the so-called "Malic enzyme" which interconverts pyruvate plus CO_2 with malate was also an abundant enzyme, and it was tightly coupled to NADP. While some difficult points remain to be clarified the overall pathway for amebal carbohydrate metabolism appears to be as presented in Fig. 3.

Some aspects of the proposed pathway which deserve special consideration are the following: First, all enzymes are present at activities far greater than required for resting cell metabolism. Second, the branched pathway between PEP and pyruvate provides for a stable balance between orthophosphate and pyrophosphate. Third, it provides a substrate-level transhydrogenation between NADH and NADP, reduced NADP being available for the later alcohol dehydrogenase reaction. Fourth, the pyruvate phosphate dikinase reaction conserves the high energy bonds of both PEP and PP_i, converting them to the two high energy bonds of ATP. Aspects not yet fully worked out are the characterization of the pyruvate decarboxylase and the steps by which hydrogen and acetate are formed. It should be noted that if NADPH rather than NADH is formed during acetate production a 1:1 flow of molecules through both branches of the split pathway could result in the observed 3:1 ratio between ethanol and acetate in the anaerobic end products.

Acknowledgements

I am indebted to my former associates, P. Bragg, F.E. Montalvo, and J. Bichoff; and to my colleagues Lionel G. Warren and T.S. Lushbaugh for their cooperation in the work described. This work supported by grants AI-02951 and GM-14,023 from the National Institutes of Health.

References

1. REEVES, R.E., MONTALVO, F. & SILLERO, A. (1967). A. Biochem. **6,** 1752.
2. BRAGG, P. & REEVES, R.E. (1962). Exptl. Parasitol. **12,** 393.
3. MONTALVO, F. & REEVES, R.E. (1968). Exptl. Parasitol. **22,** 129.
4. BAUMAN, P. & WRIGHT, B.E. (1968). Biochem. **7,** 3653.
5. HILKER, D.M. & WHITE, A.G.C. (1959). Exptl. Parasitol. **8,** 539.
6. MONTALVO, F.E., REEVES, R.E., & WARREN, L.G. Exptl. Parasitol. In the press.
7. COOPER, R.A. & KORNBERG, H.L. (1965). Biochim. Biophys. Acta **104,** 618.
8. HATCH, M.D. & SLACK, C.R. (1968). Biochem. J. **106,** 141.
9. EVANS, H.J. & WOOD, H.G. (1968). Proc. Natl. Acad. Sci. (U.S.A.) **61,** 1448.
10. REEVES, R.E. (1968). J. Biol. Chem. **243,** 3202.
11. SIU, P.M.L. & WOOD, H.G. (1962). J. Biol. Chem. **237,** 3044.
12. REEVES, R.E. (1970). Biochim. Biophys. Acta. **220,** 346.
13. REEVES, R.E., MONTALVO, F.E., & LUSHBAUGH, T.S. (1971). Internatl. J. Biochem. **2,** 55.

Table 1.

Anaerobic fermentation by resting cells of *Entamoeba histolytica.* The results given in the table are averages from data for six experiments reported by Montalvo, Reeves and Warren[6].

Glucose utilized	0.70 μmoles/min/ml cells
Products identified expressed as moles per mole glucose utilized.	
CO_2	2.02
H_2	0.60
Acetate	0.50
Ethanol	1.36
Total glucose carbon accounted for	96 %

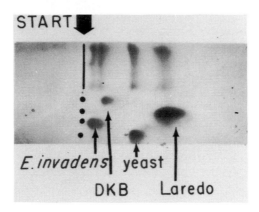

Fig. 1.

Electrophoresis of glucose-6-P isomerase from various sources on cellulose acetate strips.

In the upper half of the strip all four isomerases were superimposed upon a single starting line. In the lower half of the strip each enzyme was separately spotted along the starting line. Electrophoresis was for 1 hr in a barbital buffer, pH 8. Strips were stained for enzyme activity employing a solution containing F-6-P, glucose-6-P dehydrogenase, NADP, nitrobluetetrazolium, and phenazine methosulfate.

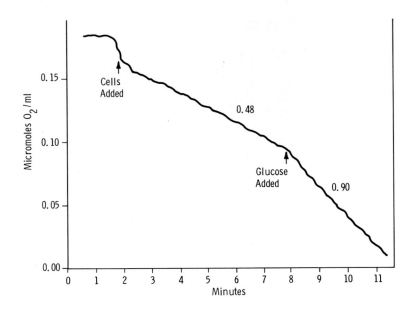

Fig. 2.
Respiration by *Entamoeba histolytica* and its stimulation by added glucose.
At the start of the experiment the oxygraph cell contained 1.3 ml of a balanced salt solution. At 1.7 min 0.04 ml ameba cells which had been preincubated at 37° for 30 min were added in a volume of 0.2 ml. At 7.6 min 5 μmoles of glucose in 0.1 ml water was added. The numbers above the two slopes refer to the rate of oxygen consumption in μmoles/min/ml cells.

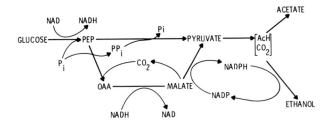

Fig. 3.
A schematic diagram of the proposed pathway for glucose metabolism in *Entamoeba histolytica*.

25

BIOCHEMISTRY OF COCCIDIA

J.F. Ryley
Imperial Chemical Industries Ltd.,
Pharmaceuticals Division,
Alderley Park,
Macclesfield, Cheshire, England

Coccidiosis biochemistry is of interest from the practical standpoint as we seek an explanation for the mode of action of anticoccidial drugs, and more particularly as we seek new methods of attacking the parasites. It is of interest from the biological viewpoint as we seek to study the processes by which these protozoa live, the functioning of their varied subcellular organelles, the reasons for their obligate parasitic mode of life and the influence of a changing environment on metabolism.

The stages in the life cycle of a typical coccidian are summarized in Fig. 1. The unsporulated oocyst voided in the faeces of an infected host must undergo the process of sporulation outside the host, and after a dormant period of months or even years, must excyst within the intestines of a new host to liberate the infective stage, the sporozoite. Within the new host the sporozoite must seek out and then penetrate a suitable cell before it can reproduce by the asexual process of schizogony. Rapid growth prior to multiple division requires an adequate supply of nutrients, and is the biochemically least understood phase of the life cycle due to difficulties inherent in studying an intracellular parasite. After two or more cycles of asexual reproduction, apparently identical merozoites give rise to sexual cells which eventually result in the discharge of oocysts.

The oocyst wall

The oocyst wall is of considerable interest, since it allows the coccidium to survive outside the host for long periods of time under unfavourable circumstances. Although physical problems may be expected in bringing a potential disinfectant into actual contact with the oocyst in say a broiler house, the most difficult problem in developing an effective disinfectant is the impermeable nature of the oocyst wall. Certain small molecules such as ammonia, methyl bromide and carbon disulphide are apparently able to penetrate the wall, and in high concentrations can bring about oocyst sterilization; on the other hand, conventional disinfectants, even in high concentration, are totally without effect, and it may be remembered that oocyst cultures in the laboratory are normally stored in potassium dichromate or sulphuric acid solutions! Gas exchanges can take place - the oocyst respires, particularly during the process of sporulation - and water can pass slowly through the wall - oocysts can be killed by dessication, and exhibit plasmolysis and bizzare distortion of the walls and contents after prolonged exposure to hypertonic salt solutions. Chemical and physical studies on the wall coupled with a study of the wall-forming bodies in the macrogametocyte have thrown some light on the nature of this remarkable structure.

The shell consists basically of two layers. This can be discerned by careful light microscopy, but its refractile nature makes study difficult. The two layers are better seen in section in the electron microscope, which reveals that they have a different consistency, and that the inner layer has a narrow dense band on its outer aspect. Monné and Hönig (1954) were able to draw some quite amazing conclusions from a combination of microscopy and a number of potentially destructive treatments. The shell is birefringent when examined in polarized light, but this birefringenoe is due chiefly to the inner layer. The outer layer is either isotropic or weakly birefringent. Because it fails to dissolve in sodium sulphide or thioglycollate it cannot be a keratin-type of protein, and because it will dissolve in hypochlorite and because it reduces ammoniacal silver nitrate it must be a quinone-tanned protein! The outer part of the inner layer is negatively birefringent in a radial direction, suggesting its protein nature, while the inner part of the inner layer scatters light and is positively birefringent, suggesting that it is mainly lipid. A radical separation of the two components of the inner layer is not however visualised; rather it is regarded as a protein matrix impregnated with lipid, with a particularly high concentration of lipid on the inner aspect.

Two types of wall-forming bodies may be seen in the developing gametocyte. These gradually assume a peripheral location, and subsequent to fertilization, fuse to form the wall. The extremely dense, osmiophilic Type I granules give a PAS positive reaction which is not influenced by amylase pretreatment, and subsequently give rise to the outer layer, while the less dense Type II granules, which give a strong reaction for protein with mercuric bromphenol blue, eventually fuse to form the inner layer. A strong reaction for β-galactosidase can be detected in the granules, and subsequently in the region of the wall, and the enzyme is evidently somehow implicated in wall formation. No differences in optical properties have been noted between unsporulated and sporulated oocysts, but it has been found that incubation of freshly passed oocysts with the alkaloid colcemid facilitates fixation and staining for cytological studies of sporulation, apparently by delaying hardening of the wall. It is desirable that the nature of possible changes in the wall and the permeability of the wall during sporulation be investigated.

We have carried out some preliminary chemical studies on the cyst wall of *E. tenella.* Oocysts were isolated from faeces by salt flotation and purified by washing on a glass bead column followed by sucrose density gradient centrifugation. The outer layer was removed by digestion with 10 % sodium hypochlorite at room temperature for 30 min, and the inner layer removed by shaking the treated oocysts with glass beads and washing the broken shells free from disintegrated oocyst contents. Dry weight determinations showed that the outer layer accounted for about 20 % of the total dry weight of the unsporulated oocyst. Difficulties have been experienced; due to the large amounts of inorganic material in the hypochlorite extracts, but it is evident that the outer layer contains carbohydrate and a protein characterized by high proline and an absence of basic amino acids. The inner layer contains a very small amount of carbohydrate, and consists mainly of protein (about 70 %) and lipid (about 30 %) extractable by hot chloroform. The protein displays no particularly unusual features, but the lipid appears to be a mixture of waxes with chain lengths up to C_{40} containing very small amounts of nitrogen and phosphorus. Table 1 summarizes an experiment carried out to determine the location of protein in the oocyst wall. The wall protein of intact oocysts was virtually inaccessible to enzyme attack, and only a small amount could be solubilized following stripping of the outer layer with hypochlorite. It was not until the inner surface of the inner layer had been exposed by fragmentation that real susceptibility to protease digestion was observed.

Sporulation

Sporulation appears to be a strictly aerobic process, although coccidia can survive and carry on some form of metabolism under anaerobic conditions. Dürr and Pellérdy (1969) find that 302-395 μl O_2 per 10^6 oocysts is required by *E. stiedae* for sporulation to occur. Wilson and Fairbairn (1961) studied the process with *E. acervulina* occysts incubated in 0.1 $N\text{-}H_2SO_4$ at $30°C$ with continuous shaking (Fig. 2). They noted a steadily declining respiratory rate which was correlated with a fall in RQ from an initial value of 1.12 to 0.8-0.9. During an initial 10 hr period, by which time 4 sporoblasts had formed, they observed an almost 50 % utilization of polysaccharide. After 10 hr, metabolism appeared to be at the expense of lipid reserves, and there was in fact some resynthesis of polysaccharide. Strout, Botero, Smith and Dunlop (1963) observed a 24 % decrease in lipid during sporulation and a halving of oocyst dry weight. Wagenbach and Burns (1969) followed the respiration of *E. tenella* and *E. stiedae* oocysts polarographically and correlated this with a variety of notable cytoplasmic events in the sporulation process (Fig. 3). A more irregular curve was obtained than by Wilson and Fairbairn, and an early temporary depression in respiratory rate was correlated with the appearance of the spindle stage. Respiration is sensitive to cyanide, and cyanide prevents sporulation taking place. The effect is however reversible, and sporulation will occur on removal of the inhibitor.

Once sporulation has taken place, oocyst respiration and metabolism fall to a barely detectable level until ingestion by a nex host causes excystation to occur. Nevertheless, over long periods of storage, depletion of polysaccharide reserves takes place, and this may explain the gradual decline in infectivity, due possibly to an inability of the sporozoite to actively excyst and penetrate a host cell. Thus Vetterling and Doran (1969) found that the polysaccharide content of *E. acervulina* fell from 33.3 μg glucose/10^6 oocysts at 3 months to 21.3 μg at 1 year, 7.8 μg at 2 years and 1.5 μg at 6 years. Oocysts can survive under anaerobic conditions, but the consequent less efficient utilization of reserves reduces the survival time. Sporulation must occur before an oocyst becomes infective to a fresh host, and is one possible point of attack when seeking to control the desease. The process, and any interference with it by speculative chemicals, can be readily followed microscopically, but the methods by which active chemicals could be used present more of a problem. It might be possible to administer an active compound in the food in the hope of preventing sporulation in the faeces and so preventing build-up of infection in the environment, or it might be possible to impregnate the litter in such a way that freshly passed oocysts would be prevented from sporulating and so break the cycle.

The reserve polysaccharide of the coccidia is particularly evident in the macrogametocyte and oocyst, although lesser amounts are found in mature merozoites. Cytochemical studies have led many people to describe the substance as glycogen, paraglycogen or coccidienglykogen, although the material is visualised in the electron microscope as large oval bodies having a high affinity for lead. Protein, and to a lesser extent lipid, can interfere with the iodine-staining properties of polysaccharides, and the specificities and limitations of the PAS reaction have not always been recognised. Isolation of material from coccidia and from gregarines by digestion with KOH and precipitation with ethanol has yielded a product which when examined by physico-chemical and enzymatic methods has shown properties typical of amylopectin rather than glycogen (Table 2). Although glycogen and amylopectin are both composed of linear chains of α-1,4-linked glucose residues which are joined by α-1,6-glucosidic interchain linkages to form a multiply branched macromolecule, amylopectin has much longer unit chains and hence a more open structure. This architectural difference between the two polysaccharides results in differing physical properties - solubility, viscosity, interaction with proteins and with iodine, physical form of intracellular deposits etc - and in tissues so far examined (unfortunately not including the coccidia as yet) arises from differences in the pathways by which the two polysaccharides are synthesised. An analogy may be made with charcoal and diamond, two rather dissimilar forms of the same basic substance! The coccidia have therefore energy reserves similar to those found in rumen ciliates and plants, rather than the animals in which they live.

Excystation

Although anomalies have been claimed from Japan, it is generally agreed that two separate stimuli are necessary before excystation will take place (Jackson, 1962). With avian coccidia, the normal first process is undoubtedly a mechanical fracture of the oocyst shell in the gizzard. All coccidia however appear to be susceptible to a more subtle change which is brought about under the influence of carbon dioxide, and in ruminants and other species where mechanical stresses generated by a gizzard are not present, carbon dioxide "triggering" would seem to be the physiological natural process. All coccidia seem to have a micropyle in the oocyst wall, although this is much more noticeable and prominent in some species than in others. Under the influence of CO_2, a lifting or splitting of the cap over the micropyle takes place and a permeability change or perforation of the wall in the micropylar region takes place. The oocyst contents will now collapse in hypertonic salt solutions, the inner wall can be stained with methylene blue and the contents can be destroyed by hypochlorite.

The optimal concentration of carbon dioxide and duration of treatment required for "triggering" varies with the species, and the process is helped by the presence of reducing agents. Body temperature is almost essential, and by studying the effect of temperature on the rate of "triggering", Hibbert and Hammond (1968) concluded that activation or production of an enzyme was involved. The origin of this presumed enzyme is an interesting matter for speculation. Nyberg, Bauer and Knapp (1968) reported "that during pretreatment with CO_2, a thinning or indentation occurred in the micropylar region of the oocyst wall. The oocyst residual body became prominently located at the peripheral central portion of the thinned area". There is however some doubt and disagreement as to whether *E. tenella* has an oocyst residual body! It is conceivable that a residual body or polar granule where present could be involved in this micropylar alteration, or the source of enzyme may possibly be the fluid which fills the oocyst around the sporocysts.

The second stage in the excystation process involves the action of trypsin and bile. These bring about an activation of the sporozoites in the sporocyst and the digestion of the Stieda body which plugs a hole in the sporocyst wall. Bile may facilitate the entry of enzymes through the altered micropyle or may alter the protein or lipoprotein surface of the Stieda body. Although not essential for the excystation of every species of coccidia, bile does seem to be necessary to initiate sporozoite motility; sporozoites can be liberated from oocysts of some species in the absence of bile, but the process is slower and the sporozoites have little motility compared with the pivoting, flexing and gliding sporozoites produced in the presence of bile. Trypsin (or chymotrypsin) seem to digest the sporocystic plug, although Doran believes that the sporozoite also secretes an enzyme which acts on the inner surface of the plug. Histochemical tests show protein to be present in the Stieda body, although the intensity of the reaction varies with species; PAS positive material stable to amylase digestion is additionally found in the sub-Stieda body where present. On addition of trypsin plus bile to "triggered" oocysts, sporozoites become motile within the sporocyst, the Stieda body becomes swollen and eventually disappears. Before actually disappearing, the Stieda body may be actively forced out of the gap it is plugging. The sporozoites then escape in turn through the small hole so formed, a process which is remarkably quick and which involves a constriction passing along the sporozoite body because of the small diameter of the hole.

Although one can visualise natural movements of the sporozoite being actively responsible for its escape, the rapidity of this escape through such a narrow hole and the popping out of the partly degraded Stieda body suggests the possibility of expulsion under pressure generated within the sporocyst. It is interesting to speculate whether this may arise from osmotic forces generated by say the hydrolysis of the amylopectin abundantly present in the sporocyst residual body; a functional explanation of these membrane-bound residual bodies, which usually disintegrate during excystation, is more attractive than imagining they are merely the consequence of bad budgeting on the part of the developing oocyst.

Initiation of excystation causes a tremendous stimulation in respiration of the hitherto dormant oocyst. Fig. 4 summarizes observations made by Vetterling (1968) on three species of poultry coccidia. When sporulated oocysts were mechanically ground and incubated in trypsin-bile at pH 7.5 and 41.5°C for 30 min, oxygen consumption followed with an electrode rapidly increased to a peak at 1.5-6 mins depending on the species, remained steady for a few minutes, decreased to a lower rate for another few minutes and then decreased to a yet lower rate for the remainder of the determination. This latter rate is presumably maintained while the sporozoite seeks out and penetrates a host cell. During excystation and cell penetration, the sporozoite utilizes most of its remaining amylopectin stores to meet energy requirements. Vetterling and Doran (1969) observed that during a 30 min period of excystation at 42.9°C, carbohydrate reserves in the sporozoites of three species of avian coccidia fell by about two thirds (Table 3). This utilization when correlated with oxygen consumption values previously obtained by Vetterling gave oxygen/glucose ratios in excess of the maximum of 6 for complete oxidation, suggesting that some other substance, possibly lipid, was additionally being oxidised. Histochemical tests with the PAS reaction show a corresponding decrease in the size and number of carbohydrate granules in the sporozoite and also in the sporocyst residual body during the process of excystation and cell penetration.

The Sporozoite

The sporozoite is an attractive stage for metabolic studies, since it can readily be obtained free from host tissues and contaminating microorganisms. Interesting problems worthy of further investigation concern the chemical and physical basis or motility in the absence of cilia or flagellae, the mechanisms by which the sporozoite seeks out and then actually penetrates a suitable host cell, and the possibility of destroying the sporozoite in the gut lumen by prophylactic drugs before infection actually takes place.

Preliminary studies show that the energy metabolism of *E. tenella* sporozoites is in many respects similar to that of toxoplasms and malaria parasites, except that it has an appreciable endogenous respiration due to the presence of amylopectin reserves. An 18 % respiratory stimulation was observed with glucose and a 46 % stimulation with fructose; oxygen uptake in the presence of fructose was 330 μl/hr/10^9 sporozoites. Respiration was sensitive to cyanide; 73 % inhibition was observed with 10^{-4}M and 93 % inhibition with 0.46 x 10^{-3}M-KCN. Homogenates of sporozoites prepared by shaking with glass beads oxidised p-phenylene diamine, and this oxidation was stimulated 30 % by added mammalian cytochrome **c**. Spectroscopic examination of thick sporozoite suspension showed cytochrome bands at 556-562 and 600-610 mμ; the cytochrome **a** band at 600-610 mμ was particularly prominent. Using diaminobenzidine, we were able to locate cytochrome oxidase at the EM level on the tubular cristae of the mitochondria (Fig. 5). Incubation was carried out without prior fixation, and this resulted in rather swollen mitochondria and poor preservation of other cellular structures. Incubation of unfixed sporozoites with DPNH and the tetrazolium salt MTT in the presence of cobalt followed by treatment with ammonium sulphide, showed oxidation had taken place and formazan was deposited in bodies which may be mitochondria or possibly structures analagous to the microbodies of the trypanosomes (Fig. 6).

Under anaerobic conditions, metabolism was maintained at the expense of amylopectin reserves with the production of organic acid and CO_2. From Table 4 it will be seen that the major metabolite was lactic acid with lesser amounts of carbon dioxide and glycerol; 97 % of the amylopectin used was accounted for in this experiment. Anaerobic acid production (Table 5) was stimulated by glucose and lactose, and to a lesser extent by fructose and mannose; amylopectin usage was spared by glucose, fructose and mannose, but not by lactose.

The conoid is probably involved in penetration of host cells by both sporozoites and merozoites, and a number of electron microscope studies have shown it in an extruded position (eg. Fig. 7). Reversible protrusion of the conoid can be obtained by suspending sporozoites in 20 % glycerol, and it is possible that such a reversible protrusion may be connected with membrane penetration. The club-shaped organelles have a gland-like appearance with ducts extending into the conoid, but so far no demonstration of lytic enzymes in these structures has been achieved. Similarly the demonstration of a secretory function for the micronomes is lacking. In view of the rapidity of cell penetration, it is conceivable that the conoid could act like a captive-bolt pistol to secure mechanical puncturing of the membrane of the prospective host cell. Although penetration of toxoplasms into tissue culture cells can be enhanced by adding lysosyme or hyaluronidase, a similar effect could not be demonstrated with the morphologically similar sporozoites of *E. adenoides,* nor could hyaluronidase be detected in sporozoites or oocysts. Thus a most interesting problem involving mechanics and/or biochemistry still awaits solution.

Intracellular stages

Several attempts were made in the late 1940's to study the metabolism of intracellular coccidia, but little more could be done than measure the rate of oxygen uptake or glycolysis of normal and infected intestinal or liver tissues. More recently the problem has been tackled by histochemistry, although here again, little more can be done than detect the presence or absence of an enzyme system. Very often - as with the phosphatases - the enzymes studied have been selected on a basis of the availability of simple techniques rather than the physiological importance of the system, and the qualitative identification of an enzyme is far removed from its quantitative evaluation in the dynamic biochemistry of the cell. In many cases, histochemical reactions have been observed as granular deposits by light microscopy, but a more rewarding approach is to investigate the location of the reaction in subcellular organelles by electron microscopy (eg. Figs 5 and 6).

Most interesting, though perhaps not entirely warranted conclusions have been drawn by Beyer (1970) from a series of studies with rabbit and chicken coccidia in which she investigated reactions for α-glycerophosphate and succinic dehydrogenases at various stages of the life cycle. Beyer considers that under the conditions of ample nutrition provided by the host cell, glycolysis can easily cover the energy requirements of the developing schizont; this is indicated by the absence of succinic dehydrogenase In the later stages of growth, as the host cell substrate becomes exhausted and as the merozoites differentiate and prepare for a temporary extracellular period of existence, transition to aerobic metabolism with the Krebs cycle takes place; this is indicated by the appearance of succinic dehydrogenase Similarly the growing macrogamete obtains its energy by glycolysis - as indicated by a high level of α-glycerophosphate dehydrogenase and the absence of succinic dehydrogenase. Immediately after fertilization, high levels of both enzymes can be found. Beyer considers this an indication of change to oxidative metabolism, which would be a more economical way of substrate utilization in an independant organism which has to provide energy to support sporulation and later excystation and cell penetration. Such an alternation of metabolic pathways with changing environment is reminiscent of the trypanosomes; what stimuli are required to bring about the necessary transformations and how they act is a mystery.

From the practical standpoint, the most important aspect of the intracellular parasite concerns nutrition and growth factor antagonism by chemotherapeutic agents. Virtually nothing is known concerning the macronutrients of the coccidia, and studies are made difficult by the obligate intracellular nature of the parasite. It may be that studies using the newly-developed tissue culture techniques will prove more rewarding, since here at least the contributions of the intestinal flora are eliminated. In spite of the difficulties, Warren (1968) was able to show the need for a number of growth factors by feeding deficient diets to chicks infected with *E. acervulina* or *E. tenella* and subsequently measuring oocyst output as a measure of coccidial growth and weight gain as a measure of host-diet interaction. Data for the six most active substances in this investigation are illustrated in Fig. 8. Pantothenate, p-amino benzoic acid, ascorbic acid and pyridoxine showed slight effects when diets deficient in them were fed, while a number of other factors could not be implicated in coccidial metablism by these methods. Although thiamine deficiency had a small effect on chick growth, it had a much more marked effect in reducing coccidial proliferation. One of the most successful coccidiostats, amprolium, exerts its action by differential thiamine antagonism.

Thiamine   Amprolium

Chicks on a normal diet are protected against *E. tenella* infection by 125 ppm amprolium, whereas more than 800 ppm dietary amprolium is necessary before chicks show the growth depression and polyneuritis characteristic of thiamine deficiency. Just as the toxic effects on the host can be overcome by increased dietary thiamine, so too the anticoccidial effects can be nullified by thiamine **in vivo** or in the chick embryo or in tissue culture. Because amprolium lacks the hydroxyethyl group of thiamine, it cannot be pyrophosphorylated, and presumably cannot therefore inhibit at the coenzyme level. It appears to reduce the uptake of thiamine from the intestine in a way which affects the parasite more than the host, but in view of the **in vitro** studies, presumably must also affect thiamine uptake by the parasite. Biotin and nicotinic acid also appear from Warren's studies to be very important in coccidial nutrition.

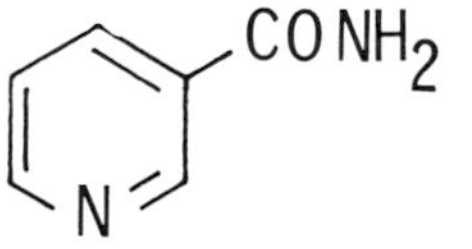

Nicotinamide

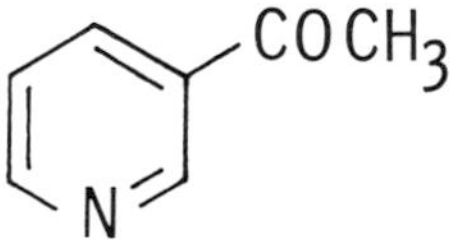

3 - acetyl pyridine

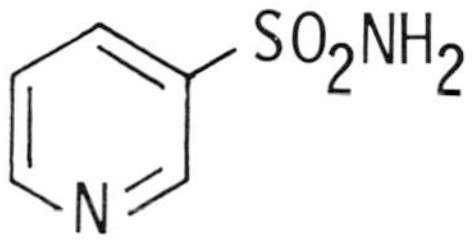

Pyridine - 3 - sulphonamide 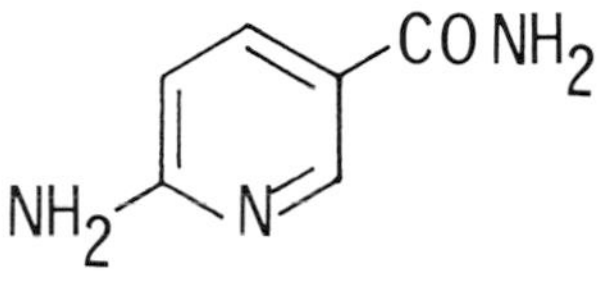   6 - amino - nicotinamide

The nicotinamide antagonists pyridine-3-sulphonamide and 6-amino nicotinamide show activity against *E. acervulina* and *E. tenella* or *E. necatrix* respectively (but not vice versa), but there is little margin between active and toxic levels. The only method of antagonising the biotin requirements of coccidia so far known is feeding diets containing 50 % egg white - hardly a practical proposition

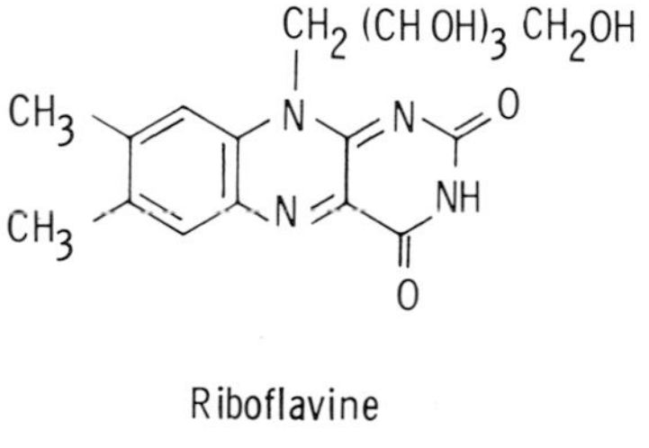

Riboflavine

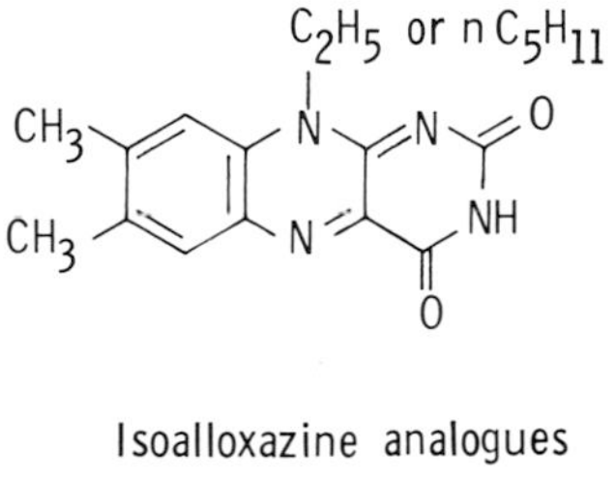

Isoalloxazine analogues

Several isoalloxazines are known which antagonise riboflavin in the case of *E. acervulina* but not *E. tenella*; the compounds however are too toxic and required at too high levels to be of practical value.

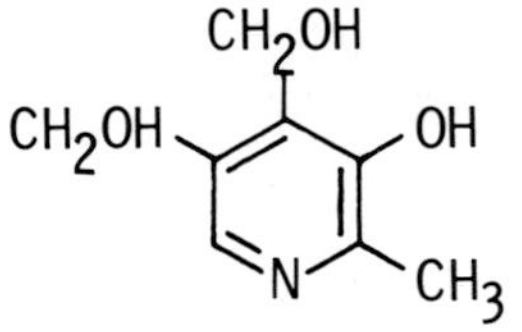

Pyridoxine

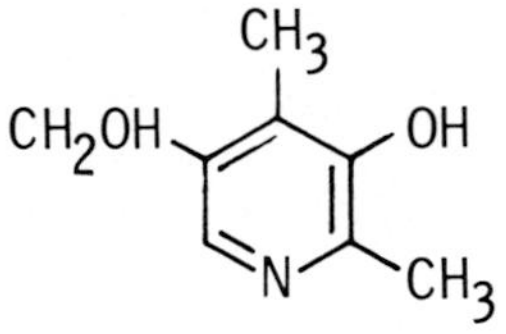

4 - desoxypyridoxine

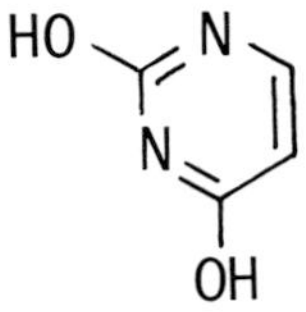

Uracil

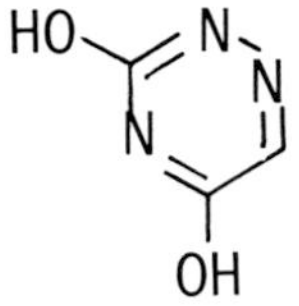

6-aza Uracil

Likewise pyridoxine in *E. acervulina* can be antagonised by 4-desoxypyridoxine, and presumably some aspect of pyrimidine metabolism by 6-azauracil, although neither compound is of value as a practical cocci-diostat. The only area of coccidial nutrition other than thiamine where antimetabolite action is of practical value is the pAB-folic acid-folinic acid pathway. The oldest synthetic anticoccidials, which inaugurated the age of anticoccidial chemotherapy, and which are still of value in treatment and prophylaxis, are the sulphonamides, the well-known pAB antagonists. More recent in discovery are the pAB-like substituted benzoic acids typified by ethopabate, which have anticoccidial activity in some strains of some species of coccidia, and apparently interfere with the conversion of dihydropteroic to difydrofolic acid.

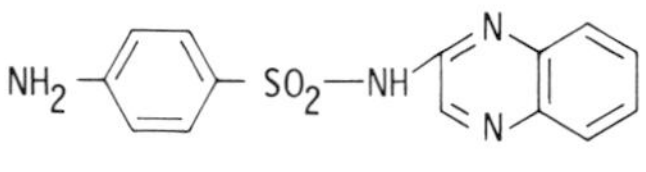

Sulphaquinoxaline

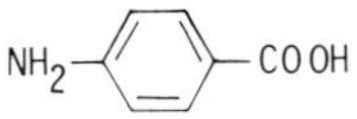

p-aminobenzoic acid

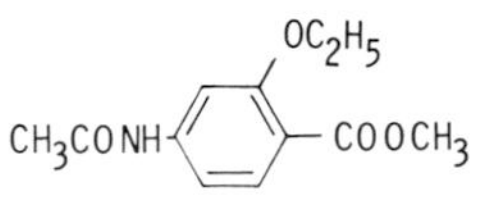

Ethopabate

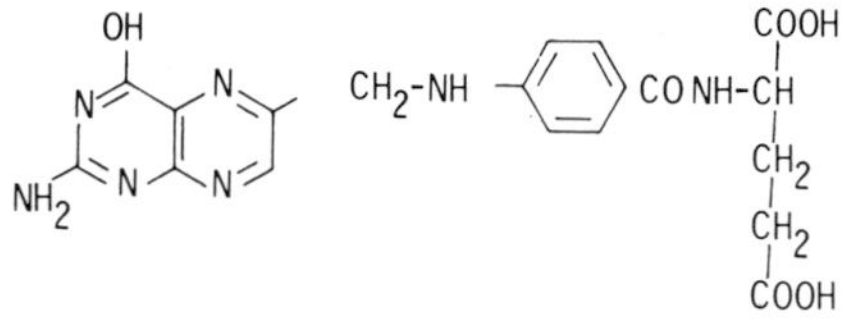

Folic acid

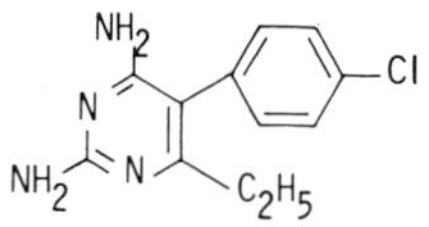

Pyrimethamine

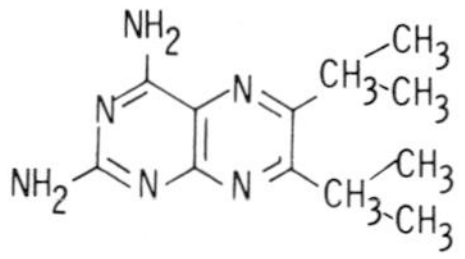

2,4-diamino - 6,7 -di isopropylpteridine

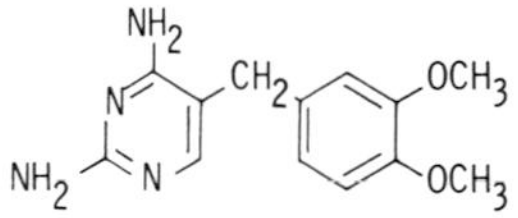

Diaveridine

Acting still later in the folic acid pathway are pyrimethamine, diaveridine and other diaminopyrimidines, dihydrotriazines and pteridines, which although they have not found usage as anticoccidials in their own right, yet are valuable as potentiators of sulphonamides. The field of growth factor antagonism is still a happy hunting ground for chemotherapy research; whether it is possible to devise further compounds which besides inhibiting a particular metabolic pathway or requirement are able to actually gain access to the parasite and exert a greater effect on parasite than host, time alone will tell.

References

BEYER, T.V. (1970). J. Parasitol. **56,** 28.
DURR, U. and PELLERDY, L. (1969). Acta vet. Hung. **19,** 307.
HIBBERT, L.E. and HAMMOND, D.M. (1968). Exp. Parasitol. **23,** 161.
JACKSON, A.R.B. (1962). Nature, Lond., **194,** 847.
MONNE, L. and HONIG, G. (1954). Ark. Zool., Stockholm, **7,** 251.
NYBERG, P.A., BAUER, D.H. and KNAPP, S.E. (1968). J. Protozool. **15,** 144.
STROUT, R.G., BOTERO, H., SMITH, S.C. and DUNLOP, W.R. (1963). J. Parasitol. **49,** (Suppl.) p. 20.
VETTERLING, J.M. (1968). J. Protozool. **15,** 520.

VETTERLING, J.M. and DORAN, D.J. (1969). J. Protozool. **16,** 772.
WAGENBACH, G.E. and BURNS, W.C. (1969). J. Protozool. **16,** 257.
WARREN, E.W. (1968). Parasitology **58,** 137.
WILSON, P.A.G. and FAIRBAIRN, D. (1961). J. Protozool. **8,** 410.

Table 1.
Susceptibility of oocyst walls to digestion.

	% amino nitrogen solubilized		
Enzyme	**intact, complete oocysts**	**unbroken, stripped oocysts**	**inner wall fragments**
Pepsin	0.49	0.4	12.5
Trypsin	0.00	3.8	91.5
Pronase	0.07	10.2	58.0

Table 2.
Characterization of coccidial reserve polysaccharides.

			E. tenella	Gregarine	Glycogen	Amylopectin
Iodine staining	in water	λ max	535	522	445	530
		ϵ max	1.05	0.90	0.1	1.1
	in half-saturated ammonium sulphate	λ max	540	528	460	560
		ϵ max	1.33	1.48	0.6	1.6
	Ratio $\dfrac{\epsilon_{max} \text{ (amm. sulph.)}}{\epsilon_{max} \text{ (water)}}$		1.27	1.65	6.00	1.45
	Chain length		21	19	12	23
Enzymatic degradation	β-amylolysis limit (%)		56	54	40-50	50-60
	α-amylolysis limit (%)		92	93	71-84	90-96
	Average chain length		20	21	10-14	20-25
	Average exterior chain length		14	14	6-9	12-17
	Average interior chain length		5	6	3-4	5-8

Table 3.
Polysaccharide utilization during excystation.

	μg glucose per 10^6 oocysts		
	E. acervulina	*E. necatrix*	*E. meleagrimitis*
Before excystation	33.3	30.1	36.7
After excystation	11.0	9.4	13.3
mols O_2/mols glucose used during excystation	6.35	5.87	7.08

Table 4.
Anaerobic endogenous metabolism of *E. tenella* sporozoites.

	μ-mols	mols/mol glucose
Amylopectin (as glucose)	- 30.05	- 1.00
Acid (from bicarbonate)	+ 50.25	+ 1.67
CO_2	+ 14.95	+ 0.50
Lactic acid	+ 45.50	+ 1.49
Pyruvic acid	0.00	0.00
Succinic acid	0.00	0.00
Glycerol	+ 8.60	+ 0.28

Sporozoites of *E. tenella* (16 mg N in total vol. of 6 ml) incubated in Ringer-bicarbonate medium for 90 min in Warburg flasks at 41°C with gas phase of 5 % CO_2 - 95 % N_2. Fermentation stopped by tipping acid from side bulb. Initial and final values based on the combined contents of 4 manometer flasks. Values given are total amounts of substrate used and metabolites recovered and also amounts of metabolite formed per mol substrate used.

Table 5.
Anaerobic carbohydrate utilization by *E. tenella* sporozoites.

	Acid (bicarbonate) (μ-equivs.)	CO_2 (μ-mols)	Amylopectin (μ-mols glucose)
Blank	19.56	1.66	- 8.05
Glucose	32.01	3.48	- 0.69
Glycerol	19.38	2.15	- 6.78
Maltose	19.60	2.64	- 7.28
Sucrose	19.40	2.54	- 4.99
Blank	16.25	2.55	-10.31
Fructose	25.45	4.38	- 5.40
Lactose	30.59	2.95	- 8.38
Galactose	15.28	3.26	- 8.50
Mannose	23.93	4.61	- 3.04

Sporozoites of *E. tenella* (3.7 and 2.7 mg N in total vol of 3 ml in two separate experiments) incubated in Ringer-bicarbonate medium in Warburg flasks at 41°C for 90 min with gas phase of 5 % CO_2 - 95 % N_2 Amylopectin determined in initial and final samples after digestion with KOH, precipitation with ethanol and hydrolysis.

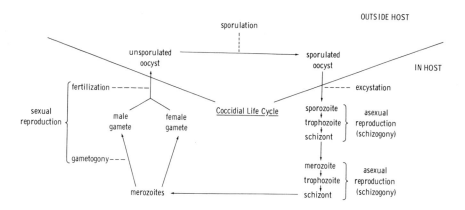

Fig. 1.
Typical coccidial life cycle

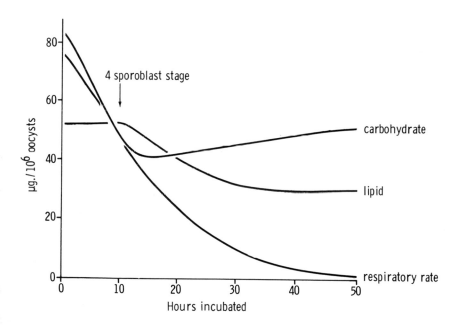

Fig. 2.
Metabolism of *E. acervulina* during sporulation (adapted from Wilson and Fairbairn, 1961).

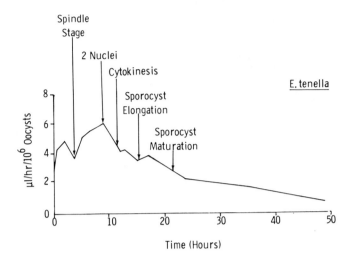

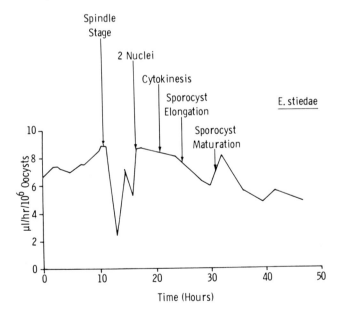

Fig. 3.
Respiration of *E. tenella* and *E. stiedae* during sporulation (adapted from Wagenbach and Burns, 1969).

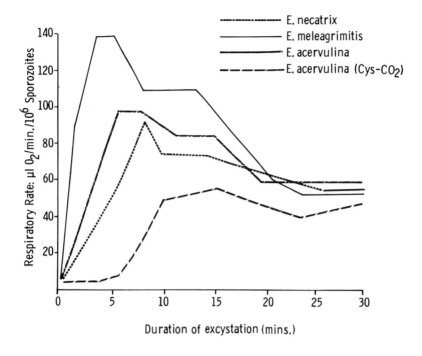

Fig. 4.

Respiration of coccidia during excystation (adapted from Vetterling, 1968).

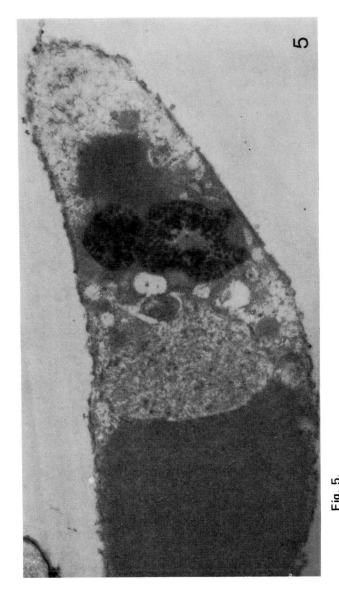

Fig. 5. Sporozoite of *E. tenella* treated with diaminobenzidine to show mitochondrial location of cytochrome oxidase (X 32,000).

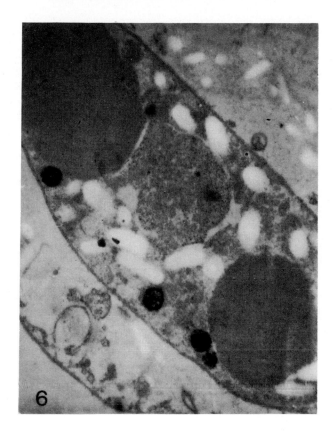

Fig. 6.
Sporozoite of *E. tenella* treated with DPNH and tetrazolium salt (X 26,000).

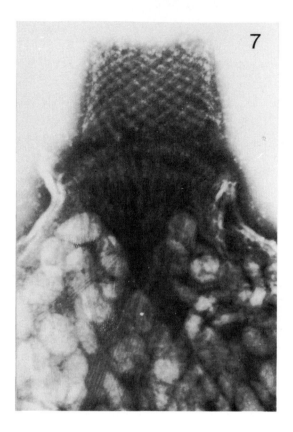

Fig. 7.
Sporozoite of *E. tenella* negatively stained with phosphotungstic acid showing protruded conoid (X 80,000).

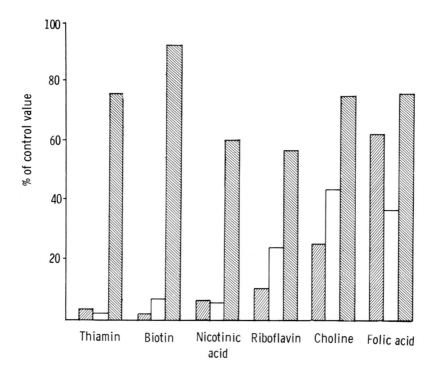

Fig. 8.
Effects of vitamin deficiency on coccidial infections (adapted from Warren, 1968). ▨ *E. acervulina* oocysts, ☐ *E. tenella* oocysts, ▧ % weight gain in 12 days.

26

LOSS OF FATTY ACID BIOSYNTHESIS IN FLATWORMS

Franz Meyer and Haruko Meyer
State University of New York
Upstate Medical Center
Syracuse, New York 13210

It is well established that animals require molecular oxygen for the biosynthesis of two classes of lipids. Oxygen is involved in the conversion of long-chain fatty acids to their corresponding unsaturated analogues[1] and participates in several steps in the biosynthesis of cholesterol: first, in the conversion of squalene to 2,3-oxidosqualene and, later, in the removal of methyl groups and in the introduction of the 5,6 double bond into the ring structure[2]. Anaerobic organisms can synthesize neither sterols nor polyunsaturated fatty acids though they can synthesize monounsaturated fatty acis **via** a novel oxygen-independent pathway[1]. Based on this knowledge, it is somewhat puzzling that cestodes which live in a predominantly anaerobic environment are rich in both sterols and polyunsaturated fatty acids[3]. These considerations prompted us to study the lipid metabolism in cestodes using the larval and adult forms of the tapeworm *Spirometra mansonoides.*

The results of these experiments have been previously published[4] and are presented here only in schematic form in Figure 1.

Figure 1.
Loss of Fatty Acid and Sterol Biosynthesis in Flatworms

Cestoda: *Spirometra mansonoides*[4], *Hymenolepis diminuta*[5, 6]
Trematoda: *Schistosoma mansoni* [7, 18, 19]
Turbellaria: *Dugesia dorotocephala* [7]

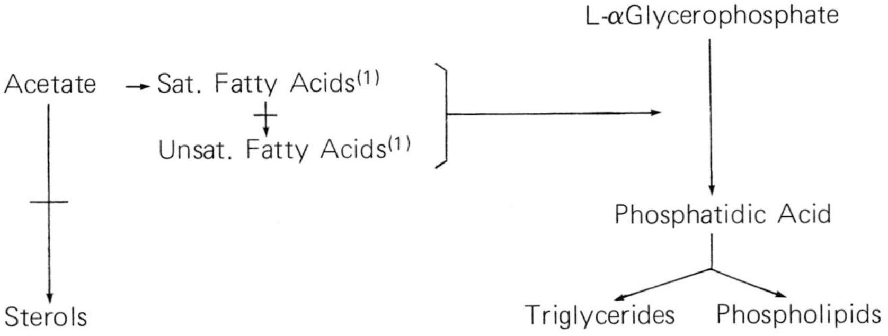

L-αGlycerophosphate

Acetate → Sat. Fatty Acids[1]
          +
         Unsat. Fatty Acids[1]

Phosphatidic Acid

Sterols

Triglycerides   Phospholipids

[1] All four organisms have a limited capacity to chain elongate preformed fatty acids by the addition of acetate.

As indicated by the crossbars, **Spirometra mansonoides** has lost the ability to synthesize sterols and unsaturated fatty acids. It has also lost the ability to synthesize long-chain saturated fatty acids, a synthesis which does not require molecular oxygen. By contrast, the parasite has the capacity to synthesize all of its complex lipids provided that an exogenous supply of fatty acids is available to it. Similar results were obtained in Fairbairn's laboratory with another cestode, *Hymenolepis diminuta* [5, 6].

From these findings two possibilities immediately presented themselves: first, that the metabolic defects are a consequence of the anaerobic environment in which the cestodes live or, second, that these defects are an adaptation to the parasitic mode of life. To explore the first point, we studied **Schistosoma mansoni**, a parasitic flatworm which lives in an aerobic environment. The results showed[7] that this parasite also lacks the ability to synthesize fatty acids and sterols (Figure 1). The absence of oxygen **per se,** therefore, cannot be the cause of this deficiency. To consider the second point - that the metabolic defect is a response to a parasitic mode of life - we investigated the free living-flatworm *Dugesia* *dorotocephala*. Again, the findings in this free-living flatworm were the same as those in the parasitic forms[7]. This unexpected result eliminated the possibility that parasitism **per se** is responsible for the loss of fatty acid and sterol biosynthesis.

We then had to consider as an alternative, the possibility that the metabolic defect is a characteristic of the entire phylum *Platyhelminthes.* For this reason we became interested in another free-living flatworm, the marine acoelous turbellarian *Convoluta roscoffensis.* Acoels are of particular significance to our problem because they are generally considered to be the most primitive flatworms. They display many of the characteristics of their hypothetical planuloid ancestor; they lack an intestine, much of their body is syncytial, and their nervous and reproductive systems are relatively simple[8]. The adult form of *Convoluta roscoffensis* harbors the green algae *Platymonas convolutae* as symbionts. The infection of the worm occurs in each generation at the larval stage by ingestion of free-living algae[9]. To characterize this symbiotic relationship in terms of lipid metabolism, we investigated in detail the lipids in 1.) the aposymbiotic larvae, 2.) the symbiotic adults and 3.) the free-living algae, all of which can be cultured in the laboratory[1].

Studies on fatty acid biosynthesis

Analysis of the fatty acids of the three organisms shows (Table 1) that the larval and adult acoels have very similar fatty acid patterns, and that 70 % of their fatty acids are structurally identical with those found in the algae. The acoels, like the algae, contain typical plant polyunsaturated fatty acids of the $\omega 3$ series such as α-linolenic acid (18:3, $\omega 3$, 6, 9) and eicosapentenoic acid (20:5, $\omega 3$, 6, 9, 12, 15) and lack the typical animal polyunsaturated fatty acids of the $\omega 6$ series such as γ-linolenic acid (18:3, $\omega 6$, 9, 12) and arachidonic acid (20:4 $\omega 6$, 9, 12, 15). The other 30 % of the acoel's fatty acids, composed largely of stearic acid and several C20 and C22 unsaturated fatty acids, are not encountered as such in the algae. They can, however, be considered as chain-elongation products of the homologous algal fatty acids, since the unsaturated acids of this group differ from the algal fatty acids with the same degree of unsaturation only in that their double bonds are separated from the carboxyl end by two or four additional carbon atoms. Thus, the fatty acids of the acoels are composed of two types, those which are structurally identical and those which are structurally homologous to the

1 The studies on **Convoluta roscoffensis** were carried out in collaboration with Dr. Luigi Provasoli of Haskins Laboratories, Yale University, New Haven.

algal fatty acids. This relation suggests that the fatty acids of the acoels are of algal origin.

This interpretation is further supported by the results of isotopic tracer experiments. When the aposymbiotic larvae are cultured with ^{14}C-acetate, their fatty acids show considerable variation in terms of specific radioactivity (Table 1). High specific radioactivity is primarily associated with those fatty acids which are structurally homologous to the algal fatty acids.

Fragmentation of these ^{14}C-acids by decarboxylation or ozonolysis and radioassay of the products shows that essentially all of the label is in the carboxyl-terminal fragment of the molecule (Table 2). Since **de novo** fatty acid biosynthesis would result in uniform labeling of the fatty acid molecule from ^{14}C-acetate, the larvae can obviously not synthesize these compounds **de novo,** but can only add ^{14}C-acetate to preformed long-chain fatty acids.

When the adult acoels are cultured with ^{14}C-acetate in the light, the specific radioactivities of their fatty acids are fairly uniform (Table 1). Furthermore, fragmentation of these acids yields both radioactive methyl-terminal and radioactive carboxyl-terminal fragments (Table 2). The distribution of label in the two fragments, however, varies somewhat with the fatty acid type. Those fatty acids which in structure are chain-elongation products of the corresponding algal fatty acids have notably more radioactivity in the carboxyl-terminal fragment (as compared to the methyl-terminal fragment) than those which in structure are identical to the algal fatty acids. The observed labeling pattern can be most readily explained by ascribing it to two processes, **de novo** fatty acid biosynthesis in the algal symbiont and chain elongation of both newly synthesized radioactive and preformed non-radioactive fatty acids by the host. Further evidence in support of this explanation comes from experiments in which adults are incubated in the dark with ^{14}C-acetate for 5 hours or 3 days. Under these conditions it might be expected that the adult acoels, like the larvae, are "functionally aposymbiotic" because their photosynthetic symbionts are metabolically inactive in the dark. And this indeed was found to be the case. The incorporation of ^{14}C-acetate in the adult fatty acids is reduced tenfold and the fatty acids vary greatly in their specific activities (Table 1) with the label being most abundant in the carboxyl-terminal fragments (Table 2).

Although the observed close dependence of fatty acid biosynthesis on the presence or absence of light is typical of photosynthetic organisms[10] and thus points to the endosymbiont as the site of fatty acid biosynthesis in the acoel, other possibilities must be considered. It can be argued, for example, that fatty acid biosynthesis does occur in the acoels, but is suppressed because the energy supplied by the symbiont is limited in the absence of light. This is unlikely, however, since the addition of a mixture of glucose and mannitol[2] to the culture medium does not enhance fatty acid biosynthesis, though both compounds are taken up by the organism. But more importantly, this interpretation cannot account for the above findings, in particular, the observation that all molecular species of fatty acids in the acoels are of the algal type or chain-elongation products of the algal fatty acids and that both fatty acid biosynthesis and desaturation are absent in the larvae.

Further evidence that the algal symbiont is the site of fatty acid biosynthesis comes from another line of investigation, employing cytochemical and auto-radiographic analysis on tissue sections of the acoels. When the acoels are cultured in the light and their lipids labeled with ^{14}C-acetate by the pulse-chase method, radioactivity is initially confined to the algal structure. Prolonging the case period results in movement of the label from the symbiont to the host tissue. In controls kept in the dark label is not accumulated at specific sites but scattered throughout the tissue in the form of weak background radiation.

We infer from these results that **de novo** fatty acid biosynthesis occurs within the endosymbiotic algae but not within the larval and adult forms of *Convoluta roscoffensis.* The acoels obtain their fatty acids from the endosymbiont either directly or, in the case of the larvae, indirectly via the parent. Once supplied, both forms are capable of chainelongating fatty acids by the addition of one or two molecules of acetate.

Studies on the biosynthesis of complex lipids.

While the fatty acid pattern of the acoel closely resembles that of its symbiotic algae, the lipid pattern of this organism is quite distinct from that of the symbiont. The acoels contain relatively large amounts of triglycerides, phosphatidyl choline, and ethanolamine plasmalogen; these compounds are either absent or present only in minute amounts in the algae (Table 3). The observed differences in the lipid patterns suggest that the acoels have the ability to synthesize their own complex lipids.

We investigated this possibility by labeling the lipids with suitable isotopic precursors such as ^{32}Pi, ^{14}C-glycerol or ^{14}C-long-chain fatty acids and by locating the label within the lipid molecules using hydrolytic procedures. Analyses of the labeled lipids and their hydrolytic

---

2 Mannitol is included, since it has been shown in other systems that polyols are frequently the principle products released by the algal symbiont[21].

degradation products shows that both the aposymbiotic larvae and the adult acoel (Table 4) incorporate ^{32}Pi into the non-fatty portion of their phospholipids and ^{14}C-glycerol into the glycerol moiety of both phospholipids and triglycerides. The two organisms, in addition, esterify long-chain fatty acids to all of their complex lipids.

Taken together these results prove that the two organisms possess the enzymes necessary for synthesizing their own complex lipids **de novo**. Most relevant here is the incorporation of glycerol into the lipid molecule since the biosynthesis of glycerol-containing lipids begins with L-α-glycerophosphate.

Both the larval and adult forms of the acoels contain sterols (Table 3), primarily phytosterols, though small amounts of cholesterol also seem to be present. Preliminary data indicate that these sterols are identical in composition to those found in the algae. In view of these results and the finding that the larvae do not incorporate label from ^{14}C-acetate into their sterols, we assume that neither the adult nor the larvae can synthesize these lipids and that they originate, like the fatty acids, from the algal symbiont either directly or indirectly via the parent.

Discussion

All platyhelminths which have so far been studied - even the most primitive free-living acoel - lack the ability to synthesize fatty acids and sterols and depend on an exogenous supply of these compounds for the biosynthesis of their complex lipids. Based on this information, it is not unlikely that the mechanisms of fatty acid and sterol biosynthesis are absent in the entire phylum **Platyhelminthes** and that these mechanisms have been lost during evolution, prior to the evolvement of present-day flatworms. The widespread occurrence of symbiosis among acoels and other lower marine invertebrates[11] might have favored this loss and in turn predisposed many of the free-living organisms to take up a parasitic mode of life.

It is tempting to speculate that whenever a close association between two organisms has been established in the form of parasitism, symbiosis, or commensalism, selective pressures will favor the elimination of duplicating metabolic pathways. Among the pathways which produce major structural and functional components of the cell, the pathways for the biosynthesis of fatty acids and sterols may be particularly susceptible to such selective pressures, since nutritional work has shown that the cell's requirement for these compounds is, in general, quite flexible and adaptable to environmental changes [2, 12, 13, 14]. In so far as the biosynthesis of fatty acids and sterols requires a large portion of the cell's energy it would evidently be of considerable selective advantage to the organism to abandon these pathways.

Whatever the real cause for the loss of these pathways may be, the observation that this phenomenon is not confined to the phylum *Platyhelminthes* but extends to other groups of organisms which contain an abundance of parasitic forms such as spirochetes [15], mycoplasma [16] and trichomonas [17] clearly indicates that a biological advantage is associated with this loss and that there is a tendency of organisms to interlink their metabolic functions when they live in close association.

References

1. ERWIN, J. and BLOCH, K. (1964), Science. **143,** 1006.
2. CLAYTON, R. Chemical Ecologoy, Ed. E. Sondheimer and J. Simeone, Academic Press, New York, 1970, p. 235.
3. von BRAND, T. Biochemistry of Parasites, Academic Press, New York, (1966).
4. MEYER, F., KIMURA, S. and MUELLER, J. (1966). J. Biol. Chem., **241,** 4224.
5. GINGER, C. and FAIRBAIRN, D. (1966). J. Parasitol., **52,** 1097.
6. JACOBSEN, N. and FAIRBAIRN, D. (1967). J. Parasitol., **53,** 355.
7. MEYER, F., MEYER, H. and BUEDING, E. (1970). Biochim. Biophys. Acta, **210,** 257.
8. HYMAN, L. (1951). The Invertebrates, McGraw-Hill Book Company, New York, vol. 2.
9. PROVASOLI, L., YAMASU, T. and MANTON, I. (1968). J. Mar. Biol. Asso. U,K, **48,** 465.
10. STUMPF, P. (1970). Comprehensive Biochemistry, Ed. M. Florkin and E. Stotz, Elsevier Publishing Company, New York, vol. 18, p. 265.
11. FLORKIN, M. and SCHEER, B. (1968). Ed., Chemical Zoology, Academic Press, New York, vol. **2.**
12. MEYER, F., LIGHT, R. and BLOCH, K. (1963). Biochemical Problems of Lipids, Ed., A. Frazer, Elsevier Publishing Company, Amsterdam, p. 415.
13. SHAW, R. (1966). Advances in Lipid Research, Ed. R. Paoletti and D. Kritchevsky, Academic Press, New York, vol. **2,** p. 107.
14. WISNIESKI, B., KEITH, A. and RESNICK, M. (1970). J. Bact., **101,** 160.
15. MEYER, H. and MEYER, F. (1971). Biochim. Biophys. Acta, **231,** 93.
16. RODWELL, A. (1968). Science, **160,** 1350.
17. SHORB, M. (1961). Progress in Protozoology, Czech. Acad. Sci., Prague, p. 153.
18. SMITH, T. and BROOKS, T. (1969). Federation Proc., **28,** 688.
19. SMITH, T., BROOKS, T. and LOCKARD, V. (1970). Lipids, **5,** 854.
20. BEROZA, M. and BIERL, B. (1967). Anal. Chem., **39,** 1131.
21. SMITH, D., MUSCATINE, L. and LEWIS, D. (1969). Biol. Rev., **44,** 17.

Table 1.

Relative Amounts and Specific Radioactivities of 1-^{14}C-Acetate Labeled Fatty Acids in Acoels and Algae

Larvae (3,000-7,000) and adults (100-300) of *Convoluta rescoffensis* were cultured for 3 days at 15°C in charcoal treated sea water supplemented with minerals and vitamins and kept under fluorescent lights set to a light-dark cycle of a 14 hour day [9]. *Platymonas convolutae was cultured* for 3 days in chemically defined medium under the same temperature and light conditions [9]. After incubating the organisms with 1-^{14}C-acetate, the fatty acids were released by saponification, extracted, methylated, fractionated according to their degree of unsaturation by thin-layer chromatography on silver nitrate-impregnated silica gel G and analyzed by gas-liquid chromatography [4]. The positions of the double bonds in the individual unsaturated fatty acids were determined through ozonolysis [20].

Fatty acid[1]	Relative amount (% of total)			Relative specific radioactivity[2]			
	Algae	Adults	Larvae	Algae	Adult-light	Adult-dark[3]	Larvae[4]
14:0	1	1	1	2	3	8	50
16:0	18	12	15	3	5	5	1
16:1 ω5	1	0	0	2	0	0	0
16:1 ω7	2	1	0	3	3	5	0
16:1 ω9	1	1	1	2	5	5	1
16:2 ω6,9	0	1	0	0	5	7	0
16:3 ω3,6,9	1	1	0	2	4	7	0
16:4 ω3,6,9,12	9	2	0	2	2	1	0
18:0	0	6	7	0	3	16	30
18:1 ω7	14	3	1	3	3	5	4
18:1 ω9	1	10	8	3	3	2	3
18:2 ω6,9	9	2	1	2	2	5	26
18:3 ω3,6,9	16	16	13	3	2	4	4
18:4 ω3,6,9,12	13	3	3	2	4	13	15
20:1 ω9	1	3	7	3	14	70	435
20:2 ω6,9	0	2	4	0	2	15	25
20:3 ω3,6,9	0	7	6	0	4	7	29
20:4 ω3,6,9,12	0	7	8	0	1	5	15
20:5 ω3,6,9,12,15	13	11	14	1	1	3	14
22:1 ω9	0	9	10	0	3	31	134
22:5 ω3,6,9,12,15	0	2	1	0	5	15	162

1 The numerals before the colon refer to the chain length of the fatty acids. The numerals following the colon refer to the number of double bonds in the molecule and the numerals following the ω indicate the positions of the double bonds counted from the methyl terminal of the fatty acid molecule.

2 Relative specific acitvity = counts per minute per area of peak; the value of 1 is arbitrarily

given to the fatty acid with the lowest specific radioactivity.

3 A shorter incubation period of 5 hours instead of 3 days yields similar relative specific radioactivities. The addition of a mixture of glucose and mannitol has no effect on the ^{14}C-acetate incorporation pattern.

4 A mixture of glucose and mannitol (0.05 % each) was routinely added to the larval culture during incubation with isotopes.

Table 2.

Distribution of Radioactivity in Fatty Acid Fragments Obtained by Decarboxylation or Ozonolysis.

The individual 1-^{14}C-acetate labeled fatty acids of larvae and adults (Table 1) were degraded. Saturated fatty acids were decarboxylated and the resulting CO_2 and alkylamine assayed for radioactivity[7]. Unsaturated fatty acids were ozonized in carbon disulfide or pentylacetate at -60°C and the ozonides reduced with triphenylphosphine. The resulting aldehydes (methyl-terminal fragments) and aldehyde esters (carboxyl-terminal fragments) were identified by gas-liquid chromatography and collected for radioassay[20].

Procedure	Fatty acid	Specific radioactivity[1]					
		Adult-light		Adult-dark		Larvae	
		CH_3-[2]	-COOH[2]	CH_3-	-COOH	CH_3-	-COOH
Decarboxylation	16:0	1165	1695	325	1821	8	2972
	18:0	287	5593	39	7160	4	2857
Ozonolysis	18:1 ω9	619	640	90	569	0	317
	18:2 ω6,9	2044	2316	56	684	0	405
	18:3 ω3,6,9[3]	1550	2305	30	441	0	210
	18:4 ω3,6,9,12	795	1252	76	993	0	387
	20:1 ω9	430	2150	10	592	0	1439
	20:2 ω6,9	40	2406	2	1247	0	209
	20:3 ω3,6,9	350	4410	0	1021	0	285
	20:4 ω3,6,9,12	77	3406	2	276	0	357
	20:5 ω3,6,9,12,15	938	2310	15	341	0	298
	22:1 ω9	137	1508	1	978	0	517
	22:5 ω3,6,9,12,15	379	7646	6	476	0	336

1 Specific radioactivity = total cpm per number of theoretical ^{14}C-carbon atoms in the fragment.

2 CH_3- refers to the methyl-terminal and -COOH to the carboxyl-terminal fragment of the fatty acid molecule.

3 The radioactivities in the methyl-terminal fragments of the ω3 fatty acid are below the expected values due to partial loss of the volatile propanal.

Table 3.

Relative Radioactivities in the Acoel and Algal Lipids
Organisms were incubated with isotopic tracers and their lipids extracted
and fractionated by a combination of silicic acid column chromatography
and thin-layer chromatography [15]. Sterols were also analyzed by gas-
liquid chromatography [4].

Lipid fraction	Relative radioactivity (% cpm)					
	Algae $^{14}C\text{-}CO_2$[1]	Adults		Larvae		
		$^{14}C\text{-}$ Acetate	$^{14}C\text{-}$ Glycerol	$^{14}C\text{-}$ Acetate	$^{14}C\text{-}$ Glycerol	$^{14}C\text{-}$ Linoleate
Neutral lipids						
Hydrocarbon	8	2	2	9	1	2
Triglyceride	2	40	27	21	35	57
Free fatty acid	2	5	1	6	0	0
Mono- and diglycerides	0	2	5	5	3	2
Phytosterol	9	1	1	0	0	0
Cholesterol (?)	1	1	0	0	0	0
Glycolipids plus pigments						
Pigment	9	3	5	5	5	9
Galactosyl diglyceride	18	3	5	2	0	1
Digalactosyl diglyceride	7	4	3	4	0	1
Sulfolipid	2	1	1	0	0	0
Phospholipids						
Phosphatidyl ethanolamine	8	7	9	9	12	5
Ethanolamine plasmalogen	0	12	11	10	14	5
Phosphatidyl choline	0	10	14	21	19	15
Phosphatidyl glycerol	29	2	7	0	1	1
Cardiolipin	1	1	1	1	2	0
Phosphatidyl inositol	4	6	8	7	8	2

1 A similar labeling pattern is observed, when $1\text{-}^{14}C$-acetate is used as tracer.

Table 4.

Identification of ^{32}P-Phospholipids as Deacylated Products
^{32}P-phospholipids were sequentially hydrolyzed by the procedure of Dawson et al. and the radioactive deacylated products identified by paper chromatography[15].

Hydrolytic condition	Hydrolytic products	Relative radioactivity (% cpm)		
		Algae	Adults[1]	Larvae[1]
Mild alkali	Glycerophosphoryl choline	1	23	35
	Glycerophosphoryl ethanolamine	17	20	21
	Glycerophosphoryl glycerol	71	7	1
	Diglycerophosphoryl glycerol	2	2	4
	Glycerophosphoryl inositol	8	8	11
Mild acid	Glycerophosphoryl choline	0	1	2
	Glycerophosphoryl ethanolamine	0	32	23
Strong acid	Unidentified	0	4	1
Residue	Unidentified	0	3	2

1 The labeling patterns with ^{14}C-glycerol labeled phospholipids were similar.

27

RECENT STUDIES ON THE CHARACTERIZATION
OF THE CYTOCHROME SYSTEM IN KINETOPLASTIDAE

George C. Hill

Department of Biology and Parasitology,
The Molteno Institute,
University of Cambridge,
Cambridge, CB2 3EE, England

Introduction

The characterization of the electron transport system in Kinetoplastidae has received some attention recently. An unusual cytochrome system exists in *Crithidia fasciculata* which possesses a mitochondrion of 40-50μ in length. Cytochromes **a** + **a**$_3$, **b**, **c**$_{555}$ and **o** are present (Hill and Anderson, 1970). Some of the unusual properties of this cytochrome system will be discussed.

Preparation of mitochondrial fraction

A mitochondrial fraction can be prepared from *C. fasciculata* after breaking the cells by grinding with neutral alumina followed by differential centrifugation (Hill and White, 1968; Hill and Anderson, 1969). As determined by electron microscopy, the final mitochondrial pellet contains mitochondria as well as kinetoplasts and pieces of flagella (Hill and Anderson, 1970). The low P/O ratios and absence of respiratory control, although the O$_2$ uptake is stimulated by ADP (Figure 1), suggest the mitochondria have been damaged. In addition, the mitochondria are smaller than those usually seen **in situ.** The mitochondrial fraction contains 8-12 % of the original cell protein and can oxidize several substrates including succinate and NADH. The cytochromes reduced by the oxidation of these substrates have been observed.

Identification of cytochromes present

No difference in spectral evidence of cytochromes in the mitochondrial fraction occurs when NADH or succinate are used as substrates. Figure 2 is an NADH reduced **minus** oxidized difference spectrum of the mitochondrial fraction. Cytochromes **a** + **a**$_3$ (602 nm) and cytochrome **b** (559 nm) are evident. A shoulder on the latter peak between 560 and 550 nm suggests the presence of cytochrome **c**. In low temperature difference spectra, a shoulder at 553 nm is quite distinct (Hill and White, 1968; Hill and Anderson, 1970). Cytochrome **c**$_{555}$ has been purified from *C. fasciculata* (Hill and Chan, 1969 and Hill **et al.**, 1971; Kusel, Suriano and Weber, 1969) and some of its properties determined.

In a CO difference spectrum (Figure 3), two cytochromes which combine with CO are evident: cytochrome **a**$_3$ with a trough at 443 nm and cytochrome **o** with absorption peaks at 569, 538 and 420 nm. The presence of cytochrome **o** makes the detection of the cytochrome **a**$_3$ - CO peaks at 590 and 432 nm difficult. These have been recorded in previous spectra (Hill and Anderson, 1969). The presence in eukaryotic cells of these two terminal oxidases together, cytochromes **aa**$_3$ and cytochrome **o**, is unusual. A similar situation exists in several parasitic helminths (Cheah and Bryant, 1966; Cheah, 1967**a**, **b**). Branched electron transport systems have been extensively studied in bacteria (see recent review by White and Sinclair, 1971) and in plants.

Properties of cytochromes c from kinetoplastidae

Cytochrome **c**$_{555}$ has been extracted and purified from intact cells of *C. fasciculata* as well as from the mitochondrial fraction (Hill and White, 1968).

It is a basic protein which reacts with mammalian cytochrome oxidase (Hill **et al.** 1971). We have now purified cytochromes **c** from two cultured forms of trypanosomes, *Trypanosoma cruzi* and *T. rhodesiense* and an insect *Leptomonas* species. All have similar spectral properties (Table 1 and Figure 4). Some of the biochemical properties of *C. fasciculata* cytochrome **c**$_{555}$ are given in Table 2. The unusual properties of Kinetoplastidae cytochromes **c** include: (1) an α-peak in the reduced form between 555-558 nm; (2) the presence of 1-2 residues of ϵ-N-tri-methyllysine (Figure 5); (3) an isoelectric point at pH 8.8 and (4) a molecular weight of 13,070. The α-peak of the reduced form of the cytochrome **c** in pyridine is at 553 nm. The home moiety present in cytochrome **c** is **not** cleaved from the protein by acid acetone (Hill and White, 1968). Recently, Pettigrew and Meyer (1971) have demonstrated that the prosthetic group of cytochrome c_{557} from *C. oncopelti* could be removed by mercuric chloride in acid, yielding a heme with a pyridine hemochrome, at 552 nm. This heme was reported to have different chromatographic properties from hematohemin and protohemin, suggesting a unique prosthetic group was present. They also suggested only one cysteine residue was present in the heme peptide.

The amino acid composition data of Kinetoplastidae cytochromes **c** has been compared with the amino acid composition data of selected yeast cytochromes **c**. These similarities can be observed on Table 3. Figure 6 is an electropherogram of *C. fasciculata* cytochrome c_{555}, *T. rhodesiense* cytochrome c_{556}, horse heart cytochrome c_{550} and *Candida krusei* cytochrome c_{550}. The isoelectric point of *C. fasciculata* cytochrome c_{555} has been determined at pH 8.8 (Hill **et al.,** 1971). The other trypanosomatid cytochromes **c** all migrated at rates similar to *C. fasciculata* cytochrome c_{555} but less than *C. krusei* or horse heart cytochromes **c**, both of which have higher isoelectric points than cytochrome c_{555}.

Position and significance of cytochrome o in the electron transport chain

Application of the "crossover theorem" of Chance and Williams (1956) has provided a tool to identifying the proper sequence of cytochromes in the electron transport system of Kinetoplastidae. As can be seen in Figure 7, the O_2 uptake of intact cells is sensitive to antimycin and KCN. This is also true of the oxidation of substrates by the mitochondrial fraction (Figure 1). However, cyanide, azide and antimycin give biphasic curves, suggesting two possible modes of inhibition with widely different values for K_i. These results would be compatible with the inhibition of two different terminal oxidases.

Attention has been devoted to the localization of cytochrome **o** in this electron transport system. This cytochrome combines with CO and is a terminal oxidase in bacterial systems. Several possibilities exist for the relationship between cytochrome **o** and the mitochondrial system:

(1) It receives electrons from only cytochrome **b**;
(2) it receives electrons from only cytochrome **c**;
(3) it receives electrons from both cytochromes **b** and **c**;
(4) it receives electrons from the oxidized substrate through perhaps a flavoprotein and is **not** mediated by other cytochromes.

(a) Evidence against cytochrome o receiving electrons from only cytochrome c.

Available experimental evidence suggests that electrons are not transferred from only cytochrome **c** to cytochrome **o** in *C. fasciculata.* This is concluded from several different types of difference spectra in the presence of antimycin and KCN.

Figure 8 is a difference spectrum of cytochromes which have been reduced in the presence of NADH and 1.0 mM KCN **minus** cytochromes reduced in the presence of NADH and 1.0 μM antimycin. In this spectrum only cytochromes localized between cytochrome **b** and cytochrome **a** would be present. No peak in the b-region (560 nm) is present. Cytochrome **c** is evident at 553 nm. In addition, the Soret peak is shifted to a lower wavelength appropriate for cytochrome **c** but not cytochrome **b**. These spectral results suggest the absence of protoheme pigments (i.e. cytochromes **b** and **o**). Similar spectra are obtained if 5 mM Na azide is used instead of KCN or if no inhibitor is present in the sample cuvette. If the forementioned samples are bubbled with CO, one would then see only CO-binding pigments localized between cytochromes **b** and **a**. In Figure 9, no CO-binding pigments are evident and the same spectrum is obtained as in Figure 8. These spectra strongly suggest cytochrome **o** does not receive electrons from only cytochrome **c**.

(b) Evidence for cytochrome o receiving electrons from cytochrome b.

Evidence does exist for the transfer of electrons from cytochrome **b** to cytochrome **o**. Figure 10 is a CO difference spectrum of cytochromes reduced with 1.0 mM NADH in the presence of 1.0 μM antimycin and then bubbled with CO **minus** cytochromes reduced with 1.0 mM NADH in the presence of 1.0 μM antimycin. Spectral evidence for cytochrome **o** is observed but cytochrome a_3-CO is **not** evident. This spectrum suggests that cytochrome **o** is mediating the antimycin-insensitive substrate oxidation. Similar spectra are obtained with either KCN or Na azide-CO difference spectra. Figure 11 is a difference spectrum of mitochondria reduced with NADH in the presence of antimycin **minus** an oxidized sample. No cytochrome **c** or cytochrome aa_3 is evident. If this sample is bubbled with CO, then the Soret peak shifts from 430 nm to 421 nm suggesting the presence of cytochrome **o** (Figure 12). These experiments strongly suggest the antimycin and KCN-insensitive O_2-uptake observed in *C. fasciculata* is due to the functioning of cytochrome **o**. In addition, the results of Figure 7 would suggest the functioning of cytochrome **o** is affected by high concentrations of antimycin and KCN.

Proposed branched electron transport system in *C. fasciculata*.

It seems clear that in *C. fasciculata* the respiratory chain bifurcates on the substrate side of the antimycin-sensitive site. We would suggest that cytochrome **o** receives electrons from cytochrome **b** and functions as the terminal oxidase, particularly during antimycin or KCN-insensitive endogenous O_2-uptake of intact cells or oxidation of substrates by the mitochondrial fraction. Figure 13 presents this proposed electron transport system. Co-enzyme Q_9 has been identified in *C. fasciculata* by several investigators (Vakertzi-Lemonias **et al.**, 1963; Kusel and Weber, 1965). Neither pathway is affected by thiocyanate, $\alpha\alpha'$-dipyridyl or 8-hydroxyquinoline, making it different from cyanide-insensitive respiration observed in plant mitochondria (Bendall and Bonner, 1971).

There are several aspects to the functioning of the electron transport system in Kinetoplastidae which need further attention:

(1) Preparation of a mitochondrial preparation with:
   (a) A single intact mitochondrion and kinetoplast
   (b) Respiratory control;
(2) Purification of cytochrome **o** and the characterization of its biochemical properties;
(3) Study of the rapid kinetics of electron transfer of this branched electron transport system;
(4) Illustration of the CO action spectrum of the two terminal oxidases (cytochrome a_3 and cytochrome **o**) in trypanosomatids to support the functioning of these terminal oxidases;
(5) Determining the signficance to the insect trypanosomatid of two terminal oxidases, particularly cytochrome **o**.

The possibility does exist that cytochrome **o** is located outside the mitochondria and is reduced by the substrate via a flavoprotein or is not mediated by other cytochromes. This possibility still exists and must be examined in light of the localization of cytochromes in bacterial membranes.

Conclusions

(1) A mitochondrial fraction can be obtained from *C. fasciculata* whose oxidation by substrates is stimulated by ADP.
(2) It is suggested that the antimycin and KCN-insensitive O_2 uptake observed is due to the functioning of cytochrome **o**. This alternate respiratory chain probably bifurcates before the antimycin-sensitive site.
(3) All the cytochromes **c** examined from Kinetoplastidae have 1-2 residues of ϵ-N-trimethyllysine and an α-peak in the reduced form between 555-558 nm.

Acknowledgements

The co-operation of Dr. B.A. Newton and Dr. G.A.M. Cross for the use of their equipment and laboratories is appreciated. The author is particularly indebted to Dr. Cross and Mrs. Sandra Ray for fruitful discussions and helpful suggestions during several phases of this research.

References

BENDALL, D.S. and BONNER, W.D. (1971). Plant Physiol. **47,** 236.

CHANCE, B. and WILLIAMS, G.R. (1956). Advan. Enzymol. **17,** 65.

CHEAH, K.S. (1967a). Comp. Biochem. Physiol. **20,** 867.

CHEAH, K.S. (1967b). Comp. Biochem. Physiol. **23,** 277.

CHEAH, K.S. and BRYANT, C. (1966). Comp. Biochem. Physiol. **19,** 197.

DE LANGE, R.J., GLAZER, A.N. and SMITH, E.L. (1969). J. Biol. Chem. **244,** 1385.

DE LANGE, R.J., GLAZER, A.N. and SMITH, E.L. (1970). J. Biol. Chem. **245,** 3325.

HELLER, J. and SMITH, E.L. (1966). J. Biol. Chem. **241,** 3158.

HILL, G.C. and ANDERSON, W. (1969). J. Cell Biol. **41,** 547.

HILL, G.C. and ANDERSON, W. (1970). Expt'l Parasitology. **28,** 356.

HILL, G.C. and CHAN, S.K. (1969). J. Protozool. (Suppl.) **16,** 13.

HILL, G.C., CHAN, S.K. and SMITH, L. (1971). Biochim. Biophys. Acta (in press).

HILL, G.C. and WHITE, D.C. (1968). J. Bacteriol. **95,** 2151.

KUSEL, J.P., SURIANO, J.R. and WEBER, M. (1969). Arch. Biochem. Biophys. **133,** 293.

KUSEL, J.P. and WEBER, M.M. (1965). Biochim. Biophys. Acta **98,** 632.

PETTIGREW, G. and MEYER, T. (1971). Proc. Biochem. Soc. 29-30 July, p. 44.

NARITA, K. and TITANI, K. (1965). Proc. Japan. Acad. **41,** 831.

NARITA, K., TITANI, K. YAOI, Y. and MURAKANI, H. (1963). Biochim. Biophys. Acta. **77,** 688.

VAKERTZI-LEMONIAS, C., KIDDER, G.W. and DEWEY, V.C. (1963). Comp. Biochem. and Physiol. **8,** 331.

WHITE, D.C. and SINCLAIR, P. (1971). Ann. Rev. Microb. Physiol. **6,** 173.

Table 1.
Spectral properties of kinetoplastidae cytochromes **c**.

Source	Peaks in Reduced Form (nm)			Soret Peak in Oxidized Form (nm)
	α	β	γ	
C. fasciculata	555	525	421	414
Leptomonas sp.	556	524	421	414
T. rhodesiense	556	524	420.5	412
T. cruzi	558	525	421	414

Table 2.
Properties of Purified *Crithidia fasciculata* Cytochrome c_{555}

Property	
Molecular weight	13,070
	(based on iron content)
Amino acid residues	117
ϵ-N-trimethyllysine	present
Isoelectric point	8.8
α-Peak of the reduced form (nm)	555
α-Peak of the reduced form in pyridine (nm)	553
Reactivity with mammalian cytochrome oxidase	Positive

For details of results, see Hill and Anderson (1970) and Hill, Chan and Smith (1971). The presence of ϵ-N-trimethyllysine in yeast and plant cytochromes **c** has been established by De Lange **et al.** (1969, 1970).

Table 3.

Amino acid composition of selected cytochromes **c**

Results for trypanosomatid cytochromes **c** are given as μ moles of amino acid per μ mole of Fe. The time of hydrolysis was 24 hours and no corrections have been made for hydrolytic losses. The amino acid compositions of *Candida krusei* cytochrome **c**, as reported by Narita and Titani (1965), of *Saccharomyces* (baker's yeast) cytochrome **c**. as reported by Narita **et al.** (1963), and of *Neurospora crassa* cytochrome **c**, as reported by Heller and Smith (1966), are included for comparison.

Amino Acid	C. fasci-culata	Leptomo-nas sp.	T. rhode-siense	T. cruzi	C. krusei	S. cere-visiae	N. crassa
Asp	10.1[a]	8.2[a]	10.3[a]	10.3[a]	8[b]	11[b]	13[b]
Thr	5.2	5.2	7.1	6.4	7	8	9
Ser	5.3	7.2	4.9	5.1	6	4	3
Glu	10.1	12.2	11.1	12.2	10	9	8
Prol	7.9	7.1	8.2	10.1	7	4	3
Gly	16.1	19.8	12.8	19.2	12	12	15
Ala	14.2	15.2	12.9	13.8	12	7	9
Val	6.8	5.8	6.1	5.1	3	3	1
Cys	2.0[c]	2.0[c]	2.0[c]	2.0[c]	2	3	2
Met	2.1	1.8	1.7	1.8	3	2	2
Isol	2.2	2.2	3.1	2.9	3	4	5
Leuc	8.9	5.1	8.2	5.8	6	8	7
Tyr	4.2	2.2	2.8	1.7	5	5	4
Phe	2.9	3.1	4.1	3.1	4	4	6
Lys $(CH_3)_3$[d]	1.3	1.4	0.8	1.4	1	1	1
Lys	10.2	11.1	13.1	8.8	11	15	13
Hist	1.8	1.7	2.1	2.1	4	4	2
Tryp	1.0[c]	1.0[c]	1.0[c]	1.0[c]	1	1	1
Arg	5.3	4.8	5.1	4.2	4	3	3
Total					109	108	107
α peak of the reduced form	555	556	556	558	550	550	550

[a] μ-moles of amino acid per μmole of Fe.

[b] Amino acid residues per molecule of protein.

[c] Assumed Value.

[d] ϵ-N-trimethyllysine.

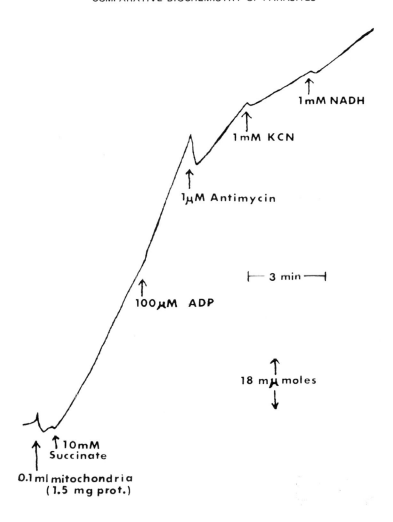

Figure 1.
Stimulation of succinate oxidation by *Crithidia fasciculata* mitochondrial fraction. The mitochondrial fraction was diluted with an isotonic buffer containing 0.25 M mannitol, 10 mM potassium phosphate, 5 mM Mg Cl$_2$, 1 mM EDTA, 20 mM Tris (hydroxymethyl) methane, pH 7.3. Additions were made as indicated. The ADP-stimulated O$_2$ uptake rate does not decrease.

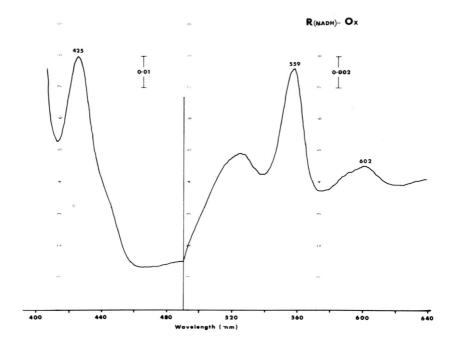

Fig. 2.

Difference spectrum of the mitochondrial fraction from *Crithidia fasciculata* with cytochromes reduced by 1.0 mM NADH **minus** mitochondria with the cytochromes oxidized. Substrate reduction of cytochromes is complete 4-5 minutes after the addition of 1.0 mM NADH or 10.0 mM succinate. The protein concentration in all spectra presented in this paper was 2.25 mg per ml. The cells were grown on 0.05 M xylose as carbon source and 1.5 μM protoheme.

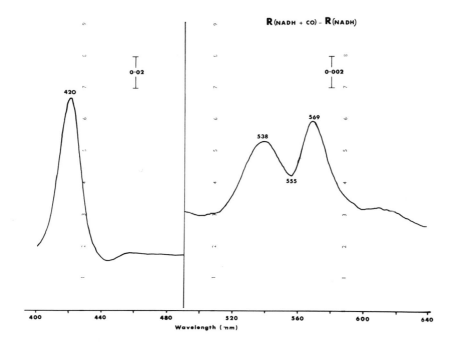

Fig. 3.
Difference spectrum of the mitochondrial fraction from *Crithidia fascicu-lata* with the cytochromes reduced with 1.0 mM NADH and then saturated with carbon monoxide for 10 minutes **minus** mitochondria reduced with 1.0 mM NADH.

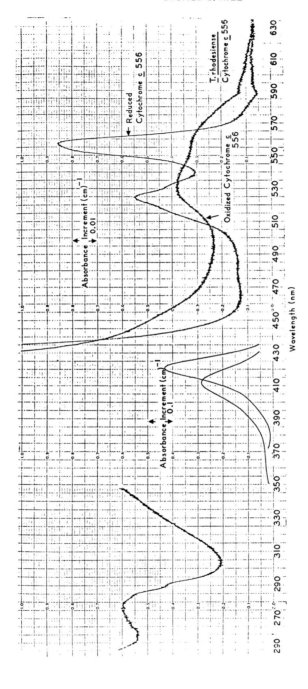

Fig. 4.

Oxidized and reduced absolute spectra of purified cytochrome **c**556 from *Trypanosoma rhodesiense.*

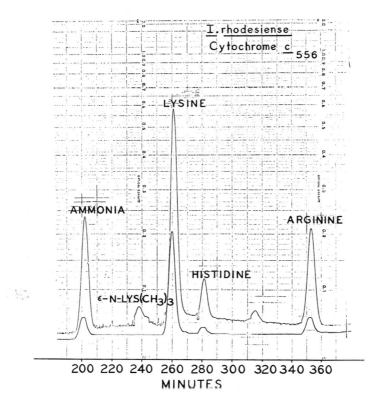

Fig. 5.

Amino acid chromatogram of the basic portion of an acid hydrolysate of *Trypanosoma rhodesiense* cytochrome c_{555}. ϵ-N-trimethyllysine is present. The cytochrome **c** was hydrolyzed for 24 hours in HCl. The Technicon autoanalyzer for amino acid analysis was employed.

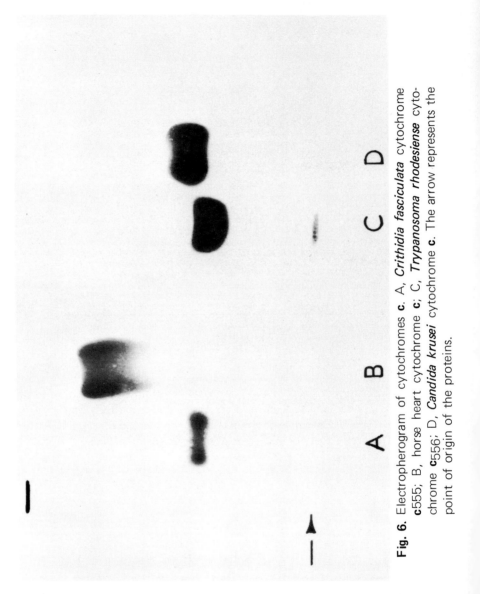

Fig. 6. Electropherogram of cytochromes **c**. A, *Crithidia fasciculata* cytochrome **c**555; B, horse heart cytochrome **c**; C, *Trypanosoma rhodesiense* cytochrome **c**556; D, *Candida krusei* cytochrome **c**. The arrow represents the point of origin of the proteins.

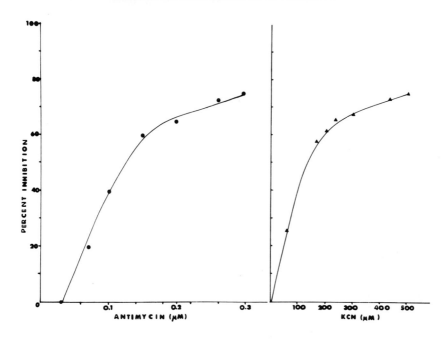

Fig. 7.
Inhibition of O_2 uptake of *Crithidia fasciculata* by KCN and antimycin.

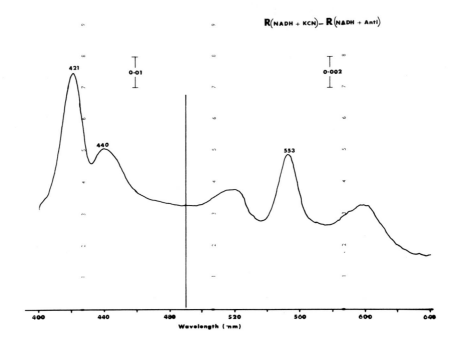

Fig. 8.
Difference spectrum of the mitochondrial fraction of *Crithidia fasciculata* with the cytochromes reduced in the presence of 1.0 mM NADH and 1.0 mM KCN **minus** mitochondria with the cytochromes reduced in the presence of 1.0 mM NADH and 1.0 μM antimycin.

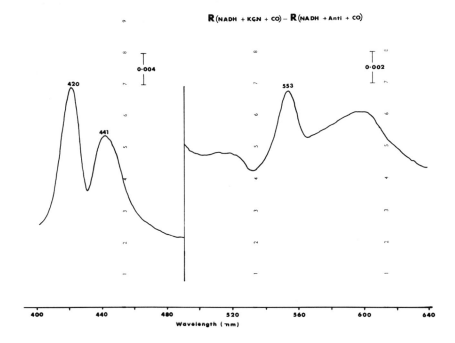

Fig. 9.
Difference spectrum of the mitochondrial fraction from *Crithidia fascicu-lata* with the cytochromes reduced in the presence of 1.0 mM NADH and 1.0 mM KCN and saturated with carbon monoxide **minus** mitochondria with the cytochromes reduced in the presence of 1.0 mM NADH and 1.0 μM antimycin and saturated with carbon monoxide.

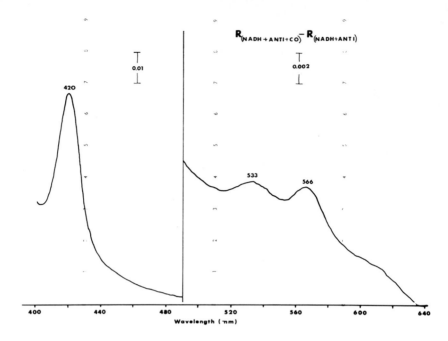

Fig. 10.

Difference spectrum of the mitochondrial fraction from *Crithidia fascicu-lata* with the cytochromes reduced by 1.0 mM NADH in the presence of 1.0 μM antimycin and saturated with carbon monoxide **minus** mitochondria with the cytochromes reduced in the presence of 1.0 mM NADH and 1.0 μM antimycin.

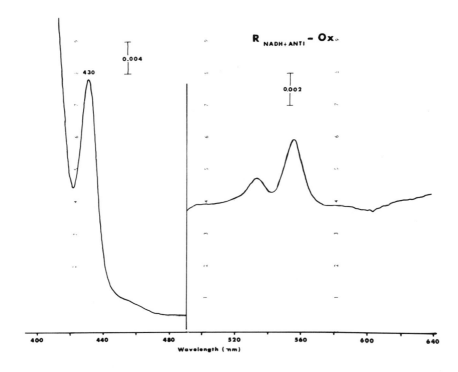

Fig. 11.
Difference spectrum of mitochondrial function from *Crithidia fasciculata* with the cytochromes reduced by 1.0 mM NADH in the presence of 1.0 μM antimycin **minus** mitochondria with the cytochromes oxidized.

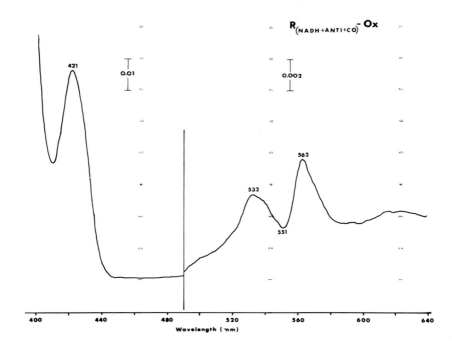

Fig. 12.
Difference spectrum of mitochondrial fraction from *Crithidia fasciculata* with cytochromes reduced by 1.0 mM NADH in the presence of 1.0 μM antimycin and then saturated with carbon monoxide **minus** mitochondria with the cytochromes oxidized.

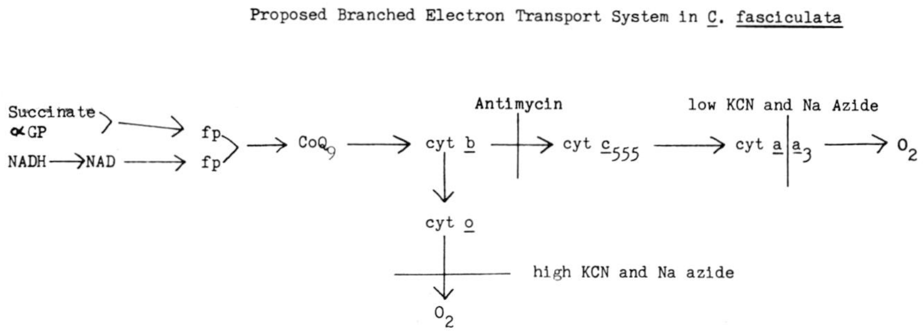

Proposed Branched Electron Transport System in $\underline{C}$. $\underline{\text{fasciculata}}$

Fig. 13.
Proposed branched electron system in *Crithidia fasciculata.*

28

CYTOCHROMES IN *ASCARIS* AND *MONIEZIA*

K.S. Cheah

Agricultural Research Council, Meat Research Institute,
Langford, Bristol BS18 7DY, U.K.

Ascaris[1] and *Moniezia*[2] have functional **b**, **c** and **a**-type cytochromes and two CO-reactive haemoproteins identified as cytochromes **o** and a_3. Besides having a higher concentration of cytochromes than *Ascaris, Moniezia* also has the classical mammalian cytochromes **b** and c_1, neither of which were observed in *Ascaris.*

Figure 1 illustrates the presence of cytochromes **b, c** and aa_3 in *Moniezia*. Cytochrome **b** (560 and 430 nm) was detected by antimycin A, and cytochromes aa_3 (443-445 nm) and **c** (550 and 420 nm) by p-phenylenediamine **plus** cyanide. The two terminal oxidases in *Moniezia* were characterized through the formation of the CO-complexes (Figure 2). The CO difference spectra recorded at 8 and 11 min with α-glycero-phosphate as substrate clearly illustrate the predominant cytochrome **o**-CO complex (419 nm) superimposed on that of cytochrome a_3, and that the overall spectrum recorded at 11 min was due to cytochromes **o** and $a3$. Moniezia cytochromes **c** (548 nm) and c_1 (552 nm) were detected only at liquid nitrogen temperature (-196°), in addition to cytochromes **b** (562 nm) and aa_3 (598 nm), using succinate **plus** cyanide (Figure 3). The autoxidizable **o**-type terminal oxidase having a split peak (552 and 556 nm) at -196° was not detected but could easily be observed under anaerobiosis in the presence of a suitable electron donor. Figure 4 shows the succinate-reduced **minus** oxidized difference spectra recorded at 20° (A) and -196° (B). The absorption band at 555-557 nm (A) is partly contributed by reduced cytochrome **o** whose α-peak at 555-557 nm is split into two sharp bands at 552 and 556 nm at -196° (B). The succinate-reducible cytochromes **b, c** and aa_3 not observed at

20° are clearly seen at -196° appearing at 562, 547 and 594-601 nm (B) respectively.

One of the most interesting observations about *Moniezia* is that its cytochrome **o** is involved in fumarate reduction. Thus fumarate in addition to oxygen could act as an electron acceptor. This is clearly demonstrated in Figure 5. The addition of fumarate to the mitochondria made anaerobic with α-glycerophosphate (A) resulted in the re-oxidation of cytochrome **o** (B), which at -196° showed its distinct split peaks at 552 and 556 nm (C), a spectrum also obtainable if CO was used instead of fumarate.

The kinetics of the reduction of cytochrome **o** by α-glycerophosphate and its subsequent re-oxidation by fumarate is shown in Figure 6. On reaching anaerobiosis, only fumarate and not crotonate or even the **cis**-isomer of fumarate, maleate (not shwon in Figure 6) could bring about the re-oxidation of cytochrome **o**. The 35 % reduction of the **b**-type cytochrome observed at the steady state at about 20 min after fumarate addition is due to other **b**-type cytochrome in *Moniezia* not associated with fumarate reduction.

Unlike *Moniezia*, *Ascaris*-muscle mitochondria appear not to have the classical mammalian cytochromes **b** and c_1, but have functional cytochrome b_{556} (-196°), cytochromes **o** and a_3 involved in electron transport. Figure 7 shows the difference spectrum (196°) of *Ascaris*-muscle mitochondria obtained with dithionite (A), α-glycerophosphate (B), malate (C) and ascorbate (D). *Ascaris* cytochrome b_{556} (-196°) and cytochrome **c** were involved with the oxidation of α-glycerophosphate and malate and cytochromes **c** and a_3 (443 nm) with ascorbate. The novel features about *Ascaris*-muscle mitochondria are the lack of any detectable α-peak of the **a**-type cytochrome (indicating that cytochrome **a** is probably missing) and the predominant maximum at about 480 nm with two corresponding minima at 460 and 502 nm in the substrate difference spectra (B and C) which is contributed by two flavin components [2].

The action spectrum of *Ascaris*-muscle mitochondria (Figure 8A) with maxima at 593 (α-peak) and 432 (γ-peak) nm is characteristic of cytochrome $a_{3,1}$ clearly indicating that respiration in *Ascaris*-muscle mitochondria involved cytochrome a_3 as its terminal oxidase [1]. An almost identical action spectrum was obtained with yeast (Figure 8B) which was used as a control. The slight discrepancy in the β-peak of the *Ascaris*-muscle mitochondria action spectrum could be due to cytochrome **o**, which is clearly observed in the CO-difference spectrum using α-glycerophosphate as the electron donor (Figure 9). Unfortunately the cytochrome a_3-CO complex (γ-peak at 427-430 nm) was not detected as it was obscured by the strong absorption band of cytochrome **o**-CO complex (417 nm) just as in *Moniezia* (Figure 2).

Table 1 summarizes the amount of cytochromes in *Moniezia, Ascaris* and the ox-neck muscle mitochondria. The concentration of cytochromes in both *Moniezia* and *Ascaris* is much lower than that of the ox-neck muscle mitochondria, with *Moniezia* having a higher content of cytochromes than *Ascaris.*

Table 2 illustrates the turnover numbers of the terminal oxidases in *Moniezia, Ascaris* and ox-neck muscle mitochondria. The turnover numbers for both *Moniezia* and *Ascaris*-muscle mitochondria based on cytochromes (**o** + **a**$_3$) are lower than those of the ox-neck muscle mitochondria based on cytochrome **a**$_3$.

The above evidence for the participation of cytochromes in *Ascaris* electron transport system is complemented by the unpublished data of Winburne and Erecinska (personal communication) who found that the re-oxidation of flavoprotein was much slower than that of cytochromes **b** and **c**. They found the half-time for the re-oxidation of *Ascaris*-muscle mitochondrial flavoprotein by ferricyanide of about 3 sec as compared with the flow time of 20 msec for 80 % re-oxidation of cytochrome **b**, both of which were reduced with malate.

Table 3 summarizes the absolute absorption peaks of purified *Ascaris* cytochrome **c**$_{550}$, *Moniezia* cytochrome **c**$_{550}$ and cytochrome **c** from the back muscle of the Large White pig. At room temperature (20°), no difference in the position of the absorption peaks was observed in the cytochrome **c** isolated from the three different tissues. However, at -196°, the α-peak of cytochrome **c** from the pig showed the characteristic satellite bands [**c**$_{\alpha_1}$ (547 nm), **c**$_{\alpha_2}$ (544.5 nm), **c**$_{\alpha_3}$ (536)] observed for other mammalian tissue while **c**$_{\alpha_1}$ appears to be missing in *Moniezia* and **c**$_{\alpha_3}$ in *Ascaris*. This is clearly shown in Figures 10 and 11 (B).

A novel haemoprotein, *Ascaris* cytochrome **b**$_{560}$, was also purified from *Ascaris* which at 20° has a single α-peak at 560 nm (absolute spectrum) but split into two sharp bands (553.0 and 560.0 nm) at -196° (Figure 11 A). This cytochrome isolated from whole muscle is most unlikely to be microsomal cytochrome **b**$_5$ or even mitochondrial cytochrome **b**$_5$, based on the positions of the two sharp symmetrical peaks observed at -196°, as microsomal cytochrome **b**$_5$ and mitochondrial cytochrome **b**$_5$ have bands at 552 and 557 nm (-196°) and at 551 and 558 nm (-196°) respectively [7].

From the substantial data just presented we can confidently say that for the two large intestinal parasites, *Ascaris* and *Moniezia*, cytochromes do participate in electron transport, as with *Taenia hydatigena*[8]. Furthermore, these parasites are aerobic as shown by the existence of cytochrome a_3. Unlike the classical mammalian respiratory chain system both *Ascaris* and *Moniezia* modify their respiratory chain systems to adapt to their environment (as do bacteria) by having more than one terminal oxidase. With *Moniezia* mitochondria, oxygen is essential for energy synthesis[9, 10] further supporting the concept[9] that *Moniezia* is aerobic.

References

1. CHEAH, K.S. and CHANCE, B. (1970). Biochim. Biophys. Acta, **223,** 55.
2. CHEAH, K.S. (1968). Biochim. Biophys. Acta, **153,** 718.
3. CHANCE, B. and SPENCER, E.L. Jr. (1959). Discussions Faraday Soc., **27,** 200.
4. CHANCE, B. and SCHOENER, B. (1966). J. Biol. Chem., **241,** 4567.
5. HYDE, T.A. (1967). M.Sc. Thesis, University of Pennsylvania, Philadelphia.
6. CASTOR, L.N. and CHANCE, B. (1959). J. Biol. Chem. **234,** 1587.
7. PARSONS, D.F., WILLIAMS, G.R., THOMPSON, W., WILSON, D.F. and CHANCE, B. (1967). in E. Quagliariello, S. Papa, E.C. Slater and J.M. Tager (Editors) Mitochondrial Structure and Compartmentation, Adriatica Editrice, Bari, Italy, p. 29.
8. CHEAH, K.S. (1967). Comp. Biochem. Physiol., **20,** 867.
9. CHEAH, K.S. Biochim. Biophys. Acta. in press.
10. CHEAH, K.S. (1972). in Comparative Biochemistry of Parasites (H. Vanden Boscche, Editor) Academic Press, N.Y. chapter 33.

Table 1.

Comparison of the concentration of respiratory pigments in *Moniezia,
Ascaris* and the ox-neck muscle mitochondria.

The concentration of respiratory pigments was estimated from difference spectra at 20° except those of *Ascaris*-muscle mitochondria. The cytochromes in *Ascaris* were calculated from difference spectra recorded at -196° with α-glycerophosphate as substrate.

—: absent in mitochondria

Respiratory components	Concentration (nmoles/mg protein)		
	Moniezia	Ascaris	Ox-neck
Cyt b_{556} (-196°)		0.073	
Cyt **b**	0.120	—	0.247
Cyt c_1		—	
Cyt **c**		0.089	
Cyt cc_1	0.139		0.582
Cyt aa_3	0.064	0.003	0.478
Cyt **o**	0.331	0.012	—

Table 2.
Comparison of the turnover numbers for cytochromes **o** and **a₃** in *Moniezia* and *Ascaris* and cytochrome **a₃** in the ox-neck muscle mitochondria.

The respiratory rates for the various substrates were measured polarographically at 25°. The turnover number was calculated from the following formula: Turnover number = [(O₂ uptake (nmoles O₂/sec/mg protein)/(concentration of cytochrome **o** or **a₃** (nmoles/mg protein)] × 4. —: not estimated. The turnover is expressed to the nearest whole number in electrons/cytochrome/sec.

	Turnover number (electrons/cytochrome/sec)						
Substrate	**o**		**a₃**			**(o + a₃)**	
	Monie-zia	*Ascaris*	*Monie-zia*	*Ascaris*	**Ox-neck**	*Monie-zia*	*Ascaris*
α-Glycerophosphate	8	3	81	13	0	7	3
Succinate	6	21	54	91	41	5	17
Ascorbate **plus** TMPD	6	31	54	136	33	5	25
Pyruvate **plus** malate	2	—	11	—	33	1	—
Malate	2	12	12	53	—	1	10
NADH	0	5	0	21	0	0	4

Table 3.

Comparison of the absorption spectra of purified *Ascaris* cytochrome c_{550}, *Moniezia* cytochrome c_{550} and cytochrome **c** from the back-muscle of the Large White pig.

The cytochrome **c** was reduced with dithionite and the absolute spectrum was recorded with an Aminco-Chance Dual-Wavelength/Split Beam spectrophotometer using 20 mM phosphate buffer (pH 7.0) as reference. The room temperature (20°) spectra were recorded using 10 mm light-path cuvettes and those at liquid-nitrogen temperature (-196°) in 2 mm light-path cells.

Absolute absorption peaks (nm)

Cyto-chrome c	α-band		β-band		γ-band	
	20°	**-196°**	**20°**	**-196°**	**20°**	**-196°**
Ascaris	550	548	521	528	417	416
		542		521		
				511		
Moniezia	550	545.5	521	524	417	416
		534		517.5		
				511		
				507		
Large White Pig	550	547	521	524		
		544.5		517.5		
		536		507		
				502		

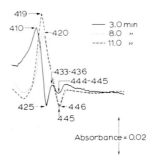

Fig. 1.
Difference spectra (20°) showing the reduced respiratory pigments of *Moniezia expansa*. (From Cheah, ref. 2: reproduced with the permission of Elsevier Publishing Company).
Both the sample and reference cuvettes (4.0 mm light-path) contained 0.48 ml preparation (4.7 mg protein/ml) in 50 mM phosphate buffer (pH 7.6). ....., antimycin A (1.0 µM) **minus** oxidized; - - - - - , p-phenyl-enediamine (1.0 mM) + cyanide (1.0 mM) **minus** oxidized.

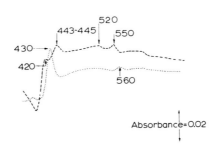

Fig. 2.
CO difference spectra (α-glycerophosphate + CO **minus** α-glycerophosphate) showing the existence of cytochromes a_3 and **o** in *Moniezia* (20°). (From Cheah, ref. 2: reproduced with the permission of Elsevier Publishing Company).

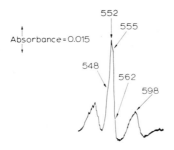

Fig. 3.

Low temperature (-196°) difference spectrum of the succinate-reducible cytochromes in *Moniezia* in the presence of cyanide **plus** oxygen.

Both the sample and reference cuvettes (2.0 mm light-path) contained 0.24 ml preparation (2.4 mg protein/ml) in 50 mM phosphate buffer (pH 7.6). The preparation in the sample cuvette was treated with succinate (10 mM) and cyanide (1 mM) after which both the contents in the sample and reference cells were oxygenated before freezing in liquid-nitrogen.

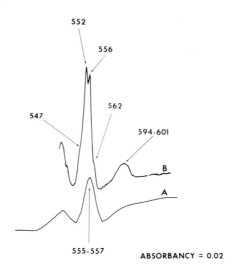

Fig. 4.

Difference spectra of the succinate-reducible cytochromes in *Moniezia* recorded at 20° (A) and at -196° (B).

425

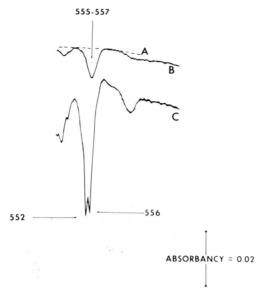

Fig. 5.
Difference spectra illustrating the re-oxidation of cytochrome **o** by fumarate in *Moniezia.*
A, α-glycerophosphate-reduced **minus** α-glycerophosphate-reduced (20°)
B, α-glycerophosphate + fumarate **minus** α-glycerophosphate (20°).
C, as B but recorded at -196°.
Final concentrations (mM): α-glycerophosphate, 10.0; fumarate, 5.0.

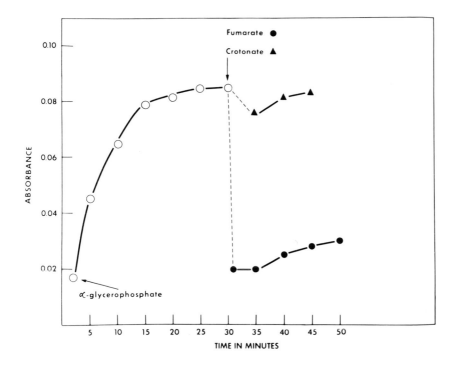

Fig. 6.

Reduction of cytochrome **o** by α-glycerophosphate and the effects on ferrocytochrome **o** following the separate subsequent addition of either fumarate, crotonate or maleate.

Cytochrome **o** reduction by α-glycerophosphate (10 mM) was measured at 425-410 nm and treated subsequently with either 5 mM fumarate (●), crotonate (▲) or maleate, the latter not shown in Figure.

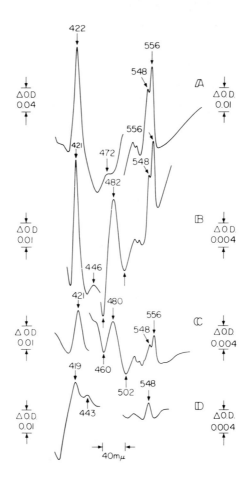

Fig. 7.
Difference spectra (-196°) of *Ascaris*-muscle mitochondria. These spectra were recorded by the rapid freezing technique[3, 4] using 2.0 mm cuvettes containing 0.35 ml *Ascaris*-muscle mitochondria (7.0 mg protein/ml) suspended in 220 mM mannitol, 50 mM sucrose and 15 mM Tris-HCl (pH 7.4).
A, dithionite-reduced **minus** oxidized;
B, α-glycerophosphate-reduced **minus** oxidized;
C, malate-reduced **minus** oxidized;
D, ascorbate + cyanide **minus** oxidized.
(From Cheah and Chance, ref. 1: reproduced with the permission of Elsevier Publishing Company).

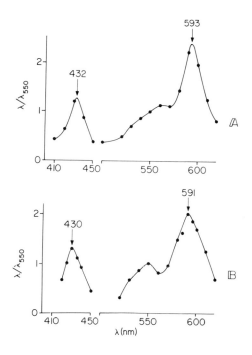

Fig. 8.

Photochemical action spectra of CO-inhibited respiration of *Ascaris*-muscle mitochondria (A) and Baker's yeast (B). The ordinate represents the efficiency of light compared with 550 nm light in reversing the CO-inhibited resipration. Gas mixture, O_2: CO: N_2 (10 : 40 : 50, by vol). *Ascaris*-muscle mitochondria (5.0 mg protein) were reduced with malate **plus** succinate in 5 mM phosphate buffer (pH 7.2). Done in collaboration with Mr. K. Olofsson using the improved technique of Hyde[5] on that of Castor and Chance[6]. (From Cheah and Chance, ref. 1: reproduced with the permission of Elsevier Publishing Company).

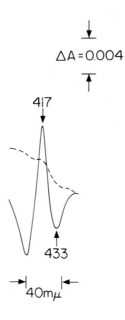

Fig. 9.
CO difference spectrum (α-glycerophosphate + CO **minus** α-glycero-phosphate) of *Ascaris*-muscle mitochondria (-196°). (From Cheah and Chance, ref. 1: reproduced with the permission of Elsevier Publishing Company).

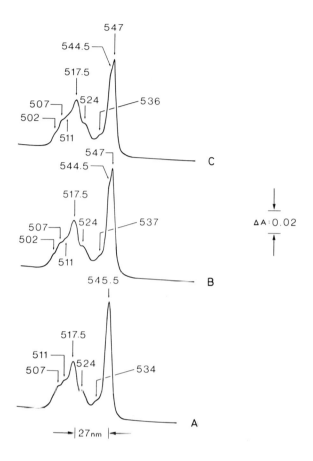

Fig. 10.
Absolute spectra (-196°) of ferrocytochrome **c** isolated from *Moniezia* (A), the back-muscle of the Pietrain (B) and Large White (C) pigs.
The purified cytochrome **c** was reduced with dithionite and then passed through Sephadex G-25 to remove the excess dithionite. Spectra were recorded in 2.0 mm light-path cuvettes using a Split-beam spectrophotometer of Dr. P.B. Garland.

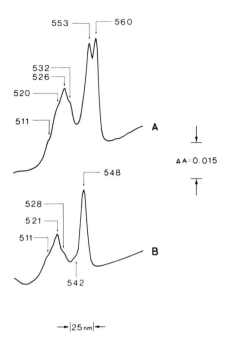

Fig. 11.
Absolute spectra (-196°) of reduced *Ascaris* cytochrome c_{550} (20°) and reduced *Ascaris* cytochrome b_{560} (20°).

The spectra were recorded in 2.0 mm light-path cells using the Aminco-Chance Dual-wavelength/Split-beam spectrophotometer. Other details are given in Figure 10.

29

ROLE OF NON-HEME IRON IN CESTODE RESPIRATION

Eugene C. Weinbach[1]

Laboratory of Parasitic Diseases
National Institute of Allergy and Infectious Diseases
National Institutes of Health
Bethesda, Maryland 20014, U.S.A.

The paramount role of iron in biological oxidations has been recognized since the pioneering studies of Warburg[2]. The monumental rediscovery of the cytochromes by Keilin in 1925 firmly established the fundamental significance of the heme proteins in cellular respiration[1].

In 1962 a new type of iron-containing protein was discovered in bacteria by Mortenson, **et al.**[2], and designated by them as **ferredoxin.** Parallel work with both plant and animal tissues soon disclosed that this new type of non-heme iron protein was widely distributed in nature. It also was shown that the non-heme iron proteins are involved in electron transport, although the picture is far from complete[3]. Green, and his colleagues[4] during the course of chemical fractionation studies of mammalian mitochondria drew attention to the relatively large quantities of iron in submitochondrial fractions, which could not be accounted for in terms of the heme groups present. It now is well documented that non-heme iron is associated with at least two enzymatically active components of the mammalian respiratory chain: succinate and NADH dehydrogenases[5]. Little evidence is available, however, for the participation of non-heme iron in the respiratory chain of parasites.

Inhibitor Studies: Several years ago, Dr. von Brand and I initiated a study of the respiratory chain of cestodes. We found that the isolated mitochondria of *Taenia taeniaeformis,* unlike their mammalian counterparts were quite restricted in their repertoire of oxidases[6]. We also established that the substrate most actively oxidized by the tapeworm

1. With the technical assistance of C. Elwood Claggett.

2. Warburg's early studies are summarized in reference (1).

mitochondria was glycerol-3-phosphate. This was of particular interest in view of von Brand's, **et al.** earlier work showing that glycerol was a preferred substrate of the intact organism[7]. During the course of these studies, we observed that although the mitochondrial respiration was only partially sensitive to cyanide, it was more sensitive to metal chelating reagents, particularly to those reagents known to bind iron (Table I).

All of the compounds listed in Table I are capable of forming coordination complexes with metals. No inhibitor, of course, is specific, and these reagents will bind metals other than iron, particularly copper. It is likely, however, that these reagents are inhibiting mitochondrial respiration by chelating iron. This point is illustrated with the substituted trifluoroacetone derivatives. 2-Thenoyltrifluoroacetone, which has a marked affinity for iron, caused complete inhibition of mitochondrial respiration at a final concentration of 3mM. In contrast, the furoyl derivative, which has a lesser affinity for iron, was much less effective as an inhibitor[3]. This phenomenon is documented further in Table II. The experiments summarized here were designed to test a series of substituted butanedione derivatives for their ability to inhibit the oxidation of glycerol-3-phosphate by *T. taeniaeformis* mitochondria. The compounds are listed in order of decreasing capacity to form coordination complexes with metals, owing to a decrease in the basicity of the enolate ion[9]. The data shows that there is good correlation between the ability of these compounds to chelate metals and their ability to inhibit mitochondrial respiration.

The site of action of the metal chelators apparently is on or closely associated with the primary dehydrogenase, and not on some other component of the respiratory chain. Phenazine methosulphate is known to transfer electrons from α-glycerophosphate dehydrogenase directly to molecular oxygen[10]. Addition of phenazine methosulphate to suspensions of *T. taeniaeformis* mitochondria enhanced the rate of respiration supported by glycerol-3-phosphate. The enhanced respiration was inhibited 70 per cent by 10 mM salicylaldoxime (Fig. I).

Iron content of mitochondria: Chemical analysis of mitochondria isolated from *T. taeniaeformis* disclosed that non-heme iron is present in these tapeworms (Table III). For comparative purposes, values for beef heart mitochondria also are shown. The values presented for the tapeworm mitochondria probably are low because the results, per convention, are expressed on the basis of protein content. The tapeworm mitochondrial fractions are not as homogeneous as the mammalian preparations and

3. Greater inhibition of mammalian electron transport systems by 2-thenoyltrifluoroacetone than by 2-furoyltrifluoroacetone has been ascribed to stronger metal coordination by the thiophene sulfur than by the furan oxygen (8).

therefore, undoubtedly, we are under-estimating the iron content of the parasite mitochondria. What is more significant than the precise values, is the ratio of non-heme to heme iron. It is clearly evident, based on these data with *T. taeniaeformis,* and on data we are obtaining with **Hymenolepis diminuta,** and **H. microstoma,** that this ratio in cestode mitochondria exceeds that obtained with mammalian mitochondria.

The results of the experiments with inhibitors, as well as the above analytical data, suggested, but certainly did not prove, that non-heme iron is participating in the respiration of cestodes. A third and final body of evidence was needed.

EPR Studies: A well-established characteristic of the non-heme iron proteins is that when the iron is in the reduced state, a specific signal can be detected by low-temperature election paramagnetic resonance spectroscopy[11]. In collaboration with Dr. Hideo Kon of the National Institute of Arthritis and Metabolic Diseases, we had the opportunity to examine cestode mitochondria by this technique.[4]

Fig. II showns the distinct and characteristic EPR signal that was obtained when *T. taeniaeformis* mitochondria were reduced with dithionite. This signal occurred at $g = 1.937$ as shown by its correspondence with the calibration signal. Beinert and his associates have observed a spin resonance signal at $g = 1.94$ on addition of substrate to mammalian mitochondria, submitochondrial particles, and in the purified succinate and NADH dehydrogenase. The signal is attributed to reduced non-heme iron in these preparations[11].

It may be seen in Fig. III that the same characteristic EPR signal of reduced non-heme iron was obtained upon the addition of glycerol-3-phosphate to mitochondria of *T. taeniaeformis.* In other words, enzymatic as well as chemical reduction of the mitochondria evoked the characteristic non-heme iron signal.

In contrast, when the mitochondrial suspension was vigorously shaken in air, the characteristic signal did not appear (Fig. IV). Under these conditions, the respiratory chain is in the oxidized state.

Fig. V presents data obtained with mitochondria isolated from *H. microstoma*. These preparations readily oxidized glycerol-3-phosphate, and upon addition of this substrate to reduce the respiratory chain, the characteristic non-heme iron signal developed. Note that the signal was abolished in the presence of 2-thenoyltrifluoroacetone, the compound shown to inhibit oxidation of glycerol-3-phosphate by *T. taeniaeformis* mitochondria (Table I).

4. These experiments were done in collaboration with Dr. James Dvorak, and we thank Dr. H. Kon for the electron paramagnetic resonance measurements and interpretations.

Analogous results were obtained with mitochondria isolated from *H. diminuta* (Fig. VI).

Concluding Remarks

Evidence has been adduced that non-heme iron proteins not only are present in cestodes but are functional in their respiration. There is little data available on the occurrence and functional significance of these important election carriers in parasites. Cheah, for example, on the basis of spectrophotometric evidence, has suggested the presence of non-heme iron in the pathway of succinate oxidation in *Monieza expansa*[12]. I believe that as we examine more species of parasites, we will find that the non-heme iron proteins are widely distributed - not only among the helminths, but also in the protozoa - particularly in those species that are deficient in the cytochromes.

Furthermore, it may be envisoned that the non-heme iron proteins are sites of chemotherapeutic activity. Model experiments done by Lovenberg and his associates[13] have shown that treatment of bacterial ferredoxin with the organic mercurial, Mersalyl, displaces the iron from its normal configuration in the protein thereby rendering it inactive as a biological catalyst. It is plausible that other heavy metals such as arsenic and antimony may act in a similar manner on the non-heme iron proteins of parasites. These are only interesting speculations now, but they should provide a fascinating field for future research.

References

1. KEILIN, D. (1966). The History of Cell Respiration and Cytochrome, Cambridge University Press.
2. MORTENSON, L.E., VALENTINE, R.C. and CARNAHAN, J.E. (1962). Biochem. Biophys. Res. Commun. **7,** 448.
3. HALL, D.O. and EVANS, M.C.W. (1969). Nature, **223,** 1342.
4. GREEN, D.E. (1956). in Enzymes: Units of Biological Structure and Function, Henry Ford Hosp. Internatl. Symp. (O.H. Gaebler, editor) p. 465, Academic Press, Inc. New York.
5. WAINIO, W.W. (1970). The Mammalian Mitochondrial Respiratory Chain. Academic Press, New York and London.
6. WEINBACH, E.C. and von BRAND, T. (1970). Internatl. J. Biochem. **1,** 39.
7. von BRAND, T., CHURCHWELL, F. and ECKERT, J. (1968). Exptl. Parasitol. **23,** 309.
8. TAPPEL, A.L. (1960). Biochem. Pharmacol. **3,** 289.
9. BAILAR, J.C. (editor), (1956). The Chemistry of the Coordination Compounds, Reinhold, New York.

10. RINGLER, R.L. and SINGER, T.P. (1962). in Methods in Enzymology (Colowick, S.P., and Kaplan, N.O., editors) vol. V, p. 432, Academic Press, New York and London.
11. BEINERT, H. (1965). in Non-heme Iron Proteins: Role in Energy Conversion (A. San Pietro, editor) p. 23, Antioch Press, Yellow Springs, Ohio.
12. CHEAH, K.S. (1967). Comp. Biochem. Physiol. **23,** 277.
13. LOVENBERG, W., BUCHANAN, B.B. and RABINOWITZ, J.C. (1963). J. Biol. Chem. **238,** 3899.
14. KING, T.E., NICKEL, K.S. and JENSEN, D.R. (1964). J. Biol. Chem. **239,** 1989.
15. KON, H. (1968). J. Biol. Chem. **243,** 4350.

Table 1.

Inhibition of Glycerol-3-phosphate Oxidation in *T. taeniaeformis* Mitochondria.

Compound	Concentration mM	Inhibition Percent
α,α-Dipyridyl	10	18
8-Hydroxyquinoline	10	40
Salicylaldoxime	10	65
2-Thenoyltrifluoroacetone	1	45
	2	60
	3	100
2-Furoyltrifluoroacetone	1	10
	2	25
	3	40

Oxidation was determined polarographically at 24°. The cuvette contained 4 mg of mitochondrial protein, and 10 μmoles of glycerol-3-phosphate in a final volume of 1.5 ml. Inhibitors were added in the final concentrations indicated. Mitochondria was isolated in 0.25 M sucrose from homogenates of *T. taeniaeformis* larvae. Details are given in reference [6].

Table 2.
Chelation and inhibition

$R\text{-CO-CH}_2\text{-CO-CF}_3$	Concentration for 50 % inhibition
R =	mM
2-Naphthyl	0.6
2-Thienyl	1.9
Phenyl	2.2
Ethyl	2.9
2-Furyl	3.1
Methyl	) 5.0

Oxidation was determined polarographically at 24°. The cuvette contained 2.7 mg of mitochondrial protein (isolated from *T. taeniaeformis* larvae) and 10 μmoles of glycerol-3-phosphate in a final volume of 1.0 ml. Details given in reference(6).

Table 3.
Iron Content of Mitochondria

Preparation	Iron Content (nanoatoms/mg protein)		Ratio Non-heme/heme
	Heme	Non-heme	
T. taeniaeformis larvae	1.0	5.6	5.6
Beef heart	2.5	6.4	2.5

Values for the iron content of beef heart mitochondria, and the analytical procedure used for determining the iron content of parasite mitochondria were as reported by King, **et al.** [14]

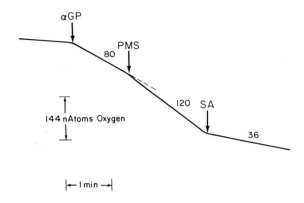

Fig. I.

Effect of salicyladoxime on the phenazine methosulfate-stimulated oxidation of glycerol-3-phosphate. The cuvette contained mitochondria (4.2 mg of protein) from *T. taeniaeformis* larvae suspended in a final volume of 1.5 ml. Additions were 10 μmoles of glycerol-3-phosphate (αGP), 0.1 μmole of phenazine methosulfate (PMS), and 15 μmoles of salicylaldoxime (SA). The numbers on the polarographic tracing express the oxygen consumption as nanoatoms per min. Other details given in reference[6].

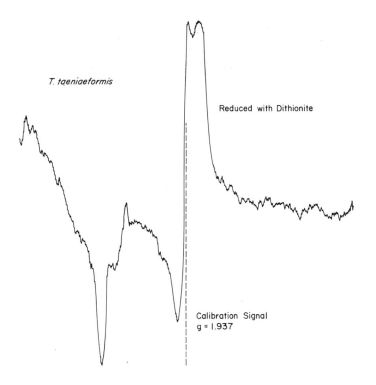

T. taeniaeformis

Reduced with Dithionite

Calibration Signal
g = 1.937

Fig. II.

The EPR spectrum of mitochondria from *T. taeniaeformis* larvae reduced with dithionite. Mitochondria (4 mg of protein) suspended in 0.25 M sucrose were reduced with dithionite at 24°, and after 3 min. were frozen in liquid nitrogen. Electron paramagnetic resonance spectra in this and subsequent experiments (Figs. III-VI) were recorded as first derivatives of the electron paramagnetic resonance absorption on a Varian V-4500 spectrometer as described by Kon[15]. The magnetic field strength was calibrated by a proton resonance gaussmeter combined with a frequency counter.

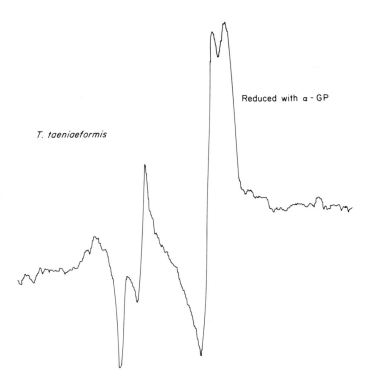

Fig. III.

The EPR spectrum of mitochondria from *T. taeniaeformis* larvae reduced with glycerol-3-phosphate. Mitochondria (4 mg of protein) suspended in 0.25 M sucrose were reduced with 10 mM glycerol-3-phosphate (α-GP) at 24°, and after 6 min. were frozen in liquid nitrogen.

T. taeniaeformis

Oxidized with Air

Fig. IV.

The EPR spectrum of mitochondria from *T. taeniaeformis* larvae oxidized in air. Mitochondria (4 mg of protein) suspended in 0.25 M sucrose were shaken in air at 24° for 3 min., and frozen in liquid nitrogen.

H. microstoma

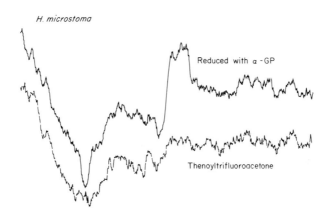

Reduced with α-GP

Thenoyltrifluoroacetone

Fig. V.

The EPR spectrum of mitochondria from adult *H. microstoma* reduced with glycerol-3-phosphate. Mitochondria (3.5 mg of protein) suspended in sucrose were reduced with 10 mM glycerol-3-phosphate (α-GP) at 24° and after 6 min. frozen in liquid nitrogen (solid line). Mitochondria also were treated with 5 mM 2-thenoyltrifluoroacetone prior to the addition of glycerol-3-phosphate.

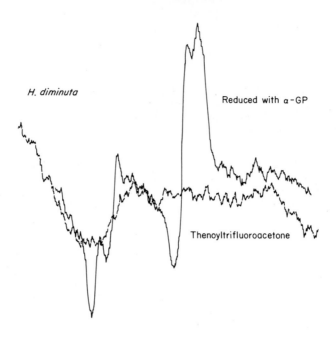

Fig. VI.

The EPR spectrum of mitochondria from adult *H. diminuta* reduced with glycerol-3-phosphate. Mitochondria (5.4 mg of protein) suspended in sucrose were reduced with 10 mM glycerol-3-phosphate (α-GP) and after 6 min. were frozen in liquid nitrogen (solid line). Mitochondria also were treated with 5 mM 2-thenoyltrifluoroacetone prior to the addition of glycerol-3-phosphate (broken line).

30

EFFECTS OF ANTHELMINTICS ON P^{32} ESTERIFICATION IN HELMINTH METABOLISM[1]

Howard J. Saz

Department of Biology
University of Notre Dame
Notre Dame, Indiana
U.S.A.

Desaspidin, one of the active principles of oleoresin of aspidium, and chlorsalicylamide (Yomesan) have high anticestodal activities **in vitro**. It has been reported by Runeberg [1, 2] and by Gönnert **et al.** [3] respectively, that both of these drugs uncouple aerobic oxidative phosphorylations in mammalian mitochondria. Subsequently, Scheibel **et al.** [4] demonstrated that desaspidin, chlorosalicylamide and a number of other compounds which possessed anticestodal activities were all capable, in low concentrations, of inhibiting the anaerobic, electron transport associated P^{32}-ATP exchange reaction in mitochondrial preparations from the cestode *Hymenolepis diminuta.*

If these agents were inhibiting the energy metabolism of *H. diminuta* under physiological conditions, then it would be expected that the same agents should also inhibit the nematode, *Ascaris lumbricoides,* which appears to have an anaerobic energy metabolism similar to that of the cestode. In spite of these similarities, the anticestodal agents are specific for the tapeworms and appear to have no inhibitory effect upon intact *Ascaris.* Saz and Lescure [5] demonstrated, however, that if the anticestodal agents are permitted to act upon isolated *Ascaris* mitochondria, they have the same inhibitory effect upon the P^{32}-ATP exchange reaction as previously demonstrated in *H. diminuta* mitochondria. These findings indicated that the mitochondria of the two organisms may be similar, but a permeability barrier is indicated to explain the failure of these agents to affect intact nematodes.

Figure 1 illustrates the pathway proposed for the utilization of carbohy-

1 These studies were supported by Grants AI-09483, and TOI-AI-00400 from the National Institutes of Health, United States Public Health Service.

drate and the mitochondrial generation of ATP in *Ascaris*[6]. Presumably, a similar mitochondrial sequence occurs in *H. diminuta*. According to the proposed pathway, one mole of the mitochondrial substrate, malate, dismutates to one-half mole of pyruvate and succinate respectively. In the process, 0.5 moles of Pi are esterified into ATP. This system has been explored further, and the effects of anthelmintics examined. Several features of this scheme should be noted. First, the incubation of *Ascaris* mitochondria with malate and Pi^{32} should result in the formation of ATP^{32}. Second, the quantity of Pi^{32} esterified into ATP should be equal to one-half the quantity of malate utilized resulting in a P^{32}/malate ratio of 0.5. Finally, the malic enzyme which catalyzes the mitochondrial formation of pyruvate from malate is DPN linked. This is required, since the energy generating fumarate reductase reaction requires DPNH.

Ascaris mitochondria were incubated anaerobically in the presence of malate, Pi^{32}, and a system for trapping the terminal phosphate of ATP formed as a more stable organic phosphate[7]. Figure 2 shows that a rapid and linear uptake of inorganic P^{32} into organic phosphate takes place. In the absence of malate, there is almost no esterification. It is of interest, that this formation of organic phosphate specifically requires either malate or fumarate. Other substrates tried resulted in little or no phosphorylation. Malonate inhibited the reaction as would be predicted. When Pi^{32} uptake was quantitated and compared to malate disappearance, results shown in Table 1 were obtained. In three separate experiments, the ratios of inorganic phosphate esterified to malate utilized were 0.43, 0.47 and 0.36 respectively; approaching the theoretical ratio of 0.5 as discussed above. These findings, then, are in accord with the pathway postulated.

Table 2 shows the effects of some known uncouplers of oxidative phosphorylation in mammalian tissues and of some anthelmintic agents on this energy yielding system. 2,4 Dinitrophenol and CCP, two uncouplers in mammalian systems, inhibit phosphorylation in the *Ascaris* mitochondria; 1.5×10^{-4} M dinitrophenol inhibiting 67 per cent, while 5×10^{-7} M CCP inhibited 44 per cent. Chlorosalicylamide was also a very effective inhibitor of phosphorylation, 1×10^{-6} M resulted in an 82 per cent reduction of P^{32} incorporation.

Desaspidin was also a very effective inhibitor of this series of reactions, 5×10^{-6} M inhibiting almost completely. Dichlorophen, on the other hand, was not very effective as an inhibitor, requiring 5×10^{-4} M to inhibit 48 per cent. Other anticestodal agents, such as SKF compound 90625 and BW compound 61-435 were also inhibitors of the anaerobic phosphorylation system, but to a lesser degree than either chlorosalicylamide or desaspidin; the SKF compound requiring a concentration of 1 x 10^{-5} M for a 45 per cent inhibition, at which concentration the BW compound inhibited only 17 per cent.

It is also of interest that the antinematodal agent dithiazanine is a very effective inhibitor in this system; 1 x 10^{-6} M inhibiting 58 per cent.

It should be stated at this point that although the inhibitory effects by these anthelmintic agents of P^{32} incorporation could explain their physiological effects, additional studies are necessary to define with certainty their primary sites of action **in vivo**.

Oligomycin, which inhibits the electron transport associated phosphorylation reaction in mammalian tissues is also a potent inhibitor of the *Ascaris* system. Similarly, rotenone, an inhibitor of site I, or flavin level phosphorylation, also inhibits the *Ascaris* anaerobic system as would be expected. In accord with the concept that the cytochrome system does not enter into these reactions, antimycin A, which inhibits phosphorylation in mammalian tissues at the cytochrome level, has very little effect on the *Ascaris* system even at relatively high concentrations.

All of the findings reported above comprise additional information which is in agreement with our current concepts concerning the mechanisms whereby *Ascaris,* and presumably a number of other helminths, obtain energy anaerobically within the mitochondrion. In addition, it appears likely that interference with the mitochondiral mechanisms for energy generation could be lethal to the worms.

Most of the enzymatic machinery required for the *Ascaris* mitochondrial energy generating system has been shown also to be present in the cestode, *Hymenolepis diminuta*. One important difference, however, has been reported by Prescott and Campbell[8]. These authors demonstrated the presence of a TPN linked malic enzyme in *H. diminuta.* As shown in Table 3, our results confirm these findings. *H. diminuta* mitochondria were assayed spectrophotometrically for malic enzyme activity. Similar to the corresponding *Ascaris* enzyme, that of the cestode is also completely dependent upon the presence of Mn^{++} for activity. Contrary to the *Ascaris* enzyme, the tapeworm system requires TPN for activity rather than DPN.

This finding must be reconciled with the fact that DPNH is the electron donor for the fumarate reductase system of *Ascaris* mitochondria. TPNH is ineffective. TPNH **per se** is generally not associated with energy yielding reactions. Therefore, the possible presence of a TPNH-DPN transhydrogenase system in *H. diminuta* mitochondria was investigated. Preparations of tapeworm mitochondria were incubated with TPNH and acetylpyridine DPN. Reduction of the DPN derivative was followed spectro-

447

photometrically at 375 mμ. Findings are illustrated in Figure 3. A non-energy dependent transhydrogenase system was demonstrable, and ATP had no effect upon the rate of this reaction. Attempts to demonstrate transhydrogenase activity in *Ascaris* muscle were negative. Whether the TPN linked malic enzyme of *H. diminuta* serves the same physiological function as the corresponding DPN linked enzyme of *Ascaris* remains to be determined. The presence of a transhydrogenase at least makes this possibility more likely. If so, then inhibition of this transhydrogenase should have an effect on the worm similar to that shown by the other anticestodal agents discussed above.

In conclusion, our increased understanding of the energy metabolisms of a number of helminths has led to the realization that many anthelmintics may act by virtue of their inhibitory effects upon these reactions. Of particular interest are those parasites which possess an anaerobic, electron transport associated mitochondrial energy generating system, since of necessity, these sequences are, in part at least, different from those of the mammalian hosts.

References

1. RUNEBERG, L. (1962). Biochem. Pharmacol., **11,** 237.
2. RUNEBERG, L. (1963). Soc. Sci. Fenn. Commentationes Biol., **26,** 1.
3. GONNERT, R., JOHANNIS, J., SCHRAUFSTATTER, E. and STRUFFE, R. (1963). Medizin und chemie, Vol. VII, Verlag chemie, GMBH, Weinheim Bergstr., p. 540.
4. SCHEIBEL, L.W., SAZ, H.J., and BUEDING, E. (1968). J. Biol. Chem., **243,** 2229.
5. SAZ, H.J. and LESCURE, O.L. (1968). Molec. Pharmacol., **4,** 407.
6. SAZ, H.J. and LESCURE, O.L. (1969). Comp. Biochem. Physiol., **30,** 49.
7. GRUNBERG-MANAGO, M., ORTIZ, P.J. and OCHOA, S. (1956). Biochim. Biophys. Acta, **20,** 269.
8. PRESCOTT, L.M. and CAMPBELL, J.W. (1965). Comp. Biochem. Physiol., **14,** 491.

Table 1.

Ratio of Pi^{32} Incorporation to Malate Utilized by Anaerobic *Ascaris* Mitochondria.

Experiment	Pi^{32} Incorporated	Malate Utilized	Ratio Pi^{32}/Malate
	(μ moles)	(μ moles)	
1	4.46	10.27	0.43
2	5.39	11.58	0.47
3	3.40	9.36	0.36

Table 2.

Inhibition of Pi^{32} Uptake by Uncouplers and Anticestodal Agents.

Inhibitor	Concentration	Pi^{32} Incorporated	Inhibition
	M	μ Moles	%
	0	10.05	—
2,4 Dinitrophenol	5×10^{-5}	6.77	33
	1.5×10^{-4}	3.27	67
	0	11.16	—
CCP (carbonyl	5×10^{-7}	6.28	44
cyanide m-chloro-	1×10^{-6}	1.39	88
phenylhydrazone)	5×10^{-6}	0.92	92
	0	10.92	—
Chlorosalicylamide	1×10^{-6}	1.97	82
	5×10^{-5}	0.54	95
	0	14.39	—
Desaspidin	1×10^{-6}	11.31	21
	5×10^{-6}	0.13	99
			continued

Inhibitor	Concentration	Pi^{32} Incorporated	Inhibition
	M	μ Moles	%
Dichlorophen	0	9.57	—
	1×10^{-5}	9.06	5
	1×10^{-4}	8.28	13
	5×10^{-4}	5.00	48
SKF Compound 90625	0	8.07	—
	1×10^{-6}	7.46	8
	5×10^{-6}	5.70	29
	1×10^{-5}	4.43	45
	5×10^{-5}	0.55	93
BW Compound 61-435	0	9.67	—
	1×10^{-5}	7.99	17
	1×10^{-4}	2.61	73
	5×10^{-4}	1.01	90
Dithiazanine	0	11.56	—
	1×10^{-7}	10.69	8
	1×10^{-6}	4.91	58
	1×10^{-5}	1.12	90
Oligomycin	0	9.43	—
	1.5×10^{-5}	1.40	85
	1×10^{-4}	1.05	89
Rotenone	0	10.10	—
	0.02 μg/ml.	8.45	16
	0.04 μg/ml.	2.53	75
	0.08 μg/ml.	0.50	95
Antimycin A	0	7.96	—
	0.4 μg/ml.	7.53	5
	1.6 μg/ml.	6.22	22

Table 3.

Cofactor Requirements of *H. diminuta* Malic Enzyme

System	Specific Activity*
Complete (TPN, Mn^{++})	124
Minus Mn^{++}	0
DPN, Mn^{++}	2

* mμ Moles substrate utilized/min./mg. Protein.

Fig. 1.
Dismutation of Malate by *Ascaris* Mitochondria.

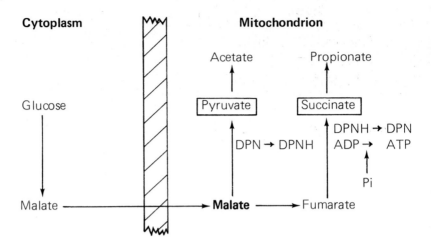

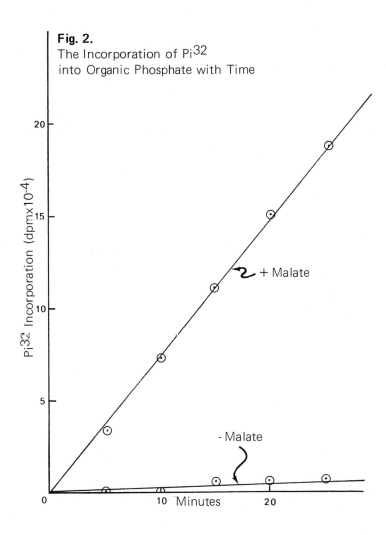

Fig. 2.
The Incorporation of Pi^{32}
into Organic Phosphate with Time

453

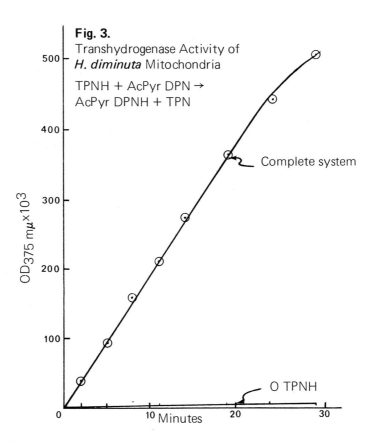

Fig. 3.
Transhydrogenase Activity of
H. diminuta Mitochondria

TPNH + AcPyr DPN →
AcPyr DPNH + TPN

Complete system

O TPNH

31

STUDIES ON THE PHOSPHORYLATION IN
ASCARIS MITOCHONDRIA

H. Van den Bossche*
Department of Comparatie Biochemistry
Janssen Pharmaceutica - Research Laboratories
2340 Beerse - Belgium.

Introduction

Evidence has recently been presented indicating that the anaerobic dismutation of malate in *Ascaris* mitochondria results in an electron transport associated phosphorylation[1]. Mitochondria isolated from the adult *Ascaris suum*[2] catalyze a ^{32}Pi - ATP exchange reaction similar to that found in mammalian mitochondria. Both the exchange reaction[2] and the anaerobic malate-induced incorporation of inorganic phosphate (Pi) into organic phosphate[1] were inhibited by 2,4 dinitrophenol, car bonyl cyanide m-chlorophenylhydrazone and chlorosalicylamide, known uncouplers of oxidative phosphorylation in rat liver mitochondria. This may indicate that the electron transport-associated phosphorylation in *Ascaris* mitochondria resembles oxidative phosphorylation in the corresponding mammalian organelles. The following studies further compare the mechanisms of phosphorylation in *Ascaris* with those in mammalian mitochondria.

Materials and Methods

Preparation of the mitochondrial fraction

Using a Potter-Elvehjem homogenizer with teflon pestle, *Ascaris* muscle was homogenized in 5 volumes of a 0.24 M sucrose solution containing 0.005M EDTA and 0.15 % albumine (pH 7). This and all subsequent procedures were carried out at 2-4° C. The homogenate was centrifuged

* This investigation was supported by Grant no. 1803 from the "Instituut tot Aanmoediging van het Wetenschappelijk Onderzoek in Nijverheid en Landbouw (IWONL)".

at 3080 **g**-min, the supernatant collected and re-centrifuged at 270,000 **g**-min in a Beckman Spinco L2-65B centrifuge. The pellet was washed with sucrose to remove the light-mitochondrial fraction and centrifuged again as before. The final pellet was resuspended in 0.25M sucrose (pH 7.4) to give an approximate concentration of 1 mg protein/ml.

^{32}Pi incorporation

The malate-induced ^{32}Pi incorporation into organic phosphate by *Ascaris* mitochondria was based on the procedure of Saz and Lescure[3]. The incubation mixture consisted of 25 μmoles $MgCl_2$, 50 μmoles glucose, 40 μmoles Tris-HCl buffer (pH 7.1), 12 μmoles K_3PO_4 containing 1.5 μC ^{32}Pi (pH 7), 20 μmoles L-malate, 2 μmoles ADP, 0.5 mg hexokinase (8 μmolar units/mg of protein) and 0.5 ml suspension of intact mitochondria. Final volume was 2 ml. Blanks were run in the absence of malate. The essays were run in Warburg vessels under nitrogen at 37°C for 30 min, and the reaction was stopped by adding 0.1 ml of ice-cold 35% perchloric acid to 1 ml of the reaction mixture. The method of Pullman[4] was followed for the extraction of the organic phosphate. The radioactivity of the sample was determined by adding 1 ml of the extracted aqueous phase to 10 ml of Scintillator fluid (Insta-Gel, Packard) and counting in a Packard 3310 Tri-Carb Liquid Scintillation Spectrometer. Correction for quenching was applied through internal standardisation.

ATPase activity

ATPase activity in *Ascaris* mitochondria was determined as described by Myers and Slater[5]. The reaction medium contained 50 mM Tris-maleate buffer of varying pH as indicated in the tables, 75 mM KCl, 0.5 mM EDTA, 2 mM ATP, 0.1 M sucrose and about 1 mg mitochondrial protein in a volume of 2 ml. When the ATPase activity was determined in the presence of Mg^{++}, EDTA was omitted from the medium. The mixture was incubated for 30 min at 37°C and the reaction terminated by the addition of 2 ml 8% (w/v) trichloroacetic acid. Inorganic phosphate determinations on the supernatants of a 12,300 **g**-min centrifugation were performed according to the method of Van Belle[6].

Uptake of pentachlorophenol

Mechanically shaken suspensions of intact mitochondria were incubated in buffered media with varying amounts of pentachlorophenol for 30 min at 37°C. The buffered media comprised: 25 μmoles $MgCl_2$, 50 μmoles glucose, 40 μmoles Tris-buffer (pH 7.1), 0.2 mmoles sucrose (pH 7.1), 3 mg mitochondrial protein and 100 μl ethanol or pentachlorophenol dissolved in ethanol. (Final volume was 2 ml). After incubation the suspensions were centrifuged in the cold at 1,200,000 **g**-min and the phenol concentration remaining in the supernatant was determined spectrophotometrically at 320 mμ.

Results and Discussion

Since the studies of De Deken[7], it has been acknowledged that the effectiveness of a phenol in uncoupling oxidative phosphorylation in yeast depends upon the degree of dissociation and upon the lipid solubility of that phenol. Hemker[8] also reached the same conclusion after his extensive investigation into the uncoupling action of a series of phenols on phosphorylation in rat liver mitochondria. He concluded that the greater the tendency of a phenol to dissociate, the higher its uncoupling activity and that increasing lipophilicity of substituents also led to higher activity.

Investigating the inhibitory action of 23 phenols on the anaerobic malate-induced ^{32}Pi incorporation into organic phosphate by isolated *Ascaris* mitochondria, we reached the same conculsion (Table 1). The results of the determinations of 50 %-uncoupling concentrations shown in this table indicate that phenols with a low pKa value and a high lipophilicity (π) are potent inhibitors of ^{32} Pi incorporation. The lower the pKa and the higher the π-value, the lower the concentration needed to obtain 50 % inhibition. The relationship between activity and lipophilic and electronic character of the substituents will be discussed further by Tollenaere in the following communication. For the moment we can assume that the dependence of activity on low pKa suggests that the phenolic anion reacts with a positively charged group of mitochondrial proteins involved in the coupling of phosphorylation to electron transport. A similar mechanism has been proposed for mammalian mitochondria[9].

It has been shown that phenylethylbiguanide is a relatively specific inhibitor of site II phosphorylation in rat liver mitochondria[10], and thus as Slater[11] pointed out, addition of this compound causes reduction of cytochrome b. Saz[1] has demonstrated that antimycin A has no effect on the ^{32}Pi incorporation in *Ascaris* mitochondria. This may indicate that no cytochrome b oxidation is involved as antimycin A is said to inhibit electron transport in mammalian mitochondria between cytochrome b and c[12]. Furthermore, it has been shown that the mitochondrial dismutation of 1 mole of malate results in the esterification of 0.5 moles of inorganic phosphate, which corresponds to 1 mole of ATP formed per mole of fumarate reduced[1]. Although phenylethylbiguanide inhibits the ^{32}Pi incorporation into organic phosphate by *Ascaris* mitochondria (Table 2), the foregoing results suggest that site II phosphorylation is not involved. A possible explanation for the inhibitory action of phenylethylbiguanide may therefore be found in the fact that Haas[10], in his study on the effects of this drug on site I phosphorylation, observed no inhibition with 38 μg of drug/mg protein. This concentration inhibited ^{32}Pi incorporation in our own system by only 20 %, and our preliminary results indicate that higher concentrations of inhibitor also affect site I phosphorylation in rat liver mitochondria.

The results presented in Table 2 indicate that the malate-induced inorganic phosphate incorporation was also inhibited by a number of benzimidazoles and salicylanilides known either as uncouplers of oxidative phosphorylation in rat liver mitochondria or as potent anthelmintics. Since oligomycin also inhibits the phosphorylation in *Ascaris* mitochondria (Table 2), it seems reasonable to assume that malate-induced phosphorylation in *Ascaris* mitochondria resembles electron transport linked phosphorylation in rat liver mitochondria.

As early as 1945, Lardy and Elvehjem had found that dinitrophenol increased the rate of hydrolysis of ATP added to minced rat muscle[13]. The same authors suggested that the ATPase might be associated with oxidative phosphorylation. Holton **et al.** [14] observed that, in the absence of 2,4-dinitrophenol, Mg^{++} stimulated the latent ATPase in heart sarcosomes but had little effect on the hydrolysis of ATP by liver mitochondria. As shown in Fig. 1, Mg^{++} also stimulates the ATPase activity in isolated *Ascaris* mitochondria. At pH 7 and 8 the optimal ratio Mg^{++}: ATP was approximately 1. However, as shown in Fig. 2, 2,4-dinitrophenol did not stimulate the ATPase activity in the absence of Mg^{++} and only a slight dinitrophenol-induced stimulation was observed when Mg^{++} were added. The fact that the latent ATPase is not stimulated by this phenol was also noted by Nomura and Obo[15].

Fig. 3 shows the effects of a few substituted phenols on the Mg^{++}-activated ATPase. 4-Cl-, 2,6-diCl- and pentachlorophenol all stimulated the hydrolysis of ATP, whereas 2,6-diCl, 4-NO$_2$ phenol seems to be totally devoid of stimulating activity. However, the concentrations needed to obtain maximal stimulation were much higher than those required for 50 % inhibition of phosphate incorporation e.g. $1.5 \times 10^{-3}M$ pentachlorophenol was required for maximal stimulation but only $6.4 \times 10^{-6}M$ for 50 % inhibition of phosphate incorporation. Further examination of the graphs presented reveals that the two substituted phenols with highest stimulating activity both have a pKa greater than 6. Pentachlorophenol with much lower stimulating activity has a pKa-value of only 4.8.

In order to determine whether or not the pKa of a phenol must be greater than 6 to obtain a significant stimulation, we measured the effects of 17 phenols on Mg^{++}-activated ATPase activity in isolated *Ascaris* mitochondria. The results presented in Table 3 are for phenols with a pKa value greater than 6, and show that with only one exception all stimulated ATPase activity. pCN-phenol was inactive, possibly due to the low lipophilic character of its substituent ($\pi = 0.14$). Table 4 presents the results obtained for phenols with pKa-values lower than 6. Only two of the investigated phenols, pentachlorophenol and 2,6 dil, 4-NO_2-phenol appeared to stimulate the Mg^{++}-activated ATPase to any degree. Both phenols have a relatively high lipid solubility.

From the foregoing it may be concluded that under experimental conditions, the phenols must have either a pKa value greater than 6 and/or a high lipid solubility before ATPase stimulation is observed. This dependence of stimulating action on high pKa is in contrast to our results for the inhibitory action of the phenols on phosphate incorporation. All our experiments to date seem to indicate that the Mg^{++}-activated ATPase of *Ascaris* differs from that of the inner membrane of mammalian mitochondria. The considerable difference between the concentrations required for 50 % inhibition of phosphate incorporation and those needed for stimulation also lend some support to this conclusion. Although a difference in optimal concentration of dinitrophenol required for inducing ATPase and stimulating oxygen uptake has also been observed in rat liver mitochondria [16], this difference does not approach the magnitude of that found in *Ascaris* mitochondria. At high concentrations however, all phenols investigated inhibited Mg^{++}-activated ATPase, and in this respect the ATPase of *Ascaris* mitochondria resembles that found in mammalian mitochondria [16]. Furthermore the concentrations necessary to obtain maximal stimulation (c_{opt}.) and 50 % inhibition of ATPase activity (I_{50}) (Tables 3 and 4) are inversely proportional to the pKa, as was shown by the effects of the phenols on phosphate incorporation.

Since it is known that small changes in the structural integrity of mitochondria lead to uncoupling and to unmasking of the ATPase, we believed that the differences observed between the ATPase of *Ascaris* and that of mammalian mitochondria may be due to the utilisation of loosely coupled mitochondria. However when phosphate incorporation and malate disappearance were determined in the same mitochondrial preparation as that used for ATPase activity determinations, Pi: malate ratios

459

between 0.34 and 0.46 were obtained. These values are close to the theoretical ratio of 0.5. Furthermore additions of albumin, which is known to restore phosphorylation in mammalian mitochondria [17] did not improve phosphorylation in isolated *Ascaris* mitochondria.

Another possible reason for the high concentrations required to stimulate ATPase is the existence of an interaction between phenol and ATP at the level of the mitochondrial membrane. This possibility was investigated by studying the uptake of pentachlorophenol by *Ascaris* mitochondria in either the absence or presence of ATP (10^{-3}M). As shown in Fig. 4, an interaction between ATP and pentachlorophenol was not observed even at an ATP concentration several times that of the phenol.

An alternative explanation may be that dinitrophenol inhibits phosphorylation without uncoupling. If this is the correct explanation, dinitrophenol must inhibit not only phosphate incorporation but also malate utilization. The results presented in Table 5 indicate, however that this particular phenol did not inhibit malate dismutation even at concentrations completely blocking phosphate incorporation.

No results to date contradict our conclusion that the ATPase of *Ascaris* mitochondria differs from that found in mammalian mitochondria. The fact that the ATPase is not stimulated by dinitrophenol may indicate that ATP does not support the reverse electron transport through site I in *Ascaris* mitochondria. However the possible errors that can be made in the study of a mechanism as complex as mitochondrial phosphorylation, necessitate further studies to establish the validity of this difference between *Ascaris* and mammalian mitochondria.

References

1. SAZ, H.J. (1971). Comp. Biochem. Physiol. **39B,** 627.
2. SAZ, H.J. and LESCURE, O.L. (1968). Molec. Pharmacol. **4,** 407.
3. SAZ, H.J. and LESCURE, O.L. (1969). Comp. Biochem. Physiol. **30,** 49.
4. PULLMAN, M.E. (1967). in Methods in Enzymology (R.W. Estabrook and M.E. Pullman, eds.) vol. 10 p. 57. Academic Press. New York.
5. MEYERS, D.K. and SLATER, E.C. (1957). Biochem. J. **67,** 558.
6. VAN BELLE, H. (1970). An. Biochem. **33,** 132.
7. DE DEKEN, R.H. (1955). Biochim. Biophys. Acta **17,** 494.
8. HEMKER, H.C. (1963). Biochim. Biophys. Acta **73,** 311.
9. WEINBACH, E.C. and GARBUS, J. (1969). Nature **221,** 1016.
10. HAAS, D.W. (1964). Biochim. Biophys. Acta **92,** 433.
11. SLATER, E.C. (1966). in Comprehensive Biochemistry (M. Florkin and E.H. Stotz, eds.) vol. 14 p. 369. Elsevier Publishing Company, Amsterdam.
12. CHANCE, B. (1958). J. Biol. Chem. **233,** 1223.
13. LARDY, H.A. and ELVEHJEM, C.A. (1945). Ann. Rev. Biochem. **14,** 16.

14. HOLTON, F.A., HULSMAN, W.C., MEYERS, D.K. and SLATER, E.C. (1957). Biochem. J. **67,** 579.
15. NOMURA, Y. and OBO, F. (1968). Acta Med. Univ. Kagoshima **10,** (2), 203.
16. HEMKER, H.C. (1964). Biochim. Biophys. Acta **81,** 1.
17. WEINBACH, E.C. and GARBUS, J. (1966). J. Biol. Chem. **241,** 169.
18. FUJITA, T., IWASA, J. and HANSCH, C. (1964). J. Am. Chem. Soc. **86,** 5175.
19. HOHORST, H.J. (1963). in Methods of Enzymatic Analysis (H.U. Bergmeyer, ed.) p. 328. Academic Press, New York.

Acknowledgements

The author wishes to thank Mrs. Horemans, Mr. Goossens and Mr. Vermeiren for their skilled assistance in these experiments; Mr. and Mrs Scott for their help in the preparation of the manuscript and Dr. Paul A.J. Janssen for his constant interest.

Table 1.

Effects of substituted phenols on the malate-induced ^{32}Pi incorporation into organic phosphate by *Ascaris* mitochondria.

Substituents	pKa	$\pi^{(a)}$	$I_{50}^{(b)}$ (x 10^{-4}M)
4-CH$_3$O	10.6	-0.12	43
4-tert.Bu	10.2	1.68	2.5
–	9.9	0	44.5
4-F	9.9	0.31	27.5
4-Cl	9.4	0.93	3.6
4-Br	9.3	1.13	2.77
4-I	9.2	1.45	1.77
4-Ph	9.2	1.89	1.04
4-(4Cl-Ph)	9.2	2.82	0.51
3-Cl	8.6	1.04	4.05
3-NO$_2$, 4-CH$_3$	8.3	1.02	2.55
3-CF$_3$	8.3	1.49	2.40
4-CN	7.9	0.14	5.85
2-Br	7.4	0.89	4.20
4-NO$_2$	7.2	0.50	1.02
2,6-diCl	6.7	0.95	5.07
2,5-diNO$_2$	5.3	0.29	0.31
2,6-diCl, 4-NO$_2$	4.9	1.88	1.15
Cl$_5$	4.8	3.27	0.06
2,4-diNO$_2$	4.1	0.83	0.17
2,6-diNO$_2$	3.7	-0.21	0.42
2,6-diI, 4-NO$_2$	3.6	2.88	0.05
2-I, 4-CN, 6-NO$_2$	3.1	1.66	0.19

(a) π = log P_x/P_H; P_H = partition coefficient of phenol between octanol and water and P_x is that of the derivative x.[18].

(b) I_{50} = concentration of substitued phenol producing 50 % inhibition.

Table 2.

Inhibition of ^{32}Pi incorporation by uncouplers and anthelmintics in *Ascaris* mitochondria.

Inhibitor	I_{50} [a] (M)
l-Phenylethylbiguanide	2.00×10^{-4}
Salicylanilide	7.20×10^{-5}
Rafoxanide [b]	1.24×10^{-6}
Niclosamide [c]	1.10×10^{-7}
Mebendazole [d]	9.20×10^{-3}
Benzimidazole	6.50×10^{-3}
Thiabendazole [e]	1.10×10^{-3}
5,6-diCl-Benzimidazole	2.00×10^{-4}
Oligomycin [f]	1.84×10^{-7}

(a) I_{50} = concentration of compound producing 50 % inhibition;
(b) 3,5-Diiodo-3'-Chloro-4'(p-Chloromethoxy)-Salicylanilide;
(c) 2',5-Dichloro-4'-Nitrosalicylanilide (Yomesan);
(d) Methyl 5(6)-Benzoyl-2-Benzimidazolecarbamate;
(e) 2-(4-Thiazolyl)-Benzimidazole;
(f) Oligomycin = 15 % A + 85 % B.

Table 3.

Effect of substituted phenols (11 > pKa > 6) on the Mg^{++}-activated (a) ATPase in *Ascaris* mitochondria (pH = 7).

Substituents	pKa	π	Copt[b] $(\times 10^{-2}M)$	Maximal[c] Stimulation %	I_{50} $(\times 10^{-2}M)$
4-F	9.9	0.31	8.0	379.5	8.9
4-Cl	9.4	0.93	2.0	1512.4	4.4
4-Br	9.3	1.13	1.4	1050.2	4.0
4-I	9.2	1.45	0.9	1824.6	1.7
3-Cl	8.6	1.04	2.2	1731.1	3.9
3-CF_3	8.3	1.49	1.4	793.9	2.1
4-CN	7.9	0.14	0.8	107.0	4.7
2-Br	7.4	0.89	2.0	1331.9	> 3
4-NO_2	7.2	0.50	0.8	166.0	2.9
2,6-diCl	6.7	0.95	1.4	1831.9	2.9

(a) $MgCl_2$: 7.5 μmoles/ml;

(b) Copt.: concentration of phenol which induced the highest ATPase activity;

(c) control = 100 %.

Table 4.

Effect of substituted phenols ($6 \rangle$ pKa $\rangle 3$) on Mg^{++}-activated ATPase in *Ascaris* mitochondria (pH = 7).

Substituents	pKa	π	Copt. $(x10^{-2}M)$	Maximal stimulation (%)	I_{50} $(x10^{-2}M)$
2,5-diNO$_2$	5.3	0.29	0.10	114.2	0.70
2,6-diCl, 4NO$_2$	4.9	1.88	$-$ (a)	100.0	0.25
Cl$_5$	4.8	3.27	0.15	310.0	0.30
2,4-diNO$_2$	4.1	0.83	0.20	111.3	0.70
2,6-diNO$_2$	3.7	-0.21	$-$	100.0	0.40
2,6-diI, 4-NO$_2$	3.6	2.88	0.06	122.1	0.15
2-I, 4-CN, 6-NO$_2$	3.1	1.66	$-$	100.0	0.26

(a) $-$: no stimulation observed.

Table 5.

Effect of 2,4-DNP on the ^{32}Pi incorporation and malate utilization by anaerobic *Ascaris* mitochondria.

DNP concen-tration (M)	^{32}Pi incorpo-rated (nmoles/min/mg protein)	Malate utilized[b] (nmoles/min/mg protein)	Ratio ^{32}Pi/malate
0[a]	38.17	111	0.34
10^{-5}	27.62	111	0.25
4×10^{-5}	25.14	122	0.21
6×10^{-5}	20.70	122	0.17
8×10^{-5}	14.93	139	0.11
10^{-4}	13.25	161	0.08
2×10^{-4}	6.23	183	0.03
4×10^{-4}	2.06	189	0.01
6×10^{-4}	1.10	183	0.00

(a) 50 μl DMSO;

(b) Malate was determined spectrophotometrically according to Ho-horst [19].

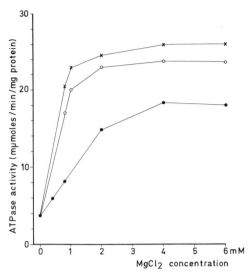

Fig. 1.
Effect of Mg^{++} on the ATPase activity in *Ascaris* mitochondria. Buffers: Tris-Maleate (0.1M) pH 6: ●; pH 7: o; pH 8: x. ATP concentration: 2mM.

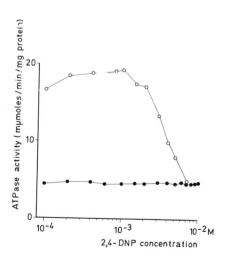

Fig. 2.
Effect of 2,4-dinitrophenol on the ATPase activity in *Ascaris* mitochondria (pH 7) in the absence (●) and presence (o) of Mg^{++} (7.5 μmoles/ml).

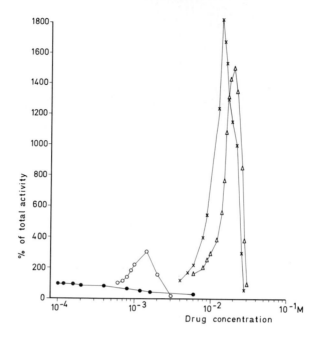

Fig. 3.
Effect of substituted phenols on the Mg^{++}-activated ATPase in *Ascaris* mitochondria. ●: 2,6-diCl, $4NO_2$-; o: pentachloro-; x: 2,6diCl-; △: 4-Cl-phenol.

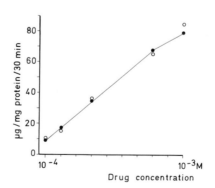

Fig. 4.
Uptake of pentachlorophenol by isolated *Ascaris* mitochondria in the absence (●) and presence (o) of 3mM ATP. (pH 7.1).

32

THE INFLUENCE OF THE LIPOPHILIC AND ELECTRONIC CHARACTER OF SUBSTITUTED PHENOLS ON THE EFFECTORS OF *ASCARIS SUUM* MITOCHONDRIA

J.P. Tollenaere

Janssen Pharmaceutica Research Laboratories
Beerse, Belgium

Introduction

One of the most complex problems in drug research is to find a systematic approach to the design of more specific and potent drugs. Due to the complexity of living organisms, the problem must be attacked with various techniques simultaneously. However, the disciplines underlying these various techniques differ widely which inevitably leads to a certain communication gap between the various workers in the field.

It is the purpose of this contribution to show how data of biochemical origin can be rationalised to a fairly high degree of success by analysing the physical nature of the compounds causing the biochemical response. The difficulty encountered in such an enterprise lies in the fact that the relationship between structure and activity is not well understood. The pioneering work of OVERTON[1] and MEYER[2] on the relationship between the narcotic action of organic molecules and their partition coefficients between oil and water may be considered as the first mile-stone in the study of the dependence of activity on structure. Subsequent studies by other authors were almost exclusively confined to the nonspecific toxic and narcotic action of various classes of compounds[3, 4].

In the early sixties HANSCH and co-workers[5] illustrated the importance of the partition coefficient of phenoxyacetic acid derivatives in explaining their activity on the growth of *Avena coleoptiles.*

A major obstacle to this sort of study, is the tremendous amount of work involved in accurately measuring partition coefficients for the various derivatives of a given series of compounds. It is the great merit of

HANSCH and co-workers in having established the additive nature of the partition coefficient. In a major paper, HANSCH and co-workers[6] introduced for 67 functional groups a new substituent constant, known as the lipohydrophilic constant π. Since the introduction of this collection of π values, structure-activity work has become feasible and has resulted in a vast amount of work being done in the area of correlation and rationalisation of biochemical and pharmacological data.

Results and Discussion

A Hansch type analysis is based on the assumption that a linear combination of free-energy related parameters reduces the variance of the biological response data. The equation to be solved is of the following form:

$$pC = \log(1/C) = aX_1 + bX_2 + cX_3 + \ldots\ldots k \tag{1}$$

where C stands for the molar concentration of a drug compound causing a standard response e.g. MIC, ED_{50} and LD_{50} etc. In this contribution, pC stands for pI_{50} i.e. the negative logarithm of the molar concentration of a substituted phenol causing a 50 % inhibition of the malate induced P^{32} incorporation into organic phosphate in *Ascaris suum* mitochondria. X_1, X_2, X_3 .... are substituent constants which may conveniently be divided into three groups:

1. The lipohydrophilic substituent constant π of Hansch is defined as

$$\pi_X = \log P_X - \log P_H \tag{2}$$

where P_H and P_X refer to the partition coefficient in the 1-octanol water system of the parent and substituted compound respectively. Thus π_X represents the contribution of the substituent to the partition coefficient of the substituted compound and $\Sigma\pi$ represents the sum of all π values of the various substituents on the parent compound. Since partitioning is a process involving molecular equilibrium, P is to be considered as an equilibrium constant which can be expressed in terms of the Gibbs free energy:

$$\Delta G = - RT \ln P \tag{3}$$

In other words, π_X is related to P and is therefore a free-energy based parameter. The 1-octanol water system has been chosen to represent the extremes of the biophase so that P should be a measure of the tendency of a drug to move either into the lipophilic or the hydrophilic phase.

2. The electronic substituent constants

Of the multitude of various types of electronic constants such as the inductive constant σ_I, the resonance constant σ_R and the polar constant σ^*, the σ constant of HAMMETT [7] is widely used. This set of σ constants is limited to meta and para substituents whereas the recently introduced δ constant of SETH-PAUL and VAN DUYSE [8], related to the Hammett constant, is also defined for ortho substituents. The use of the δ constant in a Hansch type analysis has recently been reported [9].

The Hammett constant is a measure of the electron-withdrawing or donating character of a substituent. Electron-withdrawing and donating substituents are characterised by positive and negative σ constants respectively.

3. The steric substituent constant E_s of TAFT [10].

Taft formulated the substituent constant E_s to account for intramolecular substituent effects. Naturally, it is difficult to establish whether a change in the substituents of a drug causes an inter- or intramolecular perturbation, or combination of both. KUTTER and HANSCH [11, 12] have reported excellent correlations using the E_s constant for cases apparently involving intermolecular interactions. Since large groups have negative E_s values, a positive coefficient in the regression equation indicates that small groups should result in high activities. The coefficients a,b,c,... and the intercept k in eq.1 have been calculated by means of computerised regression techniques. The values of $\Sigma\pi$, $\epsilon\sigma$, E_s and the experimental pI_{50} are presented in Table 1.

Eqs. 4-9 are the statistically significant equations correlating the experimental pI_{50} values with the various substituent constants. In these equations, n is the number of data points used in the regression analysis, r is the correlation coefficient, S is the standard error of the estimate and F the overall statistical significance. The t-test values on the significance of the coefficients have been written between brackets.

$$pI_{50} = 0.652\Sigma\Sigma\pi + 3.061$$
$$(4.94)$$

$$pI_{50} = 0.628\Sigma\sigma + 3.248$$
$$(3.96)$$

$$pI_{50} = 0.604\Sigma\Sigma\pi + 0.569\Sigma\sigma + 2.621$$
$$(11.22) \qquad (9.85)$$

$$pI_{50} = -0.106\Sigma\Sigma\pi^2 + 0.912\Sigma\Sigma\pi + 0.585\Sigma\sigma + 2.495$$
$$(-2.28) \qquad (6.36) \qquad (11.15)$$

$$pI_{50} = 0.570\Sigma\Sigma\pi - 0.159\Sigma\sigma^2 + 0.937\Sigma\sigma + 2.589$$
$$(10.49) \qquad (-1.76) \qquad (4.33)$$

$$pI_{50} = 0.681\Sigma\Sigma\pi + 0.686\Sigma\sigma + 0.398E_s^6 + 2.060$$
$$(13.88) \qquad (11.78) \qquad (3.31)$$

	n	r	s	F	
	21	0.750	0.585	24.45	(4)
	21	0.672	0.654	15.71	(5)
	21	0.965	0.237	122.53	(6)
	21	0.973	0.214	102.57	(7)
	21	0.971	0.224	92.19	(8)
	21	0.979	0.190	130.59	(9)

472

Inspection of the statistical data of eqs. 4 and 5 reveals that the lipophilic and the electronic character of the substituents alone does not account satisfactorily for the variance of pI_{50} whereas a linear combination of $\Sigma\pi$ and $\Sigma\sigma$ fits the data rather well as judged from the remarkable improvement of the correlation coefficient r and the F value accompanied by a corresponding decrease of the standard error S of the estimate.

A further analysis of the data resulted in eqs. 7-9. The t-test value on the quadratic term $\Sigma\pi^2$ in eq. 7 shows this term to be of minor importance. This result suggests that in the given series of phenols, compounds of suboptimal lipophilic character have been studied. The inclusion of a squared electronic term in eq. 8 is statistically insignificant as the coefficient of $\Sigma\sigma^2$ does only reach the $p = 0.1$ confidence level. The meaning of the insignificance of the $\Sigma\sigma^2$ term is not readily apparent since such a term can be interpreted in a number of ways [13, 14]. It has been reported that the square of the Hammett σ (i) correlates with the free radical index [15], and (ii) may also be a measure of a frontier controlled interaction [16, 17]

Applying a stepwise regression technique revealed that E_s at position 6 (or 2) contributed slightly to the reduction of the variance of pI_{50}.

Eq. 9, covering an 870-fold activity range indicates that the inhibition of the P^{32} incorporation is enhanced by increasing the lipophilic and electron-withdrawing character of the substituents. The positive sign of the coefficient of E_c suggests that small substituents at the ortho position, that is in the immediate vicinity of the hydroxy group, are beneficial for the inhibitory activity of the phenols. It is noteworthy that despite the wide structural variety of the para substituents, ranging from p-F to p-(4-Cl-phenyl), the activity is solely predictable in terms of the physical quantities appearing in eq. 9. The fact that this is true for bulky groups such as phenyl (compound 13), 4-Cl-phenyl (compound 15) and t-butyl (compound 23) indicates that steric factors at the para position are of minor importance. The calculated pI_{50} values using eq. 9 are presented in Table 1 and Fig. 1 illustrates the graphic representation of the excellent agreement between calculated and experimental pI_{50} values. Compounds 6 and 19 deviate rather seriously from the regression line. This could be attributed to the breakdown of the additivity of either π, σ or both. In a recent paper, STOCKDALE and SELWYN [18] reported on the effects of the ring substituents of 23 phenols as inhibitors and uncouplers of rat liver mitochondrial respiration. These authors noted serious deviations from the additivity of the π and σ values of the polysubstituted compounds. In order to alleviate this problem, the partition coefficients of these compounds were measured and the $\Sigma\sigma$ values were obtained from the pK_a values.

It is always tempting to use a significant regression equation as a means of elucidating some sort of reaction mechanism or to propose theories on drugreceptor interaction processes. The danger of doing so has recently been pointed out by CAMMARATA et al.[13]. Nevertheless, the dependence of the activity on the lipophilic character suggests that adsorption from an aqueous phase to the site of action is enhanced by the presence of strongly lipophilic groups. In other words, the regression equation is compatible with the view that the bioactive compounds should be able to penetrate the mitochondrial membrane. Secondly, due to the inverse proportionality between pK_a and $\Sigma\sigma$, it follows from Table 1 that those phenols having the lowest pK_a values are the most active. Thus at physiological pH, the most active members of the series are completely ionised. It is therefore conceivable that the active moieties of these phenols are their anions which could possibly react with a cationic centre at the target site thereby inducing a conformational change. The latter may activate a trigger mechanism which eventually leads to the observed behaviour.

The appearance of E_s in eq. 9 is hard to explain. If one thinks of the steric parameter in terms of being a measure of an intramolecular perturbation, then E_s might suggest that the hydroxy group should be freely accessible. On the other hand, if E_s represents an intermolecular perturbation then one might argue that the ortho substituent should be as small as possible in order to reduce steric repulsion between the substituent and some part of the receptor site. Introduction of a substituent in phenol may induce steric inhibition of resonance which affects the pK_a of the compound. In addition, the steric factor may also produce a differential change in the solvation of the phenol. These effects and others including internal hydrogen bonding frequently overlap and are therefore not easily separated. Inspection of the correlation matrix: $\Sigma\pi$, $\Sigma\sigma$, E_s:

	$\Sigma\pi$	$\Sigma\sigma$	E_s
$\Sigma\pi$	1.000	0.082	-0.277
$\Sigma\sigma$		1.000	-0.752
E_s			1.000

reveals a very strong correlation between $\Sigma\sigma$ and E_s reflected in r = -0.752. Considering all these arguments, it should be clear that with the given data, the primary role of the steric parameter E_s is not yet recognised.

A regression equation such as eq. 9 can also be used as a predictor or as a guide for further experiments (see footnotes to Table 1). For instance, this equation has predicted the activity of compounds 22 and 23 to be 3.85 and 3.56 respectively.

Conclusion

A careful selection of congeneric compounds having a large variation in the relative values of lipohydrophilic and electronic properties has resulted in a successful analysis of biochemical data. It is obivous that only under such circumstances an attempt can be made to untangle the roles of the lipohydrophilic, electronic and steric character of the substituents.

From the correlations obtained in this study it is clear that the inhibitory activity of the phenols is enhanced by increasing the lipophilic and the electronwithdrawing character of the substituents. The role of the steric factor at the ortho position could not be firmly established.

Acknowledgement.

The author is indebted to Mr. W.A. Seth Paul for many useful discussions and his help in preparing this manuscript.

References

1. MEYER, H. (1899). Arch. Exptl. Pathol. Pharmakol., **42**, 109.
2. OVERTON, E. (1899). Vierteljahrschr. Naturforsch. Ges. Zürich, **44**, 88.
3. MEYER, K.H. and HEMMI, H. (1935). Biochem. Zeit., **277**, 39.
4. BURGER, A. (1960). "Medicinal Chemistry", 2nd Ed., Wiley, New York, pp. 44-68.
5. HANSCH, C., MALONEY, P.P., FUJITA, T. and MUIR, R. (1962). Nature, **194**, 178.
6. (a) FUJITA, T., IWASA, J. and HANSCH, C. (1964). J. Am. Chem. Soc., **86**, 5175, (b) IWASA, J., FUJITA, T. and HANSCH, C. (1964). J. Med. Chem., **8**, 150. (c) HANSCH, C. and ANDERSON, S.M. (1967). J. Org. Chem., **32**, 2583.
7. McDANIEL, D.H. and BROWN, H.C. (1958). J. Org. Chem., **23**, 420.
8. SETH-PAUL, W.A. and VAN DUYSE, A. Spectrochim. Acta (in press).
9. TOLLENAERE, J.P. (1971). Chim. Thérap., **6**, 88.
10. TAFT, R.W. Jr., (1956). "Steric Effects in Organic Chemistry", M.S. Newman, Ed., Wiley, New York, N.Y., 598.
11. KUTTER, E. and HANSCH, C. (1969). J. Med. Chem., **12**, 647.

12. KUTTER, E. and HANSCH, C. (1969). Arch. Biochem. Biophys. **135,** 126.
13. CAMMARATA, A., ALLEN, R.C., SEYDEL, J.K. and WEMPE, E. (1970). J. Pharm. Sci., **59,** 1496.
14. CRAIG, P.N. (1971). J. Med. Chem., **14,** 680.
15. CAMMARATA, A., YAU, S.J., COLLETT, J.H. and MARTIN, A.N. (1970). Mol. Pharmac., **6,** 65.
16. CAMMARATA, A. (1968). J. Med. Chem., **11,** 1111.
17. CAMMARATA, A. (1969). J. Med. Chem., **12,** 314.
18. STOCKDALE, M. and SELWYN, M.J. (1971). Eur. J. Biochem., **21,** 565.

Table 1.

Substituent constants, experimental and calculated inhibition of malate induced P^{32} incorporation into organic phosphate of substituted phenols.

No	Compound	$\Sigma\pi^a$	$\Sigma\sigma^b$	E_s^{6c}	pI50 exp.[d]	pI50 calc.[e]	Δ exp. - calc.
1	H	0.00	0.000	1.24	2.35	2.55	-0.02
2	4-OMe	-0.12	-0.268	1.24	2.37	2.29	0.08
3	4-F	0.31	0.062	1.24	2.56	2.81	-0.25
4	4-CN	0.14	0.628	1.24	3.23	3.08	0.15
5	2,6-Cl$_2$	0.95[f]	0.454	0.27	3.30	3.13	0.17
6	2,6-(NO$_2$)$_2$	-0.21[g]	2.540	0.23	3.38	3.75	-0.37
7	2-Br	0.89	0.232	1.24	3.38	3.32	0.06
8	3-Cl	1.04	0.373	1.24	3.39	3.52	-0.13
9	4-Cl	0.93	0.227	1.24	3.44	3.34	0.10
10	4-Br	1.13	0.232	1.24	3.56	3.48	0.08
11	3-NO$_2$,4-Me	1.02	0.540	1.24	3.59	3.62	-0.03
12	4-I	1.45	0.276	1.24	3.75	3.73	0.02
13	4-Ph	1.89	-0.010	1.24	3.81	3.83	-0.02
14	4-NO$_2$	0.50	1.270	1.24	3.99	3.77	0.22
15	4-(4-Cl-Ph)	2.82	-0.010	1.24	4.29	4.47	-0.18
16	2,5-(NO$_2$)$_2$	0.29[g]	1.980	1.24	4.33	4.11	0.22
17	2-I,4-CN,6-NO$_2$	1.66	2.170	0.23	4.74	4.77	-0.03
18	2,4-(NO$_2$)$_2$	0.83	2.540	1.24	4.76	4.86	-0.10
19	2,6-Cl$_2$,4-NO$_2$	1.88	1.720	0.27	4.94	4.63	0.31
20	2,3,4,5,6-Cl$_5$	3.27[f]	1.427	0.27	5.19	5.38	-0.19
21	2,6-I$_2$,4-NO$_2$	2.88	1.820	-0.16	5.30	5.21	0.09
22[h]	3-CF$_3$	1.49	0.415	1.24	3.82	3.85	-0.03
23[h]	4-tBu	1.68	-0.197	1.24	3.60	3.56	0.04

(a) Ref. 6a;
(b) For ortho substituents, the σ_p values[7] have been taken.
(c) Refs. 10 and 11;
(d) Van den Bossche, H., this book chapter 31;
(e) Calculated using eq. 9;
(f) Measured in this laboratory using a Beckman UV Spectrophotometer Model DK 2A;
(g) Ref. 6c;

(h) Upon completion of the regression analysis, two additional compounds (22 and 23) became available. The activities have been
predicted by means of eq. 9.

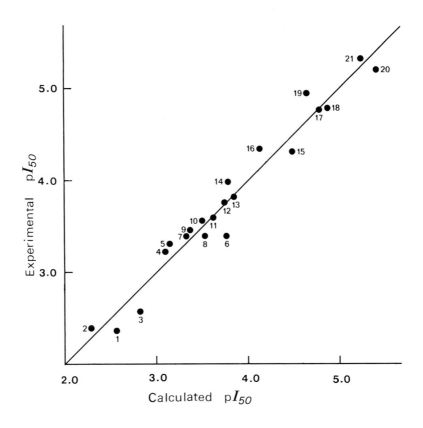

Fig. 1.
Graphic representation of experimental pI_{50} vs calculated pI_{50} (eq.9).

33

OXIDATIVE PHOSPHORYLATION IN *MONIEZIA* MITOCHONDRIA

K.S. Cheah

Agricultural Research Council, Meat Research Institute,
Langford, Bristol BS18 7DY, U.K.

The demonstration of oxidative phosphorylation in **Moniezia** mitochondria and the existence of functional cytochromes [1] are two important criteria favouring **Moniezia** to be aerobic and that oxygen is essential
for this energy synthesis.

Carefully isolated mitochondria from **Moniezia,** being impermeable to
NADH [2], have similar mitochondrial structure characteristics to those of
the mammalian tissue. Figure 1 illustrates the thin sections of mitochondria obtained from **Moniezia** (A) and the ox-neck muscle (B), the latter
tissue had previously been stored for 144 hours at 4°C prior to the
isolation of mitochondria. The ox-neck muscle mitochondria, for comparative purposes, had already been subjected to anaerobiosis **in situ**
before being isolated. The electron micrographs clearly show that **Moniezia** mitochondria have outer and inner membranes, outer compartment
and intracristal spaces, all of which were also observed in the ox-neck
muscle mitochondria. Furthermore, the existence of a double-membrane
structure (C) and elementary particles (D) are observed in the unfixed
negatively-stained samples.

Figure 2 represents typical oxygen electrode experiments demonstrating
oxidative phosphorylation in **Moniezia** mitochondria with succinate as
substrate. The classical state 3 to state 4 transition [3] could be repeated
several times (not shown) giving an average ADP/O ratio [3] value of 1.5
and a respiratory control index (RCI) of 1.6. The RCI, though low when
compared with freshly prepared mammalian mitochondria immediately
after the slaughter of the animal, corresponds very well with the value of
1.8 observed with the ox-neck muscle succinoxidase system for mito-

chondria isolated from 144 hours post-mortem tissue [4]. The state 3 rate for succinate oxidation was inhibited by oligomycin, an inhibitor of ATP synthesis by mitochondrial respiratory chain system but without affecting substrate-linked phosphorylation [5], and this could be relieved by the uncoupler p-trifluoromethoxycarbonyl-cyanidephenylhydrazone (FCCP). 50 % of the FCCP-uncoupled rate was inhibited by 24 μM CN$^-$ (Trace B). Table 1 summarizes the data on oxidative phosphorylation, oligomycin sensitivity, RCI and the FCCP-uncoupled rates of α-glycerophosphate, succinate and ascorbate **plus** tetramethyl-p-phenylenediamine (TMPD) oxidation. With ascorbate **plus** TMPD, which is used for estimating the cytochrome oxidase (EC 1.9.3.1.) activity, no clear transition of the state 3 to state 4 rate was observed. Oxidative phosphorylation was taking place since the state 3 rate of ascorbate **plus** TMPD oxidation was inhibited by oligomycin which was subsequently relieved by FCCP. The RCI, based on the ratio of ADP-induced rate divided by the rate in the presence of oligomycin, was 1.4 as compared with about 2.0 for α-glycerophosphate and succinate.

The effect of ADP concentration on succinate oxidation (state 3 rate) by *Moniezia* mitochondria is shown in Figure 3. The Michaelis constant (K_m) for ADP in accelerating the rate of succinate oxidation is about 28 μM as compared with 20-30 μM ADP for the succinoxidase system of liver and pigeon heart mitochondria [6]. The state 3 rate of α-glycerophosphate, succinate and ascorbate **plus** TMPD oxidation was inhibited by oligomycin. The sensitivity of succinate oxidation to this inhibitor is shown in Figure 4, where a 100 % inhibition of oxidative phosphorylation was obtained at a concentration of about 0.23 μg oligomycin per mg mitochondrial protein as compared with 0.18 μg oligomycin per mg rat-liver mitochondrial protein [5, 7]. 50 % inhibition of oxidative phosphorylation with succinate as substrate was achieved at 0.14 μg oligomycin per mg *Moniezia* mitochondrial protein. 92 % inhibition was obtained with 0.5 μg atractyloside, a specific inhibitor of adenine nucleotide translocator for the formation of exogenous mitochondrial ATP [8], in the presence of 1.2 mM ADP.

The oxidation of pyruvate **plus** malate was used to study the inhibitory effect of piericidin A on *Moniezia* mitochondrial NAD$^+$-linked oxidation. For a complete block for the state 3 rate of pyruvate **plus** malate oxidation about 50 pmole piericidin A per mg protein was required (Figure 5), an amount similar to that observed for the complete inhibition for the state 3 rate of pyruvate **plus** malate oxidation in mitochondria isolated from the backmuscle of the Pietrain pig [9].

The sensitivity of *Moniezia* mitochondrial respiratory chain system towards CN⁻ was tested using three different electron donors, succinate, α-glycerophosphate and ascorbate **plus** TMPD. 84 % inhibition of the state 3 rate for succinate oxidation (Figure 6) was achieved with 42 μM CN⁻, and CN⁻ in the presence of FCCP was found to be less inhibitory. Thus, 42 μM CN⁻ only blocked 63 % of the FCCP-uncoupled succinoxidase activity as compared with 84 % in the absence of FCCP. The inhibition constant (K_i) for the succinoxidase system was increased from 8 μM (without FCCP) to 24 μM (with FCCP). The same type of phenomenon, which appears to be the general property of uncouplers[10] , was also observed with rat-liver mitochondria oxidizing succinate in the presence of FCCP but azide was employed[10] instead of CN⁻. With *Moniezia* mitochondria, both the state 3 rate of α-glycerophosphate and cytochrome oxidase activities were also CN⁻ sensitive. The K_i for both these systems, in the presence of FCCP, was about 20 μM, a value almost identical to that observed for the succinoxidase system.

References

1. CHEAH, K.S. (1968). Biochim. Biophys. Acta, **153**, 718.
2. CHEAH, K.S. (1971). Biochim. Biophys. Acta, **253**, 1.
3. CHANCE, B. and WILLIAMS, G.R. (1958). in F.F. Nord, Advances in Enzymology, Vol. 17, Interscience Publishers, Inc., New York, p. 65.
4. CHEAH, K.S. and CHEAH, A.M. (1971). J. Bioenergetics, **2**, 85.
5. SLATER, E.C. (1967). in R.W. Estabrook and M.E. Pullman, Methods in Enzymology, Vol 10, Academic Press, New York, p 48
6. CHANCE, B. and HAGIHARA, B. (1961). in E.C. Slater, Intracellular Respiration: Phosphorylating and non-Phosphorylating Oxidation Reactions, Pergamon Press, London, p. 3.
7. HARDY, H.A., CONNELLY, J.L. and JOHNSON, D. (1964). Biochemistry, **3**, 1961.
8. KLINGENBERG, M., GREBE, K. and HELDT, H.W. (1970). Biochem. Biophys. Res. Commun., **39**, 344.
9. CHEAH, K.S. (1970). FEBS letters, **10**, 109.
10. WILSON, D.F. and CHANCE, B. (1966). Biochem. Biophys. Res. Commun., **23**, 751.

Table 1.

Oxidative phosphorylation in *Moniezia* mitochondria.

All the respiratory activities were measured polarographically with a Clark oxygen electrode at 25° in 2.5 ml (total volume). The data represents an average value from three separate state 3 rates induced by ADP. Reaction medium (mM): KCl, 30.0; $MgCl_2$, 6.0; sucrose, 75.0; KH_2PO_4, 20.0; EDTA, 1.0; pH 7.20. Final concentration of substrates (mM): α-glycerophosphate, 8.0; succinate, 8.0; ascorbate, 4.0; TMPD, 0.2. Rotenone (2 μM) was added prior to succinate in estimating the succinoxidase activity and 0.1 μg antimycin A per mg protein before TMPD addition for the ascorbate-TMPD oxidase activity. The final concentration of oligomycin and FCCP was 1 μg per mg protein and 1 μM respectively. The RCI values, within brackets, were calculated by dividing the state 3 rate by the oligomycin inhibited rate. (Reproduced from Biochim. Biophys. Acta **253** (1971) 1 with permission from Elsevier.

Oxidase System	Oxygen Uptake (natoms 0/min/mg protein)			ADP/O	RCI
	ADP	ADP + Oligomycin	ADP + Oligomycin + FCCP		
α-Glycerophosphate	78	42	90	1.4	1.4 (1.9)
Succinate	52	26	69	1.5	1.6 (2.0)
Ascorbate **plus** TMPD	52	32	66	–	– (1.4)

Fig. 1.

Electron micrographs showing thin sections of *Moniezia* and ox-neck muscle mitochondria and the unfixed negatively-stained *Moniezia* mitochondria.

With the exception of a few swollen mitochondria, all the intact mitochondria isolated from *Moniezia* (A) and the ox-neck muscle (B) are in the condensed configuration. The unfixed negatively-stained samples of *Moniezia* mitochondria show clear definitions of outer and inner membranes (C) and the existence of elementary particles (D). OM, outer membrane; IM, inner membrane; OC, outer compartment; IS, intracristal space; S, swollen, EP, elementary particles. Magnification: A, 64 000 X; B, 16 000 X; C and D, 112 000 X. Electron microscopy was carried out in collaboration with Mr. C.A. Voyle. (Reproduced from Biochim. Biophys. Acta **253** (1971) 1, with permission from Elsevier.).

See illustrations pages, 484-487

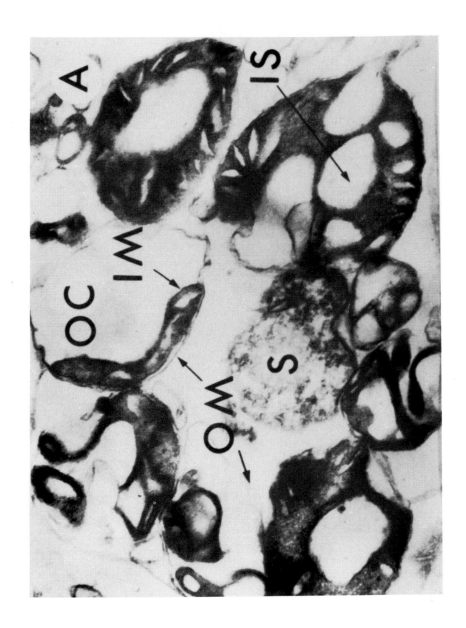

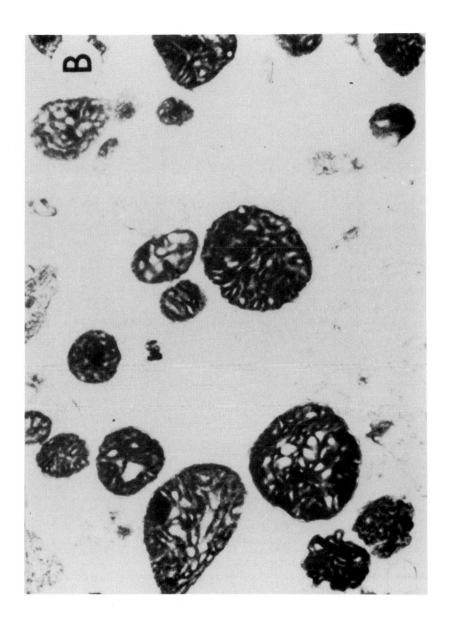

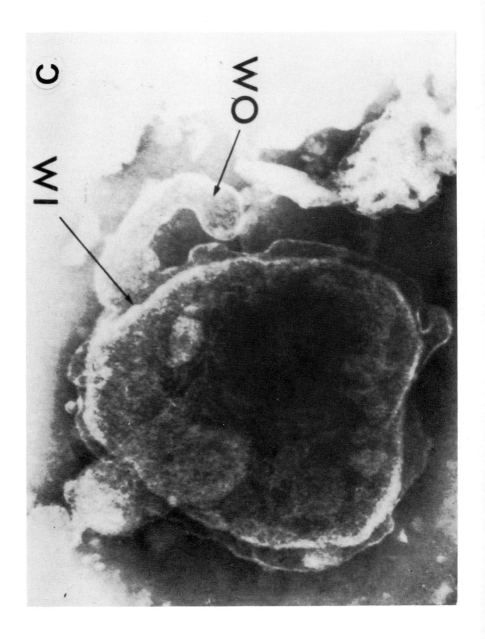

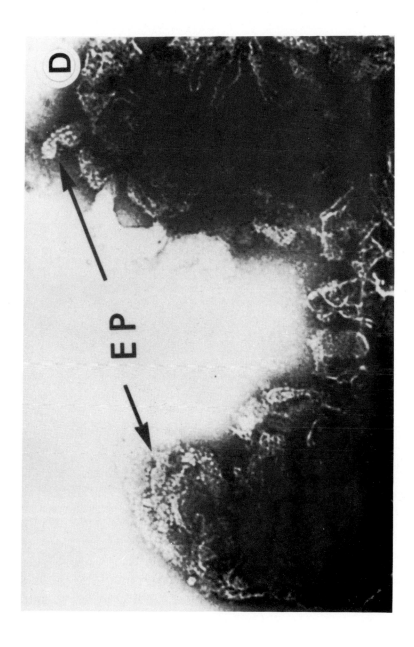

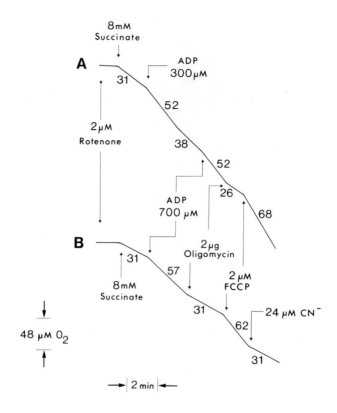

Fig. 2.
Typical oxygen electrode tracings showing respiratory control (Tracing A) and effect of oligomycin, FCCP and CN⁻ (Tracing B) on *Moniezia* mitochondrial succinoxidase system.

All respiratory rates, expressed in natoms O per min per mg protein, adjacent to the electrode traces, were estimated with a Clark oxygen electrode. Total protein, 3.07 mg. Other experimental details are given in Table 1. (Reproduced from Biochim. Biophys. Acta **253** (1971) 1, with permission of Elsevier).

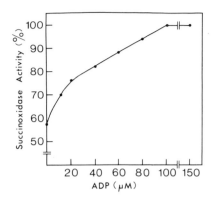

Fig. 3.
Effect of ADP on *Moniezia* mitochondrial succinate oxidation. The succinoxidase activity (state 3) was determined polarographically at 25° as described in Figure 2. (Reproduced from Biochim. Biophys. Acta **253** (1971) 1, with permission of Elsevier).

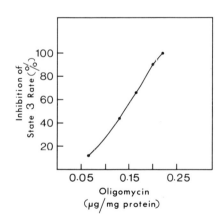

Fig. 4.
Inhibitions of oxidative phosphorylation in *Moniezia* mitochondria by oligomycin.
The effect of oligomycin on *Moniezia* oxidative phosphorylation was tested using succinate as substrate. All respiratory activities (state 3) were estimated polarographically at 25° as described in Figure 2.

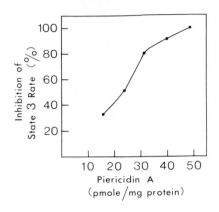

Fig. 5.
Inhibition by piericidin A on the state 3 respiratory rate of pyruvate **plus** malate oxidation by *Moniezia* mitochondria. Piericidin A was added prior to ADP (500 μM) to block the state 3 rate of pyruvate (8 mM) **plus** malate (8 mM) oxidation. Other details as described in Table 1 and Figure 2.

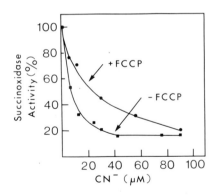

Fig. 6.
Effect of CN⁻ on the succinoxidase activity of *Moniezia* mitochondria. The % inhibition of the succinoxidase activity (state 3) was determined polarographically at 25° in a total volume of 2.5 ml. Other experimental details are given in Table 1 and Figure 2. (Reproduced from Biochim. Biophys. Acta **253** (1971) 1, with permission of Elsevier).

34

SUMMARY AND OUTLOOK

W. Trager
The Rockefeller University - New York - U.S.A.

When I agreed to this assignment I feared I might come to the place where I now stand feeling like the man who has eaten too much too quickly of a very rich diet that he can neither digest nor assimilate. - The diet has been rich and abundant but of such high quality that its assimilation has turned out not so difficult after all.

On Wednesday morning, after a few kind words of welcome by Dr. Janssen, the scientific sessions of this symposium were opened with a historical review by Theodor von Brand. Although it was clear that Dr. von Brand did not intend this, it was equally clear that the great bulk of work on physiology of parasites and parasitism has been done during the last 50 years, a period closely coinciding with von Brand's own ongoing career. This is probably not strictly coincidence.

A remarkable feature of the entire 3 days of lectures and discussions that followed was the reciprocal significance of the work on parasites to the general field of cell biology and of progress in cell biology to parasitology. This was true regardless of whether the work was of purely basic nature or was directly oriented toward a practical chemotherapeutic goal. Thus Bueding pointed out the two modes of action of antischistosomal drugs: (1) by selectively inhibiting the parasite's phosphofructokinase, the rate-limiting enzyme for its glycolytic metabolism; (2) by inhibiting neurohumoral transmitters. The latter finding has led to a fundamental study of acetylcholine and biogenic amines in neurotransmission in trematodes. An exciting byproduct of the work was the finding that hycanthone induces worms to produce progeny resistant to the drug, a trait transmitted hereditarily via the female cytoplasm. H. Saz outlined clearly the singular pathway for anaerobic phosphorylation used by the mitochondria of *Ascaris* and other intestinal helminths. This involves the reduction of fumarate to succinate with simultaneous oxidation of NADH. This method of coupled phosphorylation not requiring any of the cytochromes is of fundamental significance to our understanding of mitochondrial acitivity. In *Moniezia,* however, cytochromes are present in appreciable amount and oxidative phosphorylation does occur, as pointed out by K.S. Cheah. C. Bryant discussed the related subject of control of respiratory metabolism in the sheep tapeworm and E.G. Weinbach called attention to the presence of non-heme iron containing proteins in the mitochondria of cestodes. These compounds will probably assume considerable importance as more is learned about their occurrence and function.

Several discussions on anthelminthic drugs emphasized their pharmacological and biochemical activities with special reference to possible modes of action. The very useful drug tetramisole, of which the laevo isomer is especially active, may affect *Ascaris* primarily via its neuromuscular system, as discussed by Van den Bossche and Van Neuten, although the drug also inhibits the fumarate reductase so important to the metabolism of the worm. Evidence was provided that the anaerobic P^{32}-ATP exchange reaction in the mitochondria of cestodes is inhibited by anticestodal agents such as chlorosalicylamide and several others. Mebendazole on the other hand probably exerts its antinematodal action by inhibiting glucose uptake. The activity of substituted phenols in inhibiting malate-induced P^{32} incorporation by *Ascaris* mitochondria clearly depends on their lipophilic properties. Furthermore, Tollenaere showed how accurate predictions of structure activity relationships for these compounds could be based on the lipohydrophilic and electronic properties of the substituents.

Of great interest was the report by the Meyers that free-living as well as parasitic flatworms cannot synthesize their own fatty acids or sterols, a defect that may have encouraged so many flatworms to become parasites. Davey gave a thorough discussion of the little understood subject of humoral control of development in nematodes. Other fundamental physiological studies on helminths included reports on the phosphatases in the intestinal cells of *Ascaris* and other parasitic nematodes by Borgers and Ruitenberg. Beames showed that movement of sugars and other molecules across the intestine of *Ascaris* requires an energy source and CO_2. An especially stimulating lecture by Lee pictured the parasitic worm as a sensitive creature trying hard, like the rest of us, to find a comfortable niche, and often being unceremoniously thrown out of it by immune reactions of the host.

Despite this wealth of material on helminth parasites, the protozoan parasites were not neglected. Reeves noted some of the special aspects of carbohydrate metabolism in *Entamoeba histolytica,* a deceptively simple organism that seems to have some of its own ways of doing things. Ryley discussed the difficult subject of biochemistry of coccidia, a field as yet little worked on. Of special interest are the mechanism of emergence of sporozoites from the sporocyst and oocyst, and the vitamin requirements of the schizogonic stages. Much more work has been done with that other group of intracellular parasitic protozoa, the malaria parasites. Howells showed that mammalian malaria parasites in their insect stages have true protozoan mitochondria with succinoxidase activity, but concentric-membrane bodies lacking succinoxidase in their mammalian stages. This situation is reminiscent of the much greater development of the mitochondrion in the insect as compared to the mammalian stages of the brucei-group trypanosomes. The DNA of the primate malaria *Plasmodium knowlesi* seems to be synthesized during growth of the trophozoite, and this synthesis is not sensitive to inhibitors of dihydrofolate reductase, though the subsequent division of the nuclear material is sensitive to these agents as shown by Gutteridge. Trager presented a new hypothesis based on work with bongkrekic acid that malaria parasites require exogenous ATP for transport of essential metabolites across the outer of the 2 membranes separating the parasite from its host erythrocyte.

The hemoflagellates represent an especially active field. Jaffo discussed their dihydrofolate reductase which, like that of malaria parasites, has a much higher molecular weight than the corresponding enzymes of metazoa, including those of the parasitic helminths. As shown by Eeckhout, in the insect flagellate *Crithidia luciliae* some hydrolases are secreted into the flagellar pocket whereas others are in lysosomes. Bowman discussed the physiological differences between stumpy and slender trypomastigotes of *Trypanosoma rhodesiense* and the changes undergone in culture, especially the gradual acquisition of the electron transport chain in the mitochondrion. The details of structure of the cytochromes of hemoflagellates were presented by Hill.

Papers by Newton, Riou and Steinert provided the latest information on kinetoplast-DNA and brought forth a lively discussion. K-DNA has now been isolated in pure form from a variety of hemoflagellates and the stage seems set for studies by hybridization and other techniques which might begin to tell us something about the function of this unique DNA. In general it seems likely that for the immediate future new types of useful drugs will still be found by chance. But clearly a background of fundamental understanding is rapidly being laid down that may enable not only rational design of new chemotherapeutic agents but also other altogether new approaches to control of parasitic infections. Certainly enough progress has been made to encourage all of us to go back to work, and also enough indications of practical benefits have been obtained to justify continued support of these kinds of studies.

I know that I speak not only for myself but for all of us here when I say that this symposium was one of the liveliest and most interesting. We are most grateful to our hosts The Janssen Research Foundation, and especially to Dr. Janssen and Dr. Van den Bossche, for providing so felicitous an environment and such warm hospitality.

SUBJECT INDEX

A

495

496

I3